Metal Nanoparticles

Metal Nanoparticles
Synthesis, Characterization, and Applications

edited by

Daniel L. Feldheim
North Carolina State University, Raleigh, North Carolina

Colby A. Foss, Jr.
Georgetown University, Washington, D.C.

MARCEL DEKKER, INC.   NEW YORK · BASEL

ISBN: 0-8247-0604-8

This book is printed on acid-free paper.

Headquarters
Marcel Dekker, Inc.
270 Madison Avenue, New York, NY 10016
tel: 212-696-9000; fax: 212-685-4540

Eastern Hemisphere Distribution
Marcel Dekker AG
Hutgasse 4, Postfach 812, CH-4001 Basel, Switzerland
tel: 41-61-261-8482; fax: 41-61-261-8896

World Wide Web
http://www.dekker.com

The publisher offers discounts on this book when ordered in bulk quantities. For more information, write to Special Sales/Professional Marketing at the headquarters address above.

Copyright © 2002 by Marcel Dekker, Inc. All Rights Reserved.

Neither this book nor any part may be reproduced or transmitted in any form or by any means, electronic or mechanical, including photocopying, microfilming, and recording, or by any information storage and retrieval system, without permission in writing from the publisher.

Current printing (last digit):
10 9 8 7 6 5 4 3 2 1

PRINTED IN THE UNITED STATES OF AMERICA

Preface

Metal nanoparticles are certain to be the building blocks of the next generation of electronic, optoelectronic and chemical sensing devices. The physical limits imposed by top-down methods such as photo- and electron- beam lithography dictate that the synthesis and assembly of functional nanoscale materials will become the province of chemists. In the current literature, there are three emerging themes in nanoparticle research: (1) synthesis and assembly of metal particles of well-defined size and geometry, (2) structural and surface chemistry effects on single electron charging, and (3) size, shape, and surface chemistry effects on particle optical properties.

This book was written in order to identify and elaborate upon the unifying themes in metal nanoparticle research vis-à-vis their synthesis, characterization and applications. Specifically we have sought to: (1) compile the most up-to-date work in synthesis and characterization of nanoparticle optical and electronic properties and (2) present these topics in such a way that the volume will serve as a leading text for established researchers in the field *and* as a comprehensive primer for nonspecialists.

This volume is a particularly timely compilation of nanoparticle research because it is only within the last few years that a fundamental understanding of nanoparticle structural, optical, and electronic properties has been established. Despite these recent advances, no comprehensive treatise currently exists to tie together these three historically disparate, yet intimately related, areas of nanoparticle research.

Daniel L. Feldheim
Colby A. Foss, Jr.

Contents

	Preface	iii
	Contributors	vii
1	Overview *Daniel L. Feldheim, and Colby A. Foss, Jr.*	1
2	Transition-Metal Nanoclusters: Solution-Phase Synthesis, Then Characterization and Mechanism of Formation, of Polyoxoanion- and Tetrabutylammonium-Stabilized Nanoclusters *Richard G. Finke*	17
3	Magic Numbers in Clusters: Nucleation and Growth Sequences, Bonding, Principles, and Packing Patterns *Boon K. Teo and Hong Zhang*	55
4	Modeling Metal Nanoparticle Optical Properties *K. Lance Kelly, Traci R. Jensen, Anne A. Lazarides, and George C. Schatz*	89
5	Electrochemical Template Synthesis of Nanoscopic Metal Particles *Colby A. Foss, Jr.*	119

6	Nonlinear Optical Properties of Metal Nanoparticles Robert C. Johnson and Joseph T. Hupp	141
7	Electrochemical Synthesis and Optical Properties of Gold Nanorods Chao-Wen Shih, Wei-Cheng Lai, Chuin-Chieh Hwang, Ser-Sing Chang, and C. R. Chris Wang	163
8	Surface Plasmon Resonance Biosensing with Colloidal Au Amplification Michael J. Natan and L. Andrew Lyon	183
9	Self-Assemblies of Nanocrystals: Fabrication and Collective Properties Marie-Paule Pileni	207
10	Electrodeposition of Metal Nanoparticles on Graphite and Silicon Sasha Gorer, Hongtao Liu, Rebecca M. Stiger, Michael P. Zach, James V. Zoval, and Reginald M. Penner	237
11	Synthesis, Characterization, and Applications of Dendrimer-Encapsulated Metal and Semiconductor Nanoparticles Richard M. Crooks, Victor Chechik, Buford I. Lemon III, Li Sun, Lee K. Yeung, and Mingqi Zhao	261
12	The Electrochemistry of Monolayer Protected Au Clusters David E. Cliffel, Jocelyn F. Hicks, Allen C. Templeton, and Royce W. Murray	297
13	Nanoparticle Electronic Devices: Challenges and Opportunities Wyatt McConnell, Louis C. Brousseau III, A. Blaine House, Lisa B. Lowe, Robert C. Tenent, and Daniel L. Feldheim	319
	Index	335

Contributors

Louis C. Brousseau III North Carolina State University, Raleigh, North Carolina

Ser-Sing Chang National Chung Cheng University, Min-Hsiung, Chia-Yi, Taiwan, R.O.C.

Victor Chechik* Texas A&M University, College Station, Texas

David E. Cliffel University of North Carolina, Chapel Hill, North Carolina

Richard M. Crooks Texas A&M University, College Station, Texas

Daniel L. Feldheim North Carolina State University, Raleigh, North Carolina

Richard G. Finke Colorado State University, Fort Collins, Colorado

Colby A. Foss, Jr. Georgetown University, Washington, D.C.

Sasha Gorer University of California, Irvine, Irvine, California

Jocelyn F. Hicks University of North Carolina, Chapel Hill, North Carolina

---

*Current affiliation: University of York, Heslington, York, U.K.

Contributors

A. Blaine House   North Carolina State University, Raleigh, North Carolina

Joseph T. Hupp   Northwestern University, Evanston, Illinois

Chuin-Chieh Hwang   National Chung Cheng University, Min-Hsiung, Chia-Yi, Taiwan, R.O.C.

Traci R. Jensen   Northwestern University, Evanston, Illinois

Robert C. Johnson   Northwestern University, Evanston, Illinois

K. Lance Kelly   Northwestern University, Evanston, Illinois

Wei-Cheng Lai   National Chung Cheng University, Min-Hsiung, Chia-Yi, Taiwan, R.O.C.

Anne A. Lazarides   Northwestern University, Evanston, Illinois

Buford I. Lemon III[*]   Texas A&M University, College Station, Texas

Hongtao Liu   University of California, Irvine, Irvine, California

Lisa B. Lowe   North Carolina State University, Raleigh, North Carolina

L. Andrew Lyon   Georgia Institute of Technology, Atlanta, Georgia

Wyatt McConnell   North Carolina State University, Raleigh, North Carolina

Royce W. Murray   University of North Carolina, Chapel Hill, North Carolina

Michael J. Natan   SurroMed, Inc., Palo Alto, California

Reginald M. Penner   University of California, Irvine, Irvine, California

Marie-Paule Pileni   Université Paris et Marie Curie (Paris IV), Paris, France

George C. Schatz   Northwestern University, Evanston, Illinois

Chao-Wen Shih   National Chung Cheng University, Min-Hsiung, Chia-Yi, Taiwan, R.O.C.

---

[*]*Current affiliation:* Dow Chemical Co., Midland, Michigan

Contributors

Rebecca M. Stiger   University of California, Irvine, Irvine, California

Li Sun   Texas A&M University, College Station, Texas

Allen C. Templeton   University of North Carolina, Chapel Hill, North Carolina

Robert C. Tenent   North Carolina State University, Raleigh, North Carolina

Boon K. Teo   University of Illinois at Chicago, Chicago, Illinois

C. R. Chris Wang   National Chung Cheng University, Min-Hsiung, Chia-Yi, Taiwan, R.O.C.

Lee K. Yeung*   Texas A&M University, College Station, Texas

Michael P. Zach   University of California, Irvine, Irvine, California

Hong Zhang   Air Force Research Laboratory (AFRL/MLPO), Wright-Patterson AFB, Ohio

Mingqi Zhao[†]   Texas A&M University, College Station, Texas

James V. Zoval   University of California, Irvine, Irvine, California

---

Current affiliation: Dow Chemical Co., Freeport, Texas
[†]*Current affiliation:* ACLARA Bio Sciences, Inc., Mountain View, California

1
Overview

Daniel L. Feldheim
North Carolina State University, Raleigh, North Carolina

Colby A. Foss, Jr.
Georgetown University, Washington, D.C.

I. INTRODUCTION

Over the last decade there has been increased interest in "nanochemistry." A variety of supermolecular ensembles (1), multifunctional supermolecules (2), carbon nanotubes (3), and metal and semiconductor nanoparticles (4) have been synthesized and proposed as potential building blocks of optical and electronic devices (5). This has arisen for a variety of reasons, not the least of which is technological advance, and the promise of control over material and device structure at length scales far below conventional lithographic patterning technology.

Metal particles are particularly interesting nanoscale systems because of the ease with which they can be synthesized and modified chemically. From the standpoint of *understanding* their optical and electronic effects, metal nanoparticles also offer an advantage over other systems because their optical (or dielectric) constants resemble those of the bulk metal to exceedingly small dimensions (i.e., < 5 nm).

Perhaps the most intriguing observation is that metal particles often exhibit strong plasmon resonance extinction bands in the visible spectrum, and therefore deep colors reminiscent of molecular dyes. Yet, while the spectra of molecules (and semiconductor particles) can be understood only in terms of quantum mechanics, the plasmon resonance bands of nanoscopic metal particles can often be rationalized in terms of classical free-electron theory and simple electrostatic limit models for particle polarizability (6). Furthermore, while the composition of a metal particle may be held constant, its plasmon resonance extinction maximum can be shifted hundreds of nanometers by changing its shape and/or orientation in the incident field (7), or the number density of particles in a composite material (8). Thus, in contrast to molecular systems, the linear optical properties of nanoscopic metal particle composites can be changed significantly without a change in essential chemical composition.

The electrical properties of metal particles are also similar in form to those of their corresponding bulk metals. Surface charging and electron transport processes in individual nanoscopic metal particles and two-dimensional particle arrays may often be understood with relatively simple classical charging expressions and RC equivalent circuit diagrams (9). Again, in contrast to molecules and semiconductor nanoparticles whose electron transport properties require a quantum mechanical description, charging in metal nanoparticles only requires a knowledge of their size and the dielectric properties of the surrounding medium (9).

Recent experimental studies of metal particle optical properties and single-electron-device applications have demonstrated yet another aspect of versatility: since the surface chemistry of nanoscopic metal particles is similar to that of continuous metal surfaces, chemical surface modification (e.g., self-assembled monolayers) is straightforward and allows for particles that are soluble in a variety of media (10) or possess specific affinities for certain analyte species in solution (11).

The foregoing discussion was not meant to imply that the optical and electronic properties of metal nanoparticle systems are completely understood or that we have achieved arbitrary control over their geometry and assembly. On the contrary, relationships between particle geometry and their linear optical properties have not been established fully, except perhaps for perfect spheres. Consider, for example, that despite over 20 years of theoretical and experimental research, the optimum size and shape of a collection of metal particles for surface-enhanced Raman spectroscopy is still uncertain. Moreover, the interplay between nanoparticle surface chemistry and optical and electronic behaviors has not been addressed in detail. Finally, methods for linking particles deliberately and rationally in a manner analogous to molecular synthesis have not been developed. These issues are critically important to future device technologies such as integrated optical and electronic devices and chemical sensors.

This book reviews recent advances in nanoscopic metal particle synthesis, theory of optical properties, and applications in optical composite materials and electronic devices. Its major emphasis is on particles which are large enough to possess a well-defined conduction band and, therefore, able to manifest plasmon resonance and classical electron charging behaviors. However, we also invited contributions on the topic of smaller metal clusters because the emerging science of their synthesis and structure will almost certainly impact "nanodevice" technology. We should also note that the book does not emphasize the application of nanoscopic metal particles in catalysis, a topic extensively reviewed elsewhere (12). In the next sections, we review briefly the history of nanoscopic metal particles, including their synthesis and application until the early 1990s. We also review the very basic theories necessary for understanding plasmon resonance spectra and Coulomb blockade effects in single-electron devices.

II. HISTORICAL BACKGROUND

The first nanometal containing human artifacts predates modern science by many centuries. Perhaps the oldest object is the Lycurgus chalice from fifth-century Rome, which contains gold nanoparticles (13). The Maya Blue pigment found in the eleventh-century Chichen Itza ruins owes its color in part to nanoscopic iron and chromium particles (14). Many sources credit Johann Kunckel (1638–?) with developing the first systematic procedures for incorporating gold into molten silica, thus producing the well-known "ruby glass" (15).

As early as the sixteenth century, the darkening of silver compounds by light was known (16). The successful application of silver halide photochemistry to photography did not occur until the mid-nineteenth century, with the work of Fox-Talbot and Daguerre (16). In the early glass pigmentation and photographic plate applications, the physical basis of color in these materials was not known.

From correspondence between Michael Faraday and George Gabriel Stokes, it is clear that, by 1856, Faraday had postulated that the color of ruby glass, as well as his aqueous solutions of gold (mixed with either SO_3 or phosphorus), is due to finely divided gold particles (17). Stokes' disagreement and argument for the existence of a purple gold oxide apparently prompted Faraday's famous electrical discharge method for preparing aqueous gold colloids (18). It is noteworthy that Faraday did not have a quantitative theoretical framework, but seems to have based his postulate on an intuitive understanding of highly reflective metals and scattering processes.

The first attempt at a quantitative theoretical description of the colors of nanoscopic metal particles occurs in 1904 with the work of J. C. Maxwell-Garnett (19), who used expressions for spherical particle polarizability derived by Rayleigh and Lorenz to define effective composite optical constants. Maxwell-Garnett's theory applied only to particles whose dimensions were negligible in comparison to the wavelength of the incident light. Thus, while particle size could not be addressed in the theory, Maxwell-Garnett could attribute the different colors seen in particle systems derived from the same metal element to differences in interparticle spacing (19).

Gustav Mie's 1908 paper represents the first rigorous theoretical treatment of the optical properties of spherical metal particles (8a,20). Mie's theory yielded extinction coefficients for nanoscopic gold particles which compared well with the experimental spectra of gold sols and, unlike Maxwell-Garnett theory, was applicable to spheres of any size. Mie scattering theory is applied today to a variety of systems, including nonmetal particles. His basic approach has also been adapted to other shapes, such as cylinders (21) and ellipsoids (22).

In the first half of the twentieth century, scientific interest in metal nanoparticles was not limited to their optical properties. For example, aqueous gold particles were model systems for the study of colloidal stability and nucleation (23). The application of colloidal silver particles was also the subject of serious discussion before the advent of sulfa drugs in the 1930s (24). The use of colloidal metals as histological staining agents began in 1960 and expanded rapidly as the use of the electron microscope in cell biology became routine (25,26).

The so-called integral coloring of aluminum surfaces via anodization was first patented in the late 1950s (27). However, it was not until the late 1970s that Goad and Moskovits demonstrated that the observed colors arise from plasmon resonance extinction of metal particles embedded in the pores of the anodic aluminum oxide layer (28). In 1980, Andersson, Hunderi, and Granqvist discussed the application of anodic alumina-metal nanoparticle composite films as selective solar absorbers, outlining a generalized Maxwell-Garnett-theory-based approach to predicting spectral absorption and emissivity (29). Applications of others metal nanoparticle systems as selective solar materials were discussed in the early 1980s (30).

It was the discovery of the surface-enhanced Raman scattering effect (SERS) (31) that sparked a renewed interest in metal nanoparticle optics and physics. The discovery of the connection between electromagnetic enhancements and plasmon resonance processes (32) provided the impetus for serious experimental and theoretical investigations of particle shapes other than spheres (33). Although many workers during the mid-1970s to mid-1980s were interested primarily in the SERS effect, their work provided important insights into the fundamental linear optical properties of metal nanoparticles (34).

Largely independent of the discussions surrounding SERS phenomena, a number of groups since the late 1970s and early 1980s became interested in what Arnim Henglein has termed "the neglected dimension between atoms or molecules and bulk materials" (35). Some of the new synthetic and theoretical advances in metal nanoparticles were inspired by size quantization phenomena in semiconductor particle systems (36–39). The potential for applications in photocatalysis and electronic devices was also a driving force even a decade ago (35–38). However, it is also likely that inorganic chemists were simply interested in the challenge of preparing and crystallographically characterizing successively larger metal cluster compounds (40). In any case, in much of this work, the questions are quite fundamental: How many atoms must a metal cluster possess before it achieves metallic properties? What are the rules governing the geometry of small metal clusters?

In the 1990s, a certain confluence of perspectives seems to have commenced. For some, metal nanoparticles are interesting because of their surface properties. For others, they are simply very large molecules. The synthesis and stabilization of large structurally-well-defined metal clusters requires the presence of surface-bound moieties that are now referred to as "ligands" as opposed to "adsorbates" (38). At the same time, clusters large enough to achieve metallic properties exhibit surface-charging behavior in solution that is similar to that of bulk electrodes (41,42). Mulvaney has studied the voltammetric behavior of silver colloids in aqueous solution, demonstrating that the nanoparticles behave as redox centers in a manner analogous to molecular systems (43).

The conception of metal nanoparticles as large molecules is obviously appealing to the chemistry oriented. But the next logical step in this context, namely the use of nanoparticles as building blocks of larger structures, is still in its infancy. For example, Pileni has demonstrated the ability of metal nanoparticles to form ordered lattices (44). Schiffrin and co-workers have prepared intriguing highly ordered two-dimensional lattices composed of particles of two different diameters (45). The self-assembly of nanoparticles will undoubtedly be a key element in the maturation of the once "neglected dimension."

III. OPTICAL PROPERTIES OF METAL PARTICLES

Throughout this volume reference will be given to the so-called plasmon resonance bands of nanoscopic metal particles. We thus devote a section of this introduction to a basic discussion of this optical process, whose outward manifestation resembles the absorption of molecular systems, but is nonetheless very different in physical origin.

The polarizability of a spherical object in vacuum in either a static electric field or a time-dependent field whose wavelength is much larger than the dimensions of the sphere is given by Lorentz's well-known expression (21)

$$\alpha = 4\pi a^3 \left\{ \frac{\varepsilon_p - 1}{\varepsilon_p + 2} \right\} \tag{1}$$

where a and ε_p are the radius and complex dielectric function of the sphere, respectively.

The optical extinction cross section C_{ext} for particles which are much smaller than the incident wavelength can be related to their polarizability via (21)

$$C_{\text{ext}} = k \, \text{Im}\{\alpha\} + \frac{k^4}{6\pi} |\alpha|^2 \tag{2}$$

where k is the wavevector ($= 2\pi/\lambda$), Im denotes the imaginary part of α, and $|\alpha|^2$ denotes the square modulus of α. The first term on the r.h.s. of Eq. (2) is associated with absorption losses. The second term describes losses due to scattering. The complex dielectric function of a material capable of undergoing photon-induced electronic transitions can be described by the general Lorentz dispersion equation (46)

$$\varepsilon = 1 + \frac{N_e e^2}{m_e \varepsilon^\circ} \sum_j^n \frac{f_j}{\omega_{0,j}^2 - \omega^2 - i\delta_j \omega} \tag{3}$$

where N_e and m_e are the number density and mass of an electron, e and ε° are the electronic charge and permittivity of vacuum, respectively, and f_j is the oscillator strength of a given electronic transition. The spectral frequency and bandwidth (FWHM) of the jth electronic transition are given by ω_{0j} and δ_j. The frequency of the incident light is given by ω.

In the classical mechanical interpretation of Eq. (3), the resonance frequency ω_{0j} is equal to the square root of K/m_e, where K is the oscillator restoring force constant. For materials that contain free electrons (i.e., for which $K = 0$), one of the n resonance frequencies ω_0 is equal to zero. Thus Eq. (3) can be recast as

$$\varepsilon = \left(1 + \frac{N_e e^2}{m_e \varepsilon^\circ} \sum_j^{n-1} \frac{f_j}{\omega_{0,j}^2 - \omega^2 - i\delta_j \omega}\right) - \frac{(N_e e^2/m_e \varepsilon^\circ) f_F}{\omega(\omega + i\delta_F)} \tag{4}$$

Equation (4) describes well the frequency dependence of the complex dielectric function of metals. The first two terms on the r.h.s. describe the contribution of bound electrons to the dielectric function. The third term is identical to the frequency-dependent term in Drude's free-electron model (47) if we equate the numerator term $(N_e e^2/m_e \varepsilon^\circ) f_F$ with the square of the plasma frequency ω_p^2, and the damping factor δ_F with the reciprocal of the electron mean-free lifetime τ^{-1}.

For the present discussion, the key result of Eq. (4) is that the real dielectric function of metals takes on negative values above a certain wavelength. Figure 1, from Johnson and Christy (48), shows plots of the real and imaginary parts of the dielectric function of gold. In the case of gold and many other metals, the real component (ε') is negative, and the imaginary component (ε'') is small in the visible region of the spectrum. Considering now the polarizability function (Eq. 1), it is clear that α can become very large when the denominator is close to zero (i.e., when $\varepsilon_p = \varepsilon_m = -2$). The minimization of the denominator is often referred to as the *plasmon resonance condition*. The first curve in Fig. 2 shows the extinction cross section for a gold particle (radius = 5 nm) in vacuum.

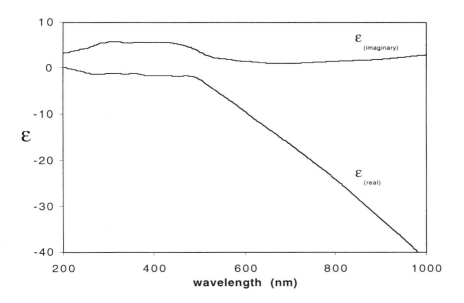

Fig. 1. Complex dielectric function of gold as a function of wavelength (based on Ref. 48).

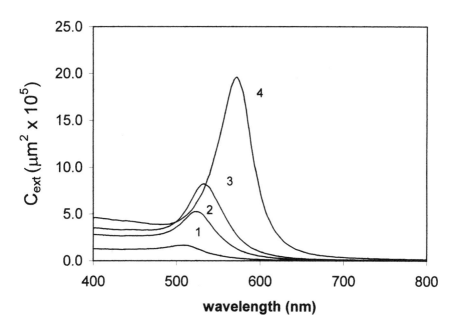

Fig. 2. Extinction spectra for gold particles calculated using Eqs. (1) and (2) and experimental optical constants (from Johnson and Christy). Curves 1–3: 5-nm radius Au sphere in vacuum (1), host dielectric = 1.8 (2), and host dielectric = 2.8 (3). Curve 4: oblate Au spheroid, rotational axis = 2 nm, radius = 8 nm, in host dielectric = 1.8. Electric field perpendicular to rotational axis.

While Eq. (1) pertains to the specific case of a sphere in vacuum, it can be generalized to particles of other shapes embedded in other host media (21,46):

$$\alpha = \frac{4\pi ab^2}{3q}\left(\frac{\varepsilon_m - \varepsilon_h}{\varepsilon_m + \kappa\varepsilon_h}\right) \tag{5}$$

In Eq. (5), a and b are the semiaxes of an ellipsoid of revolution, q and κ are shape factors, and ϵ_h is the dielectric function of the host medium.

Curves 2 and 3 in Fig. 2 are the extinction spectra calculated for a 5-nm Au sphere embedded in hosts with dielectric functions 1.8 and 2.8, respectively. As the host dielectric function is increased, the plasmon resonance condition is shifted to longer wavelengths. Curve 4 is a spectrum calculated for a nonspherical particle (in this case a squat disk with its rotational axis parallel to the propagation vector of the incident light). The plasmon resonance maximum can shift with changes in particle shape.

The spectra calculated in Fig. 2 represent the simplest case of isolated particles whose dimensions are very small relative to the incident wavelength. The polarizability expression (Eq. 5) used in these calculations is also the foundation for many theoretical treatments, such as Maxwell-Garnett theory, that attempt to model interacting ensembles of metal nanoparticles (49).

Note that Eq. (5) describes only electric dipole induction, not higher-order electric and magnetic induction modes, which become important as the particle dimensions increase relative to the incident wavelength (21,46). Nearly a century ago, Mie developed a theory to address higher multipoles in isolated spheres. However, particles of other shapes are more difficult to treat within the rigorous Mie context, and interparticle interactions for systems that involve anything beyond an electric dipole require very sophisticated treatments. Needless to say, the relevance of such theoretical treatments increases as more complex structures are achieved in experiment.

IV. ELECTRON TRANSPORT IN METAL NANOPARTICLES

More recently, the electronic properties of metal particles have been investigated within the context of decreasing electronic device size features to the nanoscopic level (5). Applications of individual particles as computer transistors, electrometers, chemical sensors, and in wireless electronic logic and memory schemes have been described and in some cases demonstrated (50), albeit somewhat crudely at this point.

Many of these studies have revealed that electronic devices based on nanoscopic objects (e.g., metal and semiconductor nanoparticles, molecules, carbon nanotubes, etc.) will not function analogously to their macroscopic counterparts. Thus, a conventional MOSFET (metal oxide semiconductor field effect transistor) will no longer be able to control the flow of electrons as its size reaches the sub-50-nm regime. At these dimensions, electron transport in n- and p-doped contacts is affected by the quantum mechanical probability that electrons simply tunnel through the interface. These tunneling processes will begin to dominate in the nanometer size regime, causing errors in electronic data storage and manipulation.

A second problem inherent in any nanoscale device is that chemical heterogeneities will influence device properties such as turn-on voltage. Defects, size dispersity, and variable dopant densities, normally of little concern in macroscale devices, can cause fluctuations in electronic function and make device reproducibility unlikely on the nanoscale. In fact, even a single pentagon-heptagon defect in a single-walled carbon tube can change I–V response (51). These seemingly insurmountable obstacles to fabricating nanoscale electrical devices in many ways form the genesis of research into new methods for synthesizing size monodisperse and chemically tailorable metal nanoparticles. Establishing basic nanoparticle size and surface chemistry-electronic function relationships in these materials is at the forefront of current nanoscale electronics research. The identification of novel electronic behaviors and device applications which capitalize on quantum effects is expected to follow from fundamental structure-function determinations. These are discussed in more detail below.

One electronic behavior observed in nanoscale objects is single-electron tunneling—the correlated transfer of electrons one-by-one through the object. Single-electron tunneling was first hypothesized in the early 1950s (52), a time when many physicists pondered how the electronic properties of a material (e.g., a metal wire) would change as material dimensions were reduced to the micron or nanometer scale. Gorter and others argued that, provided the energy to charge a metal with a single electron, $e/2C$ (e is electron charge, C is metal capacitance), was larger than kT, electrons would be forced to flow through the metal in discrete integer amounts rather than in fluid-like quantities normally associated with transport in macroscopic materials. Further reasoning led to the prediction that current-voltage *(I–V)* curves of a nanoscopic metal should be distinctly nonohmic; that is, current steps should appear corresponding to the transport of $1e^-$, $2e^-$, $3e^-$, etc., currents through the metal (Fig. 3A).

In fact, these predictions turned out to be true, although it was not until the late 1980s that well-defined single-electron tunneling steps were observed experimentally. Even then, enthusiasm for single-electron devices was tempered by the fact that these initial experiments were performed on relatively large metal islands (micron sized) prepared with photolithography or metal evaporation (Fig. 3B) (53). Thus, in order to satisfy the requirement $e/2C > kT$, it was necessary to cool the microstructures to below 1 K. Herein lies perhaps the greatest obstacle to implementing single-electron devices: to avoid thermally induced tunneling

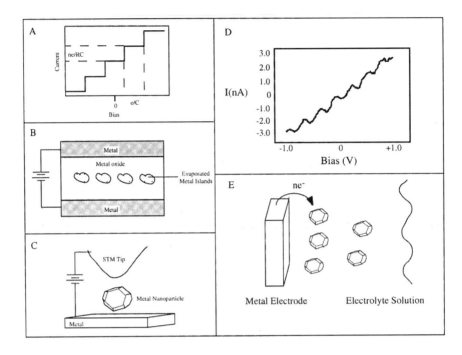

Fig. 3. Idealized single-electron tunneling *I–V* curve (A); sandwich metal/insulator/nanoscopic metal (particle)/insulator/metal (substrate) double-tunnel junction configuration (B); single-metal-particle configuration (C); typical *I–V* curve for configuration shown in C (D); and solution-phase configuration (E).

processes at room temperature, the metal island of any single-electron device must be less than 10 nm in diameter. This dimension is difficult to reach with electron beam lithography or scanning probe microscopies, but is now easily attained by chemists using solution-phase nanoparticle synthesis methods.

The realization that chemical synthesis is an ideal way to obtain large numbers of potential nanoscale device components prompted chemists and physicists to initiate research programs aimed at elucidating the electronic properties of metal particles. Much of this work has focused on gold and silver particles because synthetic methods for producing these particles of virtually any size are well developed. In addition, gold and silver surfaces (even surface areas afforded by a particle as small as 1.4 nm) can be modified with polymers (54), ceramics (55), alkythiols (56), enzymes (57), proteins (58), etc., to tune particle solubility, reactivity, optical extinctions, refractive index, and electron-hopping barriers. Electrical behaviors have been measured for individual gold particles (5) and in two-dimensional nanoparticle arrays (57).

One concern in characterizing electron transport in individual nanoparticles is particle size dispersity. Since electrical charging behaviors of metal particles depend on size, any size dispersity will tend to "smear" out individual particle properties. Monodisperse collections of gold particles have been isolated and addressed electronically primarily via (i) an STM tip to contact a single particle, or (ii) fractional crystallization to isolate highly pure samples of size monodisperse particles, followed by an ensemble average electronic measurement (e.g., electrochemistry, solid-state current-voltage measurements). In STM experiments, ligand-capped nanoparticles are cast onto metallic substrates and the tip is positioned directly over a single particle to form a metal (tip)/insulator (ligand)/nanoscopic metal (particle)/insulator (ligand)/metal (substrate) double-tunnel junction (Fig. 3C). Because gold particles with diameters as small as ca. 2 nm behave as free-electron metals (e.g., contain a continuum of electronic states), this system can be treated as a simple series RC circuit. Staircase-shaped I–V curves are then expected with voltage plateau widths of

$$V = \frac{(Q_0 - 1/2)e}{C_2 + V_{\text{offset}}} \tag{6}$$

and current steps of

$$I = \frac{e}{2R_2 C_T} \tag{7}$$

where Q_0 is the charge on the particle, C_2 and R_2 are the capacitance and resistance, respectively, of the most resistive junction (typically the particle-substrate junction), C_T is the total particle capacitance, and V_{offset} accounts for any initial misalignment in tip-particle or particle-substrate Fermi levels and any charged impurities residing near the particle. Sample data of the I–V behavior of the system are shown in Fig. 3D. Similar data have been reported by other groups.

Single-electron tunneling may also be observed in parallel circuits of gold nanoparticles, provided particle size dispersity is small. This has been demonstrated in the solid state by Murray's group, Heath's group, and others. In parallel particle systems, Eqs. (6) and (7) hold, except that the current scales by the number of particles in the array. Murray's group has observed Coulomb staircase behavior in parallel arrays, using solution-phase electrochemical experiments (Fig. 3E). Using differential pulse voltammetry, ca. 10 electron charging waves were detected for 1.64-nm-diameter clusters over a 1-V window in 2:1 toluene:acetonitrile. In solution, charging waves appear at formal potentials given by

$$E^0_{Q,Q-1} = E_{\text{PZC}} + \frac{(Q \pm 1/2)e}{C_T} \tag{8}$$

where $E^0_{Q,Q-1}$ is the formal potential of the $Q/(Q \pm 1)$ charge state, E_{PZC} is the potential of zero charge of the cluster, and Q and C_T were described earlier. [Note Eqs. (6) and (8) are essentially identical, with E_{PZC} being the electrochemical equivalent of V_{offset}].

Equations (6)–(8) have been used to determine particle capacitance and junction resistance experimentally as a function of ligand shell, solvent, and pH. These studies are better defining the sensitivity of electron transport in metal particles to a variety of environmental factors; an important consideration given the fact that wiring up and integrating particles together to form more complex architectures will likely involve chemical assembly. In one recent STM experiment, the Coulomb staircase was used to calculate the resistance of only a few p-xylene-$\alpha\alpha'$-dithiol molecules bound between the particle and substrate (5). A similar STM experiment on single particles was also recently performed in solution, where reagents were used to manipulate the charge state of pH-responsive ligands bound to the particle surface (58). Neutral to anionic conversion of the ligands was found to shift the Coulomb staircase and change particle capacitance predictably through the V_{offset} term in Eq. (6).

Solvent effects on single-electron charging have been explored by using differential pulse voltammetry. Murray found that formal potentials for successive single-electron charging events are solvent independent when the ligand shell on gold nanoclusters was a tightly packed monolayer of hexanethiolate. However, Feldheim and co-workers have found a strong solvent dependence when the capping ligand is a less densely packed layer of triphenylphosphine (see Chapter 13). These results suggest that particle capacitance (charging energies) is influenced strongly by the ability of surrounding molecules (solvent) to penetrate the ligand shell.

V. OVERVIEW OF THE FOLLOWING CHAPTERS

We attempted to assemble a broad sampling of research in metal nanoparticles which would cover synthesis, physical properties, and applications. We begin with two contributions that come from the *metal nanoparticle as atomic cluster* context. In Chapter 2, Richard Finke discusses the bulk solution phase synthesis of metal clusters and the influence of anionic ligands on their size and properties. In Chapter 3, Boon Teo and Hong Zhang introduce the concept of magic numbers, which pertains to the number of atoms in a cluster that nature often seems to prefer. In Chapter 4, George Schatz and colleagues review the theory of the optical properties of metal nanoparticles and present spectral simulations of complex systems not amenable to the simple theory introduced in this chapter. In Chapter 5, Colby Foss describes the electrochemical template synthesis method for metal nanoparticles, including noncentrosymmetric nanoparticle pairs that show second harmonic generation (SHG) activity. In Chapter 6, Robert Johnson and Joseph Hupp review their recent work on hyper-Rayleigh scattering, which is another second-order nonlinear optical technique that provides insight into the symmetry of nanoparticle assemblies in solution. Chris Wang, in Chapter 7, discusses the electrochemical synthesis of rodlike gold nanoparticles in surfactant solutions and rationalizes the optical spectra of these rods at the level of theory described in Sec. III. In Chapter 8, Andrew Lyon and Michael Natan describe applications of self-assembled colloidal gold films and surface plasmon effects to bioanalytical chemistry. Marie Pileni then provides a very detailed look at nanocrystal synthesis and assembly on a variety of length scales from isolated crystals to 2D and 3D nanocrystal superlattices (Chapter 9). These extended nanocrystal solids are important in establishing collective electrical and optical phenomena on the nanoscale. In Chapter 10, Reg Penner and colleagues take a quantitative look into the electrodeposition of metal nanostructures on graphite and silicon surfaces. Penner's group has shown very elegantly how important are proximity effects in determining the growth of nanostructures by electrodeposition. Richard Crooks and his group describe the synthesis of metal nanoparticles in dendrimer hosts in Chapter 11. This new class of encapsulated nanoscale materials has potential applications in various fields, including nanoelectronics to heterogeneous catalysis. Finally, Chapters 12 and 13 pertain to the electronic properties of ligand-capped gold nanoclusters. Royce Murray and co-workers provide a detailed analysis of electron charging of gold nanoclusters by solution-phase electrochemical techniques in Chapter 12. In Chapter 13, Dan Feldheim and colleagues review how the capping ligand can affect particle charging energies in experiments performed on individual clusters.

REFERENCES

1. JM Lehn. Supramolecular Ensembles. New York: VCH, 1995.
2. CB Gorman, BL Parkhurst, K-Y Chen, WY Su. J. Am. Chem. Soc. 199:1141–1142, 1997.
3. TW Ebbesen, HJ Lezec, H Hiura, JW Bennett, HF Ghaemi, T Thio. Nature 382:54–56, 1996.
4. B Alperson, S Cohen, I Rubinstein, G Hodes. Phys. Rev. B 52: R17017-R17020, 1995.
5. RP Andres, T Bein, M Dorogi, S Feng, JI Henderson, CP Kubiak, W Mahoney, RG Osifchin, R Reifenberger. Science 272:1323–1325, 1996.
6. C Flytzanis, F Hache, MC Klein, D Richard, Ph Roussignol. Nonlinear optics in composite materials. In: E Wolf, ed. Progress in Optics. Amsterdam: North-Holland, 1991, Vol. 29, pp 32–411.

7. (a) TR Jensen, GC Schatz, RP Van Duyne. J. Phys. Chem. B 103:2934, 1999. (b) BMI van der Zande, GJM Koper, HNW Lekkerkerker. J. Phys. Chem. B 103:5754, 1999. (c) NAF Al-Rawashdeh, ML Sandrock, CJ Seugling, CA Foss, Jr. J. Phys. Chem. B 102:361–371, 1998.
8. (a) U Kreibig, M Vollmer. Optical Properties of Small Metal Clusters. Berlin: Springer, 1995. (b) RW Cohen, GD Cody, MD Coutts, B Abeles. Phys. Rev. B 15:3689, 1973.
9. S Chen, RW Murray, SW Feldberg. J. Phys. Chem. B 102:9898, 1998.
10. (a) M Brust, M Walker, D Bethell, D Schiffrin, R Whyman. J. Chem. Soc. Chem. Commun. 801, 1994. (b) DV Leff, PC Ohara, JR Heath, WM Gelbart, J. Phys. Chem. 99:7036, 1995.
11. (a) CA Mirkin, RL Letsinger, RC Mucic, JJ Storhoff. Nature 382:607–609, 1996. (b) CD Keating, KM Kovaleski, MJ Natan. J. Phys. Chem. B 102:940, 1998.
12. JA Moulijn, PWNM van Leeuwen, RA van Santen. Catalysis: An Integrated Approach to Homogeneous, Heterogeneous and Industrial Catalysis. Amsterdam: Elsevier, 1993.
13. L Lee, G Seddon, F Stephens. Stained Glass. New York: Crown, 1976.
14. M Jose-Yacaman, L Rendon, J Arenas, MCS Puche. Science 273:223, 1996.
15. (a) F Mehlman. Phaidon Guide to Glass. Englewood Cliffs: Prentice-Hall, 1983. (b) NH Moore, Old Glass. New York: Tudor, 1935.
16. P Turner. History of Photography. New York: Exeter, 1987.
17. LP Willimas, ed. The Selected Correspondence of Michael Faraday. London: Cambridge University Press, 1971, Vol. 2.
18. M Faraday. Philos. Trans. 147:145, 1857.
19. JC Maxwell-Garnett. Philos. Trans. R. Soc. A 203:385, 1904.
20. G Mie. Ann Phys. 25:377, 1908.
21. HC van de Hulst. Light Scattering by Small Particles. New York: Dover, 1981.
22. S Asano, M Sato. Appl. Optics 19:962–974, 1980.
23. HB Weiser. Inorganic Colloid Chemistry. New York: Wiley, 1933, Vol. 1.
24. AB Searle. The Use of Colloids in Health and Disease. New York: E.P. Dutton, New York, 1919.
25. DA Handley. In MA Hayat, ed. Colloidal Gold. San Diego: Academic Press, 1989, Vol. 1.
26. (a) JE Beesley. Proc. R. Microscop. Soc. 20:187, 1985. (b) MA Hayat, ed. Colloidal Gold: Principles, Methods, and Applications. San Diego: Academic Press, 1989, Vol. 1.
27. DR Gabe. Principles of Metal Surface Treatment and Protection. 2nd ed. Oxford: Pergamon, 1978.
28. DGW Goad, M Moskovits. J. Appl. Phys. 49:2929, 1978.
29. A Andersson, O Hunderi, CG Granqvist. J. Appl Phys. 51:754, 1980.
30. (a) GA Niklasson. Solar Energy Mater 17:217, 1988. (b) TS Sathiaraj, R Thangaraj, H Al-Shabarty, OP Agnihotri. Thin Solid Films 195:33, 1991.
31. (a) M Fleischmann, PJ Hendra, AJ McQuillan. Chem. Phys. Lett. 26:163, 1974. (b) AJ McQuillan, PJ Hendra, M Fleischmann. J. Electroanal. Chem. 65:933, 1975. (c) JP Jeanmarie, RP van Duyne, J. Electroanal. Chem. 84:1, 1977.
32. M Moskovits. J. Chem. Phys. 69:4159, 1978.

33. (a) RK Chang, TE Furtak, eds. Surface Enhanced Raman Scattering. New York: Plenum Press, 1982. (b) M Kerker, ed. Selected Papers on Surface-Enhanced Raman Scattering. Bellingham: SPIE Optical Engineering, 1990. (c) PF Liao. In Ref. 32(a), pp 379–390. (d) MC Buncick, RJ Warmack, TC Ferrell. J. Opt. Soc. Am. B 4:927?, 1988.
34. (a) M Meier, A Wokaun. Optics Lett. 8:851, 1983. (b) EJ Zeman, GC Schatz. J. Phys. Chem. 91:634, 1987. (c) PW Barber, RK Chang, H Massoudi. Phys. Rev. B 27:7251, 1983.
35. A Henglein. Chem. Rev. 89:1861, 1989.
36. A Henglein. Top. Curr. Chem. 143:113, 1988.
37. M Gratzel. Energy Resources Through Photochemistry and Catalysis. New York: Academic Press, 1983.
38. G Schmid. Chem. Rev. 92:1709, 1992.
39. Size quantization phenomena in metal particles were first considered four decades ago. See, for example, M Doyle. Phys. Rev. 111:1067, 1958.
40. (a) GA Ozin, SA Mitchell. Angew. Chem. Int. Ed. Engl. 22:674, 1983. (b) MD Morse. Chem. Rev. 86:109, 1986. (c) BK Teo, KA Keating, YH Kao. J. Am. Chem. Soc. 109:3494, 1987.
41. P Mulvaney. In Electrochemistry in Colloids and Dispersions. RA Mackay, J Texter, eds. New York: VCH, 1992.
42. T Ung, M Giersig, D Dunstan, p Mulvaney. Langmuir 13:1773, 1997.
43. S Chen, RW Murray. Langmuir 15:682, 1999.
44. CJ Keily, J Fink, M Brust, D Bethell, DJ Schiffrin. Nature 396:444, 1998.
45. MP Pileni. Langmuir 13:3266, 1997.
46. CF Bohren, DR Huffman. Absorption and Scattering of Light by Small Particles. New York: Wiley, 1983.
47. RE Hummel. Electronic Properties of Materials. Berlin: Springer-Verlag, 1993.
48. PB Johnson, RW Christy. Phys. Rev. B Condens. Matter 6:4370, 1972.
49. D Aspnes. Thin Solid Films 89:249, 1982.
50. DL Feldheim, CD Keating. Chem. Soc. Rev. 27:1–12, 1998.
51. PG Collins, A Zettl, H Bando, A Thess, RE Smalley. Science 278:100–103, 1997.
52. CJ Gorter. Physica 17:777, 1951.
53. (a) BJ Barner, ST Ruggiero. Phys. Rev. Lett. 59:807, 1987. (b) M Amman, R Wilkins, E Ben-Jacob, PD Maker, RC Jaklevic. Phys. Rev. B 43:1146, 1991. (c) TA Fulton, G Dolan. Phys. Rev. Lett. 59:109, 1987.
54. SM Marinakos, JP Novak, LC Brousseau, J Feldhaus, AB House, DL Feldheim, J. Am. Chem. Soc. 121:8518, 1999.
55. M Giersig, T Ung, LM Liz-Marzan, P Mulvaney. Adv. Mater. 9:575, 1997.
56. M Brust, M Walker, D Bethell, DJ Schiffrin, RJ Whyman. J. Chem. Comm. Chem. Commun. 801, 1994.
57. (a) RS Ingram, MJ Hostetler, RW Murray, TG Schaaff, T Khoury, RL Whetten, TP Bigioni, DK Guthrie, PN First. J. Am. Chem. Soc. 119:9272, 1997. (b) G Markovich, DV Leff, S-W Chung, JR Heath. Appl. Phys. Lett. 70:3107, 1997. (c) H Wohitjen, AW Snow. Anal. Chem. 70:2856, 1998.
58. LC Brousseau, Q Zhao, DA Shultz, DL Feldheim. J. Am. Chem. Soc. 120: 7645–7646, 1998.

2
Transition-Metal Nanoclusters

Solution-Phase Synthesis,
Then Characterization and Mechanism
of Formation, of Polyoxoanion- and
Tetrabutylammonium-Stabilized Nanoclusters

Richard G. Finke
Colorado State University, Fort Collins, Colorado

I. INTRODUCTION

A. General Introduction and Key Definitions

Nanoclusters (1) are those "strange morsels of matter" (2) about 1–10 nm (10–100 Å) in size. They are of considerable current interest, both fundamentally and for their possible applications in catalysis, in nanobased chemical sensors, as light-emitting diodes, in "quantum computers," or other molecular electronic devices. Additional possible applications of nanoclusters are in ferrofluids for cell separations or in optical, electronic, or magnetic devices constructed via a building-block, "bottoms-up" approach [for lead references to these topics see Weller's recent review (1n), Refs. 1–13 elsewhere (3), as well as Chapter 1 and the other chapters in this volume). Our own main interest has been in transition-metal nanoclusters and their applications in catalysis (3,4).

It is important to distinguish *modern nanoclusters* from at least *traditional colloids*, and this is done in Figs. 1 and 2. As these figures summarize, it is the control over the composition, size, surface-ligating anions and other ligands, and, therefore, control over the desired properties that distinguish modern nanoclusters from their older, colloidal congeners.

Nanoclusters:

By definition should be or have:

- 1-10 nm (10-100 Å)
- Well defined composition (best examples)
- ≤ 15% size dispersion ("near-monodisperse")
- Reproducible syntheses (control over size, shape and composition)
- Reproducible (≤ ±15-20%) catalytic activity or other physical properties
- Isolable, redissolvable ("bottleable")
- Organic solvent soluble (or other control over solubility)
- Clean surfaces (no X^-, O^{2-}, OH^-, H_2O, or polymers)

Examples [4]:

Moiseev's $[Pd_{\sim570}(phen)_{\sim69}(OAc)_{\sim190}]^{x-}$

Finke's $Ir(0)_{\sim300}(P_2W_{15}Nb_3O_{62})_{\sim66}(Bu_4N)_{\sim300}Na_{\sim238}$

Reetz's $[Pd_bBr_a]^{x-}[N(octyl)_4]_x^{x+}$

Fig. 1. Definition of nanoclusters plus three prototype examples.

Some additional, important definitions for what follows are *near-monodisperse nanoclusters* (4) [those with ≤±15% size distributions, typically determined by transmission electron microscopy (TEM)] and *magic number nanoclusters* (i.e., full-shell, and thus extra-stability, nanoclusters), M_{13}, M_{55}, M_{147}, M_{309}, M_{561}, M_{923} as discussed in Teo's work on this topic (see the references in Refs. 3 and 4).

B. Key Research Goals in Modern Nanocluster Science

Some of the initial, key research goals in modern nanocluster science are summarized in Fig. 3, with some bias toward our own interests in nanocluster-based catalysts. However, general to all of nanocluster science are the key goals of rational, reproducible nanocluster syntheses with control over their size, shape, size and shape dispersity, surface and overall composition, and, therefore, control over their resultant optical, electronic, magnetic, catalytic, and other physi-

Traditional Colloids

- Are typically larger, > 100 Å

- Broader, >> 15% size dispersion

- Poorly defined molecular composition, e.g., $[Ir_aH_bO_c(OR)_dCl_e(poly(vinyl\ alcohol))_f]$ (R=R, H)

- Are not isolable, redissolvable

- Not reproducibly prepared

- Irreproducible catalytic activity, often ≥ 500% [a]

- Contain surface-bound, catalysis-rate-inhibiting X^-, O^{2-}, OH^-, H_2O, etc.

- Historically are H_2O, but not organic solvent, soluble

(a) For a 500% rate variation for the photoreduction of CO_2 by 10 different batches of Pd_n colloids, see: I Wilner, D Mendler. J. Am. Chem. Soc. 111: 1330, 1989.

Fig. 2. Definition of traditional (nano and other) colloids, as opposed to nanoclusters.

cal properties. Another main goal, with profound implications for the ease of use and reproducible physical properties of nanoclusters, is that of obtaining "bottleable" nanoclusters. That is, one needs to be able to isolate, then redissolve and otherwise use at will, identical samples of the resultant nanoclusters, all with the convenience and reproducibility that any bottleable chemical species offers. (Imagine, for a moment, the added time and complexity chemistry would engender if one had to freshly synthesize and purify each and every solvent and other reagent used in each and every synthesis rather than simply "cracking open," when needed, a bottle of fresh, certified reagent.) Having the nanoclusters available on a reasonably large, convenient scale—the current "scale-up issue"—is another important, general goal in nanocluster and other materials science.

Key Research Goals in Modern Nanocluster Science Include:

- Rational, reproducible syntheses

- Syntheses yielding near monodisperse, bottleable nanoclusters (± ≤15% size range)

- Compositionally well-defined nanoclusters

- Nanoclusters with well-defined *surface* composition

- Shape as well as size controlled nanoclusters

- Isolable, redissolvable, well-characterized nanoclusters

- Scaled-up syntheses of nanoclusters (≥ 1 gram scales)

- Reproducible (± ≤15%) catalytic activity or other physical properties

- High catalytic activity in solution, with long lifetimes (or long device lifetime with reproducible performance)

- High—"single-site"—catalytic selectivity

- Mechanistic information (e.g., mechanisms of nanocluster formation, catalytic reactions, or device operation)

- Understanding of why "magic number" (= full shell) nanoclusters form (M_{13}, M_{55}, M_{147}, M_{309}, M_{561}, M_{923}...)

- Hetero bi-, tri- and higher-metallic nanoclusters of known composition and structure

- 1-D (wires), 2-D and 3-D assemblies, superstructures and devices based on nanocluster building blocks

Fig. 3. A list of some key research goals in modern nanocluster science.

C. Nanocluster Electrostatic (charge, or "inorganic") and Steric ("organic") Stabilization Analogous to That Well Known for Colloids

The classical literature of colloid stabilization teaches that colloids are stabilized against agglomeration by the adsorption of anions (e.g., Cl^-, citrate^{3-}, others) and polymers [e.g., polyvinylpyrrolidone (PVP), polyvinylalcohol (PVA), others] onto their surfaces resulting in electrostatic (Coulombic repulsion) or steric (polymeric, or other organic "overcoat") stabilization effects (5), as illustrated schematically in Fig. 4.

A feature of the above literature and Fig. 4a that causes confusion is the +++ surface charge shown on the metal particle. At least in the case of $M(0)_n$ nanoclusters with an uncharged central core, it is *not* a full positive charge but rather a $\delta^+\delta^+\delta^+$ partial charge from an *electrostatic charge mirror* produced by the adsorption of X^- anions (6) to the coordinatively unsaturated, electron-deficient metal surface. The resulting particles are rendered anionic, $M(0)_n \cdot X_m^{m-}$. Hence, similarly charged colloidal particles electrostatically repel each other via an anionic, charge-based *kinetic stabilization* toward aggregation. The countercations necessary for charge balance, plus more anions, are also present in what is closely analogous to the electrical "double layer" (actually a multilayer) at an electrode surface (6).

Also worth noting at this point is that such stabilization of at least transition-metal colloids and nanoclusters is a *totally kinetic phenomenon*. That this is true can be seen by realizing that the conversion of, for example, $Ir(0)_n$ to n Ir(0) atoms requires a heat of vaporization of 159 kcal/mol. Hence (see Fig. 5), the most stable form of the metal is as bulk metal with its lowest possible surface area.

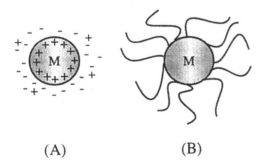

(A)   (B)

Fig. 4. Schematic for (A) an electrostatically stabilized metal (M) particle (i.e., one stabilized by the adsorption of ions and the resultant electrical double layer) and (B) a sterically stabilized metal particle (i.e., one stabilized by the adsorption of polymer chains).

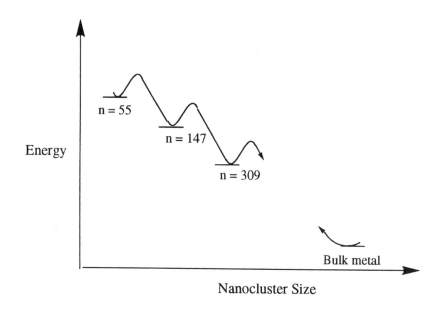

Fig. 5. Schematic of the expected continuous decrease in energy as a function of nanocluster size out to the limit of bulk metal. Local minima are shown at each closed shell, magic-number size, and thus extra-stability nanocluster.

Figure 5 also illustrates that one expects local minima at full-shell, "magic number" structures where each metal atom is maximally coordinated to other metals via its maximum number of metal-metal bonds, each Ir-Ir bond being worth ca. 159/(12/2) or 28 kcal/mol [see footnote 10 elsewhere for further discussion and derivation of this approximate Ir-Ir bond dissociation energy (4)].

The preceding discussion also makes apparent the importance of other nanocluster stabilization schemes, for example capping nanoclusters with groups such as RS^- or completely covering and thereby encapsulating them with SiO_2. Although these are not suitable strategies for nanocluster stabilization in cases where the surface needs to be accessible or readily modifiable (e.g., as in our own case of using nanoclusters in catalysis), they are very important strategies for other applications of nanoclusters.

D. Introduction to Polyoxoanions and the Prototype Nanocluster-Stabilizing Polyoxoanion, $P_2W_{15}Nb_3O_{62}^{9-}$

Of central importance to the reproducibly prepared, compositionally well-defined, isolable, yet redissolvable, and high stability—plus record solution cat-

Transition-Metal Nanoclusters

alytic lifetime (7)—nanoclusters in this chapter is the custom-made (8,9) polyoxoanion (10,11), $P_2W_{15}Nb_3O_{62}^{9-}$. Polyoxoanions can be defined as those species of compositions $M_bO_c^{n-}$ or $X_aM_bO_c^{m-}$ where, for example, X = P^V, Si^{IV}; M = W^{VI}, Mo^{VI}, Nb^V, V^V, and so on (from among many other possible elements and combinations of higher-valent metals and bridging plus terminal O^{2-} ligands) (10). While polyoxoanions perhaps appear esoteric to the reader not aware of them, polyoxoanions are actually a broad subclass of inorganic oxides that are discrete, readily made, readily modifiable, thermally robust, and oxidation resistant. Moreover, polyoxoanions (or, equivalently, polyoxometalates) have a broad generality: they are Nature's products from placing high-valent metals in Nature's solvent, H_2O, changing the pH, and then observing the myriad of possible polyoxoanion products that result. Polyoxoanions also have a broad a range of applications, Müller and Pope having noted that "polyoxometalates form a class of inorganic compounds that is unmatched in terms of molecular and electronic structural versatility, reactivity, and relevance to analytical chemistry, catalysis, biology, medicine, geochemistry, materials science, and topology" (11).

The polyoxoanion of significance to this chapter is the novel, custom-made $P_2W_{15}Nb_3O_{62}^{9-}$ polyoxoanion. It is available in ca. 116 g quantities from a couple of weeks of work and following an experimentally checked, *Inorganic Syntheses* procedure (9). Key to the synthesis is the preparation of the underlying, metastable synthon, $P_2W_{15}O_{56}^{12-}$, into which 3 Nb^{5+} are inserted in the presence of H^+ and H_2O_2; we have made recent improvements in the synthesis of the $P_2W_{15}O_{56}^{12-}$ which need to be followed in order to reproducibly obtain the purest $P_2W_{15}O_{56}^{12-}$ (9c). The resultant $(Bu_4N)_9[P_2W_{15}Nb_3O_{62}]$ then is readily converted in a single step involving the addition of $(1,5\text{-COD})Ir^+$ to the desired, discrete, fully characterized, "polyoxoanion-supported organometallic" nanocluster precursor, $(1,5\text{-COD})Ir \cdot P_2W_{15}Nb_3O_{62}^{8-}$ (9,12) (Fig. 6) (oxide-"supported" by analogy to solid-oxide-supported transition-metal catalysts). Of central importance for what follows is that the $[Bu_4N]_5Na_3[(1,5\text{-COD})Ir \cdot P_2W_{15}Nb_3O_{62}]$ *is a compositionally well-defined, pure, reproducibly prepared, structurally characterized, and thus atomically reproducible precursor for the (reproducible) preparation of polyoxoanion stabilized $Ir(0)_{\sim 300}$ nanoclusters*. Hence, much of the success of the story which follows can be traced back to our adherence, insofar as possible, to the rigorous standards and principles of smaller-molecule chemistry. For this reason, those standards and principles will be a dominant theme in this account of our work in making, characterizing, and studying the mechanism of formation of polyoxoanion-stabilized transition-metal nanoclusters.

The reader may be wondering at this point: "OK, but why polyoxoanions? Why the $P_2W_{15}Nb_3O_{62}^{9-}$ polyoxoanion?" These questions are answered in additional detail elsewhere (3,4,13), but briefly: (a) $P_2W_{15}M'_3O_{62}^{9-}$ (M' = Nb^V, V^V) is an uncommon type of polyoxoanion with *basic surface oxygens* (i.e., with anionic surface charge density), one deliberately designed and then custom-made to

Polyoxoanion-Supported Nanocluster Precursor, $[(1,5\text{-COD})M \cdot P_2W_{15}Nb_3O_{62}{}^{8-}]$ (M = Ir^I, Rh^I)

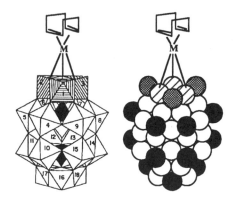

Synthesis: 9 Steps [9a]

Improved Synthesis [9b,c]
[2 Steps, 4 days (30%) shorter, 59% higher yield (116g vs. previous 73g)].

Characterization [12a-d]

Analysis (all elements: 100.18% total); ^{31}P, ^{17}O, ^{183}W, ^{13}C, ^1H NMR; MW (FAB-MS; solution MW); X-ray diffraction ($P_2W_{15}Nb_3O_{62}{}^{9-}$).

Fig. 6. The polyoxoanion-supported nanocluster precursor, $[(1,5\text{-COD})M \cdot P_2W_{15}Nb_3O_{62}{}^{8-}]$ (M = Ir^I, Rh^I), and its synthesis and full, atomic-level characterization.

be a good ligand for coordinating to, and thereby stabilizing, the $(1,5\text{-COD})Ir^+$ cation in the nanocluster precursor (and then, as it turns out, also the coordinatively unsaturated surface of clean transition-metal nanoclusters); and (b) the selection and use of $P_2W_{15}Nb_3O_{62}{}^{9-}$ was the result of up-front catalysis survey experiments plus five years of painstaking mechanistic studies solving the classic problem "is it homogeneous or heterogeneous catalysis?" (13), studies which led to the discovery of the true polyoxoanion- (and tetrabutylammonium-) stabilized nanocluster catalyst when beginning from supported organometallic precursors such as $(1,5\text{-COD})Ir \cdot P_2W_{15}Nb_3O_{62}{}^{8-}$ (13).

An important point for what follows is that $P_2W_{15}Nb_3O_{62}{}^{9-}$ is really a 12 × 15 Å size, surface 3− charge containing, six surface oxygen and thus chelating type of ligand. Hence, this polyoxoanion can be seen to be something like a "giant citrate^{3-} trianion equivalent" with respect to its more basic $Nb_3O_9^{3-}$-containing end and from the perspective of how nanoclusters are stabilized (vide supra). That there is formally only a 3− surface charge on the $P_2W_{15}Nb_3O_{62}^{9-}$ polyoxoanion can be recognized by rewriting this polyoxoanion as it structurally exists (Fig. 6) with its two central, templating phosphate trianion tetrahedra (in the leftmost polyhedral representation in Fig. 6), $\{(PO_4)_2^{6-}[W_{15}Nb_3O_{54}]^{3-}\}^{9-}$, a representation which allows one to see this polyoxoanion's trianionic, "$[W_{15}Nb_3O_{54}]^{3-}$" oxide surface. The "$[W_{15}Nb_3O_{54}]^{3-}$" fragment can be further deconvoluted conceptually into a formally trianionic "$[Nb_3O_9]^{3-}$" minisurface onto which $(1,5-COD)M^+$ (Ir, Rh) (12), $Ru(benzene)^{2+}$ (14a), $Rh(C_5Me_5)^+$ (14a), $Re(CO)_3^+$ (14b), and other organometallic cations bind. The structures of these complexes have these bound organometallic cations centered about the threefold axis of $P_2W_{15}Nb_3O_{62}^{9-}$, structures proven by X-ray crystallography [in the case of $Rh(C_5Me_5)_2^+$ (14a)] or by a combination of ^{17}O NMR, ^{183}W NMR, and IR for the other polyoxoanion-supported organometallic cations [e.g., see the M = Ir and Rh $(1,5-COD)M^+$ complexes in Ref. 12]. Note that the polyoxoanion is, nevertheless, a very large 9- anion, so two nanoclusters with multiple $P_2W_{15}Nb_3O_{62}^{9-}$ polyoxoanions affixed to their surfaces should experience a sizable electrostatic (Coulombic) repulsion and thus stabilization toward agglomeration; the significant steric repulsion of the associated bulky Bu_4N^+ (plus Na^+) countercations associated with the polyoxoanion are undoubtedly another reason for the relatively high stability toward agglomeration in solution of the resultant nanoclusters (vide infra, Fig. 9).

Note that, while also important for the nanocluster stabilization, the R_4N^+ countercations are not novel, having been introduced to the nanocluster area by Grätzel in a classic paper in 1979 (15a) and then expanded significantly subsequently by several workers, for example through Reetz's or Bönnemann's important efforts [see the references summarized elsewhere (3)]. Furthermore, there is no compelling evidence for a chemically implausible[1,2] direct adsorption of the

---

[1] The source of this apparent myth appears to be primarily threefold (and is discussed here since a referee raised this issue). First, there appears to be misunderstanding and miscitation of a 1988 paper studying SERS on very poorly compositionally characterized "$[(Ag(0))_a(Ag\ (surface)^+_b(X^-)_c(EDTA)_d)]^{b-c}$ $[Me_3NR^+]_{b-c}$" (X^- = an ill-defined, apparent mixture of Br^- and NO_3^- or ClO_4^- or deprotonated EDTA) (15b). This paper implies that the long-chain Me_3NR^+ (R = cetyl) is adsorbed directly to the Ag surface (see the misleading Fig. 3 in Ref. 15b). However, these authors actually say that the binding of a cationic (i.e., Me_3NR^+) surfactant to a cationic (Ag^+) surface "*must imply the intermediacy of. . . the Br^- counteranion*"; that is, they do not believe the implausible direct coordination of a cationic Me_3NR^+ to the cationic Ag^+ surface as, however, their Fig. 3 shows. The authors do provide zeta potential data showing that before $Me_3NR^+\ Br^-$ addition the nanocluster is anionic ($\zeta = -94$ mV), but is cationic after the addition of $Me_3NR^+Br^-$ ($\zeta = \geq +100$ mV). However, this evidence does *not* demand direct coordination of Me_3NR^+ to the metal's surface (something that is Coulombically uphill in any event); instead, it only requires that the sum of $Ag(surface)^+$ and Me_3NR^+ exceed the amount of surface-bound, anion present, X^-, in the species being detected by the zeta potential measurements.

R_4N^+ countercations to the electrophilic metal surfaces as others have written, at least for nonanionic, neutral-core, $M(0)_n$, transition-metal nanoclusters (see the discussion and references on this important point on pp. 21, 26, 36, and 37 of Ref. 4.) Hence, propagation of this apparent myth should be stopped until, and unless, direct, compelling evidence for a non-anion-mediated, direct R_4N^+ nanocluster interaction is forthcoming.

Returning to the $P_2W_{15}Nb_3O_{62}^{9-}$ polyoxoanion, the important points, then, for the present chapter are that (a) the nanoclusters are synthesized from a carefully prepared, atomically very precise precursor; and (b) the polyoxoanion can be viewed as a "giant citrate" type of nanocluster-stabilizing ligand on its $Nb_3O_9^{3-}$ end, an unusual ligand that binds to transition-metal nanocluster surfaces via its up to six total, basic, chelating three Nb—O—Nb bridging, and three Nb—O terminal, oxygen atoms. In a moment we will see that another valuable feature of the nanocluster precursor $(1,5-COD)Ir \cdot P_2W_{15}Nb_3O_{62}^{8-}$ is (c) that this precursor plus H_2 produces the desired nanoclusters, plus their stabilizing polyoxoanion and associated Bu_4N^+ and Na^+ countercations *only* (save the relatively inert by-products cyclooctane and cyclohexane; see Sec. II, vide infra).

II. THE FIRST STEP IN NANOPARTICLE RESEARCH. THE SYNTHESIS OF COMPOSITIONALLY-WELL-DEFINED, SIZE- AND SHAPE-SELECTED, NEAR-MONODISPERSE, AND "BOTTLEABLE" NANOCLUSTERS

Nanochemistry [one subdivision of "supramolecular chemistry" in the organic genre (16)] presents one sizeable, up-front, and ultimately all-important challenge

---

Second, few authors measure the overall charge on their nanocluster cores as we have (13,27) (and which include the coordinated anions, for example), so they are guessing about the true overall charge on their nanoclusters. Moreover, even if they had that overall charge, they would not be able to interpret it correctly since the complete anion (and cation) composition of their nanoclusters is typically unknown as well. Third, and ultimately, many authors continue to work with *compositionally poorly characterized nanoclusters* made by unbalanced reactions leading to nanoclusters of unknown compositions. This is a poor, dangerous practice, one which should be avoided in the future—a main message of the present chapter.

[2] One, of course, does not expect a R_4N^+ to coordinate to a coordinatively unsaturated, electrophilic metal in solution where anionic ligands and coordinating solvent are present as competing ligands. This follows since the most coordinating component of R_4N^+ (i.e., with its absence of any Lewis nonbonding pairs) is the poorly donating C—C and C—H σ bonds with their M ← RH BDEs (bond dissociation energies) of ≤ ca. 8–10 kcal/mol (32). On the other hand, solvent or anionic ligand BDEs are generally ≥15 to 30–40 kcal/mol or more, respectively (32). The solvent is also generally at a much higher concentration than the R_4N^+ as well. In short, it is hard to understand why some authors have implied that R_4N^+ are directly coordinated to electrophilic metal surfaces since this defies chemical logic (see the discussion and references provided on pp. 21, 26, 36, and 37 of Ref. 4).

to chemists and other materials scientists: the synthesis of the requisite nanoclusters or other materials that are as pure, as controlled compositionally and structurally, and as otherwise well defined as possible. Ideally, and insofar as possible, one would like to adhere to the rigorous standards of purity, composition, and structure held by small-molecule inorganic, organic, and organometallic chemists. This not easily achieved—and in some cases impossible—goal is, nevertheless, key: only with compositionally and otherwise well-defined nanoclusters or other materials can one avoid the insidious "garbage in, garbage out" syndrome; only with well-defined materials can one avoid composite results or misinterpretations of composite (and possibly artifactual) data. Moreover, any material that will eventually have commercial applications will very likely need to rise to the highest standards of purity, performance, and easy inexpensive synthesis (and in the highest yield in an environmentally "green" fashion). This leaves little room for inexact science along the way.[3]

Of course, as one's molecules or materials become of higher and higher molecular weight or exist in extended, noncrystalline structures, one must accept that some standards of small-molecule chemistry must be abandoned.[4] An illustrative example here is polyoxoanion fast-atom-bombardment mass spectra (FAB-MS). The FAB-MS of $(Bu_4N)_9P_2W_{15}Nb_3O_{62}$ with its MW of 6272 has an envelope width of ca. 40 m/z (see Fig. 3 in Ref. 17a) due primarily to the presence of 15 W atoms along with the 5 naturally occurring isotopes of W (17). [The case is only slightly improved with a ~15 m/z inherent envelope width in the mass spectrum of 10,000 MW polystyrene, In-$(C_8H_8)_{\sim 100}$-H (In-H initiator), with its ca. 800 carbons magnifying the effect of the relatively low, 1.1% natural abundance of ^{13}C(17c).] Clearly in these nano- or larger molecules (recall the $P_2W_{15}Nb_3O_{62}^{9-}$ polyoxoanion itself is 1.2 × 1.5 nm), a mass spectrum of even quite low, ±1.0 m/z resolution by small-molecule standards is *physically impossible*. Nevertheless

---

[3] Interestingly, Weller makes essentially the same point in his recent minireview (1n) covering nanosemiconductor materials (CdS, CdSe, InAs, InP, GaAs, or the CdS/HgS/CdS "nano-onions"), nanoclusters under intensive investigation, via an exploding literature, as building blocks for use in, eventually, molecular electronics. See especially the references cited therein to atomically precisely defined nanoclusters such as $[Cd_{17}S_4SPh_{28}]^{2-}$ or $[Cd_{32}S_{14}SPh_{36}DMF_4]$ (e.g., Refs. 10–14); to the use of micellar or reversed micellar block copolymers as molecular reactors to obtain some of the best, near-monodisperse nanoclusters known (ca. ±3–5% size dispersity); to the synthesis of gram quantities of good materials in the best cases; and to, for example, Whetten's impressive *crystalline* 2-D array of Ag or Au nanoclusters [Ref. 33 in Weller's review (1n)]. Of special interest is Weller's comment that the synthesis of large amounts of high-quality nanoclusters with a narrow size range ". . . is basically the prerequisite for large scale applications in modern materials science" (1n, p. 198).

[4] Another obvious but illustrative example is a comparison of the state-of-the-art standards in small-molecule versus protein crystallography: the R factors of ca. 0.02–0.08 in the former in comparison to the inherently larger, ca. 0.15–0.25, R factors in the case of the larger, more complex proteins.

FAB-MS of ≥6,000 MW polyoxoanions in comparison to calculated envelopes are quite valuable (17). In the case of mass spectrometry as applied to nanoclusters, the literature data presented elsewhere "demonstrate the power of mass spectrometry in nanocluster science and argues for its use in every applicable case" (3, p. 13). Hence, one still can, and must, strive toward rigorous standards in nanocluster science, standards that include adherence to the principles of complete stoichiometry (i.e., full mass and charge balance) in nanocluster syntheses and complete compositional characterization of the resultant materials.

A case in point here illustrates the difference between nanoclusters and the historically better known (nano)colloids. In careful work, Bradley has shown that attempts to make multiple batches of PVP-protected Pd_n nanocolloids from the following, typical (nano)colloid-synthesis reaction yielded compositionally-ill-defined $Pd_n \cdot Cl^-_a \cdot H_3O^+_b \cdot PVP_c \cdot H_2O_d (\cdot MeOH_e \cdot HCHO_f)$ in which the initial rates of catalytic hydrogenation varied by up to ca. 670% (18):

$$n\, H_2Pt^{IV}Cl_6 + 2n\, CH_3OH + \text{excess PVP}$$
$$\rightarrow 1/n Pt_n \cdot PVP_c + 2n\, \text{"HCHO"} \;(\text{not identified/quantitated}) \quad (1)$$
$$+ 6n\, H^+ + 6n\, Cl^-$$

Bradley notes that the formation of each ca. 35 Å, $Pt_{\sim 1500}$ particle is accompanied by the production of a large excess of ~9000 equiv. of Cl^- (i.e., in addition to that required to stabilize the nanocluster) as well as a large excess of ~9000 equiv. of H^+. These workers further showed that dialysis yielded a set of (nano)colloids that now gave *indistinguishable rates of hydrogenation catalysis* (with, by implication, very similar amounts of surface-modifying Cl^- or other ligands) (18); in a sense, via their careful efforts these workers have converted their nanocolloids into nano*clusters*. The case cited back in Fig. 2, in which 10 batches of Pd_n nanocolloids made in other work showed rates of photoreduction of CO_2 that varied by 500%, and the data in Lewis's review on colloidal catalysis (19), all show that the example in Eq. (1) is not the exception but is the more general rule: one needs to carefully control the synthesis and resultant composition of one's nanoclusters in order to obtain reproducible physical properties. Bradley's to-the-point remark in his scholarly review that "perhaps the most important irritant in colloid synthesis is irreproducibility" (20) further underscores the need for reproducible syntheses of compositionally-well-defined nanoclusters if one is to avoid irreproducibility problems of the type noted above for nano*colloids*.

To summarize, the important conceptual points of this section are fourfold (and apply more generally in the writer's opinion to other areas of materials chemistry): (a) only with carefully controlled syntheses of compositionally and structurally-well-defined nanoclusters can one avoid the insidious "garbage in, garbage out" syndrome and its associated problems; (b) the use in one's syntheses of the very simple, yet fundamentally powerful, concept of full stoichiometry with complete mass and charge balance is one of the main weapons toward avoiding subsequent problems; and (c) as the compositional complexity and difficulty in characterizing the resultant material increases, it becomes increasingly important *to use synthesis as the foremost tool to (pre)determine the composition and structure of the (synthesized) material*. This last point is, on one hand, profound, yet, on the other hand, is just a restatement of an old lesson of organic chemistry (i.e., organic chemistry before modern NMR and before the other methods for the rapid determination of purity and structure): in the old days, synthesis was among the main composition- and structure-determining tools (plus of course degradative chemistry). In addition, (d) mechanistic chemists know that a balanced reaction is the first requirement for reliable kinetic and mechanistic studies, a point that follows unequivocally since the underlying elementary steps must add up to the observed, net reaction. Hence, the absence of a balanced nanocluster formation reaction automatically precludes any type of reliable kinetic and mechanistic work. It is nano*molecular* chemistry that is the main goal (as opposed to nanocolloidal or nanomaterials chemistry)!

III. AN ILLUSTRATIVE CASE HISTORY: THE SYNTHESIS AND CHARACTERIZATION, THEN KINETICS AND MECHANISM OF FORMATION OF POLYOXOANION- AND TETRABUTYLAMMONIUM-STABILIZED Ir(0)$_{\sim 300}$ NANOCLUSTERS

A. The Synthesis and Full Characterization of the Nanocluster Precursor, $(Bu_4N)_5Na_3[(1,5\text{-}COD)Ir \cdot P_2W_{15}Nb_3O_{62}]$.

As introduced in Sec. I.D, the precursor for polyoxoanion-stablized Ir(0)$_n$ nanoclusters is the organometallic-polyoxoanion complex, $(Bu_4N)_5Na_3[(1,5\text{-}COD)Ir \cdot P_2W_{15}Nb_3O_{62}]$. It is reliably obtained on 4–12 g scales (9,10) [and potentially up to the 116 g scale of the underlying $P_2W_{15}Nb_3O_{62}^{9-}$ (9) if desired], so long as recent improvements in the synthesis of the underlying $P_2W_{15}O_{56}^{12-}$ (9c) and other details of the synthesis (9) are closely followed. The use of ^{31}P NMR handles deliberately built into this precursor proved absolutely essential to the success of the work, allowing, for example, the purity of the organometallic-polyoxoanion complex to be rapidly and *directly* determined each time it is prepared (9).

Note that this precursor is fully characterized at the atomic level by an all-elements elemental analysis, ^{31}P, ^{183}W, ^{17}O, ^{1}H and ^{13}C NMR, and IR, as well as FAB-MS and X-ray structural analysis for the underlying $P_2W_{15}Nb_3O_{62}^{9-}$. Our work synthesizing and fully characterizing this precursor complex goes back more than 15 years now (21), the original synthesis of $P_2W_{15}Nb_3O_{62}^{9-}$ appearing in 1988 (8). Those interested in inorganic synthesis might wish to read about the "six-month problem areas" that had to be overcome along the way (see footnote

15 in Ref. 21a)—that is, challenges in the synthesis and characterization of polyoxoanions of MWs $\geq$ 6000, and, as we did our best to adhere to the rigorous standards of smaller-molecule chemistry, challenges overcome only after the commitment of $\geq$six months of a postdoctoral fellow's research effort. To put these challenges into a different perspective, they are perhaps of a similar magnitude to the challenges involved in the full characterization of a small protein of similar, ~6000 molecular weight.

B. The Evolution under H_2 of the $(Bu_4N)_5Na_3[(1,5\text{-}COD)Ir \cdot P_2W_{15}Nb_3O_{62}]$ Precursor to Yield $Ir(0)_{\sim 300}$ Nanoclusters: An Example of a Well-Defined Nanocluster Synthesis Stoichiometry

Under 3.5 atm of H_2, in acetone solvent and in the presence of cyclohexene cohydrogenation to cyclohexane [not shown in Eq. (2) below], near monodisperse, 20 $\pm$ 3 Å $Ir(0)_{\sim 190-450}$ (hereafter $Ir(0)_{\sim 300}$)[5] nanoclusters are reproducibly formed according to the stoichiometry in Eq. (2) (13). Note that this synthesis and its balanced reaction produce only hydrocarbon by-products (cyclooctane plus cyclohexane) and avoid the possibility of Cl^-, polymer, and (since the reaction is done under dry N_2) O_2 or O ligands which could otherwise poison the catalytic surface of the nanocluster. A small amount of H_2O is produced by anhydride formation of the polyoxoanion, 2Nb—O^- (i.e., $2P_2W_{15}Nb_3O_{62}{}^{9-}$) + $2H^+$ $\rightarrow$ Nb¬O¬Nb (i.e., $P_4W_{30}Nb_6O_{123}{}^{16-}$) + H_2O, a reaction which serves the useful purpose of scavenging the H^+ side product. The presence of H_2O does not inhibit the catalytic activity; in fact, small amounts of water have a favorable, accelerating effect on catalysis as demonstrated elsewhere (13,22). The resulting nanoclusters can be isolated and then bottled for future use and in ca. 60% yield, although the synthesis of these particular $Ir(0)_{\sim 300}$ nanoclusters has not been scaled up and, hence, re-

---

[5] Note we have labeled our nanoclusters here and previously (3,4,13) with the indicated short-hand labels $Ir(0)_{\sim 300}$ [and $Ir(0)_{\sim 900}$ (27)] for convenience only; elsewhere the interested reader will find a discussion of the problems in the literature when others labeled, as atomically precise, what prove not to be exactly or only Au_{55} or Pd_{561} nanoclusters (3,4). Hence, the presence of only the atomically precise $Ir(0)_{300}$ or $Ir(0)_{900}$ is *not* what we mean with the use of the convenient labels $Ir(0)_{\sim 300}$ and $Ir(0)_{\sim 900}$. This point is perhaps obvious when one considers more closely the details of the production of even these near-monodisperse particles, a reaction that must have a mechanism with many more than 300 discrete steps, vide infra; that is, there is no precedent for a >300-step nanocluster-producing reaction yielding anything even close to an *exactly monodisperse* $Ir(0)_{300}$.

Moreover, as nanoclusters become larger, one expects less of a distinction in the thermodynamic stability of nanoclusters differing by even hundreds of atoms (see Refs. 20 and 21 in Ref. 28 for more on this point); hence, in the case of the 40 $\pm$ 6 Å $Rh(0)_{\sim 1500}$ to $Rh(0)_{\sim 3700}$ nanoclusters, we prefer not to use the above short-hand nomenclature [i.e., to not call these $Rh(0)_{\sim 2200}$], and have not used such a short-hand nomenclature in our publication describing polyoxoanion-stabilized $Rh(0)$ nanoclusters (28).

mains relatively small scale (13). We have, however, scaled up to 1 g the preparation of 38 ± 6 Å $Ir(0)_{\sim 2000}$ nanoclusters in propylene carbonate in some of our most recent work (23), a synthesis which also is performed without the need for co-hydrogenated cyclohexene.

$$300\ (1,5\text{-COD})Ir \cdot P_2W_{15}Nb_3O_{62}{}^{8-} + 750\ H_2 \rightarrow 300\ \bigcirc$$
$$+\ Ir(0)_{\sim 300} + 300\ [\underbrace{P_2W_{15}Nb_3O_{62}{}^{9-}}] + 300\ H^+ \quad (2)$$
$$150\{H_2O + [(P_2W_{15}Nb_3O_{61})_2\text{-O}]^{16-}\}$$

The $Ir(0)_{\sim 300}$ nanoclusters are among the best compositionally characterized nanoclusters in the literature, having been characterized by elemental analysis as well as TEM (Fig. 7), HRTEM, STM, electron diffraction, ultracentrifugation solution molecular-weight measurements (which also show the presence of the otherwise very difficult to detect Nb¬O¬Nb anhydride form of the polyoxoanion, $P_4W_{30}Nb_6O_{123}{}^{16-}$), electrophoresis, ion-exchange chromatography, IR spectroscopy, FAB-MS, and by H_2 uptake stoichiometry studies. Elemental analysis and solution molecular-weight measurements yield an average molecular formula of $[Ir(0)_{\sim 300}(P_4W_{30}Nb_6O_{123}{}^{16-})_{\sim 33}](Bu_4N)_{\sim 300}Na_{\sim 228}$. FAB-MS and IR spectroscopy confirm that the polyoxoanion stabilizing agent is intact in the nanocluster product. The larger $Ir(0)_{\sim 900}$ nanoclusters formed in the absence of cyclohexene (but still in acetone and under H_2) have been similarly well characterized (27).

The most important aspect of the complete mass balance is the H_2 uptake stoichiometry; it proves that the $Ir(0)_{\sim 300}$ core is *uncharged* (13,27), a point rarely examined, much less unequivocally demonstrated, in the nanocluster literature. Electron diffraction confirms that the metal core of the nanoclusters is composed of ccp Ir(0) metal. Significantly, electrophoresis and ion-exchange chromatography show that, in solution, the clusters *contain an overall negative charge*, despite the overall neutral Ir(0) core. This observation, when coupled with the evidence for the neutral Ir(0) core, demands that the negatively charged stabilizing polyoxoanion—the only anion present—*must be binding to, and thus is crucial in stabilizing, the Ir(0) nanoclusters in solution*. This result alone made the effort to establish the complete stoichiometry in Eq. (2) worthwhile.

A schematic view of an idealized, ccp $Ir(0)_{\sim 300}$ nanocluster and a portion of its stabilizing polyoxoanions is shown in Fig. 8a in which the polyoxoanion is represented in its monomeric, $P_2W_{15}Nb_3O_{62}{}^{9-}$ form, that is, the expected form when 1.0 equiv. of $Bu_4N^+OH^-$ is added to scavenge the H^+ produced [Eq. (2)] and to cleave any $P_4W_{30}Nb_6O_{123}{}^{16-}$ formed, cleavage which occurs via the known reaction (8) $\frac{1}{2}[\text{Nb-O-Nb} + 2OH^-] \rightarrow \frac{1}{2}[2\text{Nb-O}^- + H_2O]$ thereby reverting any $P_4W_{30}Nb_6O_{123}{}^{16-}$ to its monomeric, $P_2W_{15}Nb_3O_{62}{}^{9-}$ form. A portion of the diffuse layer of polyoxoanions that must be present is shown schematically in Fig. 8b. This diffuse layer of polyoxoanions and their associated

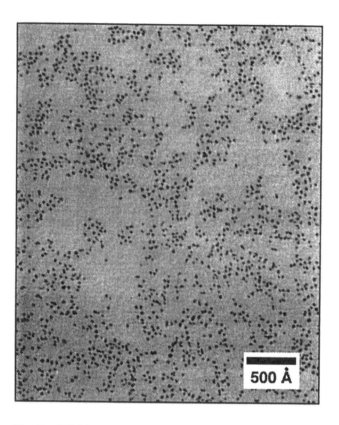

Fig. 7. TEM image (100K magnification) of near-monodisperse Ir(0)$_{\sim 300}$ nanoclusters grown by the reduction of 1.2 mM **1** in acetone, 3.5 atm H$_2$, room temperature, and our other, "standard conditions" detailed elsewhere (13) (reproduced from Ref. 27). Statistics on the nanoclusters in this image reveal 20 ± 3 Å (i.e., ±15%) diameter, and hence Ir(0)$_{\sim 300}$, nanoclusters. Although not visible in this image, the polyoxoanion component is clearly clustered around the Ir(0) nanoclusters as seen in Fig. 8 in Ref. 27.

countercations must exist since ca. 66 P$_2$W$_{15}$Nb$_3$O$_{62}$$^{9-}$ polyoxoanions are present in the isolated nanoclusters, yet only ca. 17 polyoxoanions maximum can fit about the surface of the 20 Å Ir(0)$_{\sim 300}$ nanocluster; see the calculations in footnote 43 in Ref. 27. This is yet another illustration of the value of knowing even the average molecular formula of the nanoclusters and their associated, stabilizing polyoxoanions, in this case from elemental analysis plus ultracentrifugation MW measurements.

The finding that the catalytic activity of the nanoclusters is inhibited by additional P$_2$W$_{15}$Nb$_3$O$_{62}$$^{9-}$ (13,22) confirms (a) that the polyoxoanion binds to

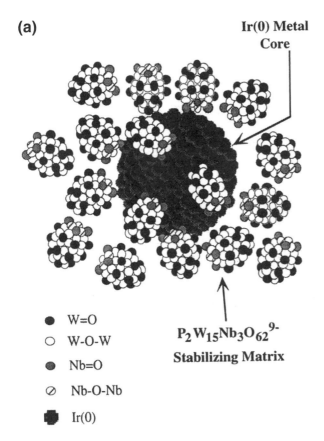

Fig. 8a. $P_2W_{15}Nb_3O_{62}^{9-}$ polyoxoanion inner and outer (Ir Coordination) sphere. (a) An idealized, roughly-to-scale representation of a $P_2W_{15}Nb_3O_{62}^{9-}$ polyoxoanion and Bu_4N^+ stabilized 20 ± 3 Å $Ir(0)_{\sim300}$ nanocluster, $[Ir(0)_{\sim300}(P_4W_{30}Nb_6O_{123}^{16-})_{\sim33}](Bu_4N)_{\sim300}Na_{\sim228}$. For the sake of clarity, only 17 of the polyoxoanions are shown, and the polyoxoanion is shown in its monomeric, $P_2W_{15}Nb_3O_{62}^{9-}$ form (and not as its Nb-O-Nb bridged, anhydride, $P_4W_{30}Nb_6O_{123}^{16-}$ form). The ~330 Bu_4N^+ and ~228 Na^+ cations have also been omitted, again for the sake of clarity.

the surface of the nanocluster (i.e., where the catalysis, and thus the inhibition, takes place), and (b) that the surface of the isolated nanoclusters is not completely packed with polyoxoanions. This latter finding is further fortified by (c) our demonstration that the $Ir(0)_{\sim300}$ nanoclusters are as active per exposed Ir(0) as even the most highly (80%) dispersed $Ir(0)/Al_2O_3$ heterogeneous catalyst known (13), an interesting finding given the belief that the surface metal in such

(b)

$P_2W_{15}Nb_3O_{62}{}^{9-}$ Polyoxoanion Inner and Outer (Ir Coordination) Sphere

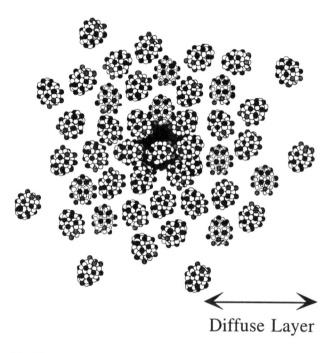

Diffuse Layer

Fig. 8b. *continued* (b) A more expanded, schematic view of a Ir(0)$_{\sim 300}$ nanocluster surrounded by a significant portion of the additional, ~66 polyoxoanions known to be present and showing both the polyoxoanions coordinated inner sphere to Ir(0) and the outersphere polyoxoanions in the diffuse or multilayer. The colloidal literature teaches that as the thickness of the diffuse layer goes up (i.e., with bulky polyoxoanions, $R_{4-x}NMe_x{}^+$ cations, or higher-dielectric solvents such as propylene carbonate) the nanocluster thermal stability toward agglomeration is expected to increase concomitantly (5,6), and this has been verified at least in part (23). Again, the Bu$_4$N$^+$ and Na$^+$ cations have been omitted from this simplified representation for the sake of clarity of the remaining components shown.

heterogeneous catalysts is largely "naked." Hence, a very important aspect of the polyoxoanion's stabilization is that it occurs while still allowing sufficient room at the nanocluster surface for facile catalysis to proceed, catalysis that is as fast as even that of "naked surface" Ir(0) heterogeneous catalysts examined under otherwise identical conditions.

C. Source of the Record Stability and Solution Catalytic Lifetime for the Ir(0) and Rh(0) Nanoclusters

An important finding resulting from the demonstration that the polyoxoanions are coordinated to the neutral $Ir(0)_{\sim 300}$ core is that it confirms the nanocluster stabilization is of the colloidal inorganic (polyoxoanion; charge or electrostatic) and organic (Bu_4N^+; steric) types shown in Fig. 4. A pictorial view of this high charge and high steric bulk stabilization due to the $P_2W_{15}Nb_3O_{62}^{9-}$ and its associated, multiple Bu_4N^+ is provided in Fig. 9. As we first remarked elsewhere (27), it is the "combined high charge plus significant steric bulk present intrinsically within the polyanion and poly-Bu_4N^+ cation components of $(Bu_4N^+)_9(P_2W_{15}Nb_3O_{62}^{9-})$" that appears to be special in providing the high stabilization. Worth noting here is that the use of polyoxoanions as a nanoparticle stabilizing agent was unprecedented prior to the work described in this chapter.

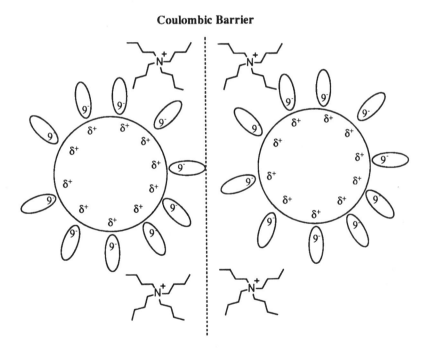

Fig. 9. Rationalization of the high stability of polyoxoanion-stabilized nanoclusters in terms of well-established colloidal electrostatic and steric stabilization theory. Unusual, however, is the *combined, intrinsic* high polyoxoanion charge and associated high tetrabutylammonium and polyoxoanion steric stabilization components.

Recently we have shown that in situ generated, polyoxoanion-stabilized Rh(0) nanoclusters exhibit record catalytic lifetime compared to any other modern transition-metal nanocluster undergoing sustained catalysis in solution, exhibiting 190,000 total turnovers (TTOs) of olefin hydrogenation before they become deactivated. This TTO lifetime in solution is better than all other Rh catalysts examined under identical conditions [e.g., Wilkinson's catalyst $Rh(PPh_3)_3Cl$ deactivated after only 19,000 TTOs], save the single case of a 5% Rh/Al_2O_3 heterogeneous catalyst which underwent 350,000 TTOs before it, too, deactivated (7). The low concentration of in situ generated nanoclusters was chosen deliberately (see footnote 19 in Ref. 7) to help stabilize the nanoclusters kinetically against precedented bimolecular agglomeration (28). TEM of the Rh(0) nanocluster catalyst after 140,000 TTOs reveals that the nanoclusters are ca. 26 ± 6 Å (7), smaller, as expected, than the normally ca. 40 Å Rh(0) nanoclusters we prepare under the "standard conditions" detailed elsewhere (28).

So far the $P_2W_{15}Nb_3O_{62}^{9-}$ polyoxoanion is the best we have found at providing both high stabilization and high catalytic activity versus other X^- or polyanions (X^- = halides, $P_3O_9^{3-}$, $SiW_9Nb_3O_{40}^{7-}$) (13,24), although additional comparisons are in progress of even higher charge polyoxoanions, other polyanions, plus different, longer-chain $R_{4-x}NMe_x^+$ cations (24). One of our goals here is to beat our own record of stability and record lifetime nanocluster catalysis (7).

We have also followed a lead from Reetz's lab (25) and found, as the theories of colloidal stabilization predict (5,6), that our nanoclusters are considerably more stable in higher–dielectric solvents, such as propylene carbonate, which expand the thickness of the diffusion- or multilayer shown in Fig. 8b (agglomeration is not seen in propylene carbonate even at ca. 150°C, whereas agglomeration begins at 50–60°C in the less polar acetone). In fact, propylene carbonate has been used to scale up our nanocluster syntheses to 1 g of the precursor, $(Bu_4N)_5Na_3[(1,5\text{-COD})Ir \cdot P_2W_{15}Nb_3O_{62}]$ (23). A catalysis-inhibiting, surface-deactivating phenomenon does appear to occur at these higher temperatures and in propylene carbonate, however (i.e., the catalytic activity is lost, yet TEM shows the nanoclusters to be unaggregated). The needed additional studies of the deactivation process and of other, high-dielectric-constant solvents as potential nanocluster-stabililizing media (24) are also in progress.

D. Reproducibility of the Nanocluster Size Distributions and Catalytic Activities

An important point is that, as first noted in our original paper (13, p. 4906), the small-scale syntheses of the $Ir(0)_{\sim 300}$ nanoclusters had been repeated more than 100 times by two different researchers as of 1994 (13), and have now been repeated more than 200 times by at least six different researchers over an eight-year period and with >30 batches of the precursor, all with a reproducibility of better than

≤±15% in size distribution and catalytic activity. This level of reproducibility does require that the purities of the $P_2W_{15}Nb_3O_{62}^{9-}$ and $(1,5\text{-COD})Ir \cdot P_2W_{15}Nb_3O_{62}^{8-}$ be carefully controlled; see also the effects due to water and to old, less pure acetone presented elsewhere (13,22), variables which must also be controlled to yield such high levels of reproducibility in the syntheses. So far, the catalytic rate appears to be a sensitive measure of the surface and its composition; the ≤±15% reproducibility of the catalytic activity is another pleasing result from all the effort that went into (a) the synthesis and characterization of the well-defined precursor $(Bu_4N)_5 Na_3[(1,5\text{-COD})Ir \cdot P_2W_{15}Nb_3O_{62}]$ (8,9), (b) the effort to carefully control all the variables in the nanocluster synthesis and its stoichiometry (13,22), and (c) the effort to follow the nanocluster formation reaction in real time (22,26).[6]

In fact, another bonus came from the ≤±15% reproducibility which resulted: this high level of "small-molecule-like" catalytic reproducibility disproved the widely held belief that only "homogeneous" (single metal or other, small organometallic), but not nanocluster or other "heterogeneous," catalysts could exhibit such reproducibility (13)! Nanoclusters *can* have small-molecule-like reproducibility in their physical properties, at least if one invests sufficient care in controlling all the variables involved in their synthesis and composition.[7]

The presence of cyclohexene has a significant effect on the resultant size of the nanoclusters (13,27) [as do H_2O or H^+ (13,22), impurities in old or improperly purified acetone (13), and as does changing the solvent to propylene carbonate (23)]—indeed, each and every variable seems to matter, a not unexpected finding given the mechanism of nanocluster growth that we will consider in a moment must consist of ≥300 steps. In the absence of cyclohexene and in acetone alone

---

[6] Another important point here, one which helped to yield the reproducible nanoclusters described in this chapter, is that we followed another rule well known to synthetic chemists: *follow all your reactions directly if at all possible*. While this is much easier using, for example, ^1H NMR in small-molecule organic chemistry than with many other methods in nanocluster and other materials chemistries, we commonly use the hydrogenation activity of the nanoclusters to follow their nucleation and growth in real time, albeit indirectly. We also followed the nanocluster evolution reaction a bit more directly, monitoring the cyclooctane evolution by GLC. Following the growing nanoclusters these two ways has proved invaluable, allowing kinetic and mechanistic studies, vide infra, and allowing us to know immediately whether a given nanocluster synthesis reaction produced the same nanoclusters as expected and as seen before [i.e., from the k_1 and k_2 rate constants one gets out of the kinetic fits and which characterize, respectively, the induction period (and its associated nucleation phase, k_1) and the autocatalytic growth phase (k_2), vide infra]; the ratio of k_2/k_1 is of further use as a good predictor for the size of the resultant nanoclusters (26).

[7] The implication here is that the best, carefully controlled nanocluster syntheses for other physical properties (i.e., electronic, optical, or magnetic properties) should also be able to exhibit high reproducibility in those properties as well, although it is not clear how rigorously this implication has been tested to date.

under H_2, 30 ± 4 Å, near-monodisperse $Ir(0)_{\sim 900}$ nanoclusters result (27). This significant effect of cyclohexene, while not yet studied in detail, can be nicely rationalized in terms of the mechanistic insights we now have (22): it is almost surely due to the coordination of cyclohexene to the Ir(0) surface of the nanocluster and, hence, its effects on either enhancing nucleation or retarding growth (or both), thereby leading to smaller—but more numerous—$Ir(0)_{\sim 300}$ [as opposed to $Ir(0)_{\sim 900}$] nanoclusters.

In the case of the less-well-investigated near-monodisperse, 40 ± 6 Å $Rh(0)_{\sim 1500}$ to $Rh(0)_{\sim 3700}$ nanoclusters (28), we have been able to control the syntheses and resultant catalytic activity to ± 15–20% for *intra*batch kinetic runs using the analogous rhodium, $(Bu_4N)_5Na_3[(1,5\text{-COD})Rh \cdot P_2W_{15}Nb_3O_{62}]$ nanocluster precursor. However, the *inter*batch variability of the catalytic activity was this good (± 15–20%) in only five of seven batches of precursor; in two of the seven batches the rate varied by a (colloidal-like) 500–600%. In one of those batches we now know that a new postdoc inadvisedly changed the synthesis[8] of the underlying $P_2W_{15}Nb_3O_{62}^{9-}$ and obtained impure $P_2W_{15}Nb_3O_{62}^{9-}$; hence, the results with that batch of precursor can be discarded. We also did controls ruling out five other conceivable explanations for the rate variations seen in the remaining one of six samples (28), and now have evidence that the previous failure to control a couple of now-understood, important variables in the synthesis of the $P_2W_{15}O_{56}^{12-}$ precursor (9c) is the source of the occasional variability in the catalytic activity of the Rh(0) nanoclusters. The important point for our purpose here, however, is that one can achieve a $\pm 15\%$, "small-molecule" level of reproducibility in the catalytic, and thus presumably in other, desired physical properties of nanoclusters, but only after exquisite care to control all the variables of these complex, multistep, nanoscale syntheses.[9]

---

[8] If the author could offer one piece of advice to young students doing a given synthesis for the first time, a piece of advice garnered from his experience of working in the synthetically demanding area of polyoxoanion synthesis and its often difficult, associated characterization issues, it would be this: do not change anything in the synthesis until, and unless, you have repeated the synthesis at least twice; then, and only after you have some idea of the amount of effort and insight that went into the original synthesis, should you try to "improve" it. In the PI's experience, most inexperienced synthetic chemists seem to underestimate badly the complexity of, or work and insight behind, a given synthesis, naively viewing themselves as "great chefs in the kitchen of chemistry" only, in the end, to find themselves—and their poor advisor!—"eating" their fallen, sometimes burnt to a crisp, "chemical soufflé."

[9] See for example our precautions, from "day one" (13,27), using a new glass reaction vessel (a borosilicate test tube plus a new stir bar) in our nanocluster syntheses, precautions designed to guard against *inhomogeneous*, as opposed to the desired *homogeneous*, nucleation of the nanocluster formation reaction.

E. Nanocluster Formation Kinetic and Mechanistic Studies: Elucidation of a Novel Mechanism Consisting of Slow, Continuous Nucleation Followed by Fast, Autocatalytic Surface Growth

1. Literature Background

A look in the literature at what is known about the mechanism of nanoparticle formation reveals that the available information was rather primitive when we began our kinetic and mechanistic studies. This of course makes sense: one needs to first have samples of modern, stable, bottleable, and reproducibly prepared and compositionally-well-characterized nanoclusters—as well as a complete stoichiometry for their formation reaction—before one can do meaningful kinetic and mechanistic studies. Such samples simply did not exist until recently. A comprehensive summary of the 19 papers which constituted the nanoparticle formation mechanistic literature we located prior to our 1997 paper are summarized in Ref. 22. Four relevant Ag_n nanoparticle formation mechanistic papers we subsequently found, and which are listed as Refs. 180–183 in footnote 10 in Ref. 3, plus the above-mentioned 19 papers constitute the 23 main references which are available on the mechanism(s) of particle formation, even when considering particles as diverse as sulfur sols, S_n, to Ag_m nanoparticles related to the silver photographic process.

A study of that literature reveals two key points: first, (i) LaMer's mechanism of S_n sol formation done in the 1950–1952 period, and consisting of (a) rapid, *burst nucleation* from supersaturated solution followed by (b) *diffusive, agglomerative growth* [Eqs. (3a), (3b)], has dominated mechanistic thinking about particle growth for the intervening nearly 50 years; second, (ii) there has been little new in terms of kinetic and mechanistic studies of nanoparticle formation since LaMer's original and influential work. Further discussion of these points, plus additional references and comments on important work by Turkevitch and others, is available in our initial kinetic and mechanistic paper (22, pp. 10382–10384, references, footnotes 1–19). Some recent work by Alivisatos's group on the kinetics and diffusional focusing and defocusing of CdSe and InAs semiconductor nanocrystals is also available (29). The important point for the present summary is the one we first noted elsewhere (22): "if one adds that any postulated mechanism needs to be clearly expressed in terms of the usual elementary (or pseudoelementary, vide infra) chemical equations and not just words and phenomenology, so that others can test, use and adapt the otherwise only "implied" mechanisms, then there has been no truly new mechanistic paradigm governing nanocluster formation reactions since the 1950s."

LaMer's Sulfur Sol Particle Formation Mechanism from the 1950s

(a) Nucleation from supersaturated solution $\quad nS \rightleftharpoons S_n \quad$ (3a)

(b) Diffusion-controlled growth $\quad S_n + S \rightleftharpoons S_{n+1} \quad$ (3b)

A list of the questions that we identified and hoped to answer about how modern transition-metal nanoparticles are formed is summarized in Fig. 10.

2. Kinetic and Mechanistic Studies of $Ir(0)_{\sim 300}$ Nanocluster Formation

The mechanistic studies were conducted along the same standards and "rules" for optimally proceeding that the author has constructed during his now 22 years of kinetic and mechanistic studies leading to, among our other publications, 48 kinetic

(1) The lack of *any* demonstrated kinetic and mechanistic scheme for transition-metal nanocluster formation (e.g., does LaMer's 1951 S_n sol-formation mechanism apply?).

(2) How to follow nanocluster formation? (TEM? Light scattering? *Catalytic activity*? Other methods?)

(3) How does a $Ir(0)_{\sim 300}$ "self assemble" from 300 $(1,5\text{-COD})Ir \cdot P_2W_{15}Nb_3O_{62}^{8-}$ precursors?

(4) How to control nanocluster size distributions (and why is a near-monodisperse, $\leq \pm 15\%$, size distribution seen for $Ir(0)_{\sim 300}$?).

(5) More generally, how can one control size, size-distribution, shape (and bi-, tri- or higher multi-metallic composition *and structure*)?

(6) Why have $Ir(0)_{\sim 300}$ and $Ir(0)_{\sim 900}$ "magic number" nanoclusters formed in our initial syntheses and without any effort on our part to achieve magic number nanoclusters?

(7) Why do added $P_2W_{15}Nb_3O_{62}^{9-}$ polyoxoanion, H_2O and H^+ have their observed dramatic effects on nanocluster formation?

(8) Does a more general, "universal" mechanism for transition metal nanocluster formation (e.g., under H_2) exist, or will each new nanocluster synthesis proceed by a different mechanism?

Fig. 10. A list of some nanocluster formation mechanistic issues that were unsolved when we began our kinetic and mechanistic studies.

and mechanistic papers at present.[10] Specifically, first we ensured that a complete stoichiometry was available, recall Eq. (2).

Next we devised a method to follow the nanocluster growth in real time, one of the most challenging problems for most nanoclusters (i.e., those other than some cases, such as Cu, Ag, and Au, which have a very convenient, size- and surface-composition-sensitive plasmon band in their visible spectrum). We managed to follow the nanocluster evolution reaction by two methods and to "document" the methods by showing that each kinetic method yielded equivalent kinetics and rate constants within experimental error. Figure 11 shows the first, and primary, kinetic method that we use, namely following the H_2 uptake using a computer-interfaced pressure transducer (which, from the reaction stoichiometry, can be equivalently expressed as the olefin loss as is done Fig. 11). The sigmoidal shape of the curves with their induction period, then fast, near-linear rate after the induction period, are characteristic of autocatalysis (by definition the elementary step A + B → 2B, where the product, B, is also a reactant; hence, the reaction goes faster and faster past the induction period and as more and more B is produced in the reaction). A in this case is just the nanocluster precursor, $(Bu_4N)_5Na_3[(1,5\text{-COD})Ir \cdot P_2W_{15}Nb_3O_{62}]$, while B is just the resultant Ir(0) which does the hydrogenation catalysis, first the Ir(0) on the surface of the critical-sized nucleus, and then the Ir(0) on the surface of the growing nanocluster.

The most novel aspect of the treatment of the kinetics is probably our use (Fig. 12) of Noyes's little known pseudoelementary step concept (33) to allow us to

---

[10] A couple of interesting statistics, and a resultant insight, arise here: first, and despite the fact that the PI considers himself primarily a mechanistic chemist, but probably because we have worked with larger, custom-made molecules (e.g., nanosized polyoxoanions, and now nanoclusters), only ca. 38% of our publications are in kinetics and mechanistic studies; the other 62% are dominated by synthesis and characterization work. Second, some time ago we noted that even for the nanosized molecules polyoxoanions, and even though our primary interests in polyoxoanions is their applications in catalysis, our polyoxoanion publications up to 1988 (as well as the talks at a 1988 Toronto Symposium ostensibly on "Heteropolyoxoanions and Their Catalytic Properties") were ca. 75% (and 70%), respectively, *on synthesis and characterization*, not catalysis (21b). The perhaps obvious insight, which may, however, be of use to younger researchers (which is why it is included here), is that one should expect to wind up spending 60–75% of one's research time in synthesis and characterization in nanocluster and other materials science areas, at least if one is making truly new materials and in the earlier, "pioneering days" of the work. This fact of scientific life is true even if one's "scientific heart" is ostensibly in other areas such as mechanism, the applications, or the physical property measurements of the new materials. This of course makes sense given the complexity of "nanomaterials." However, the failure to recognize the amount of synthesis and characterization work required up front means that one may fall into the trap of making sophisticated, time- and resource-consuming measurements on ill-defined or impure materials—the "garbage in, gargbage out" syndrome.

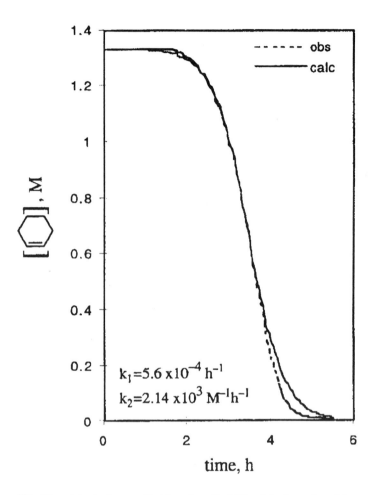

Fig. 11. A typical curve fit, taken from Ref. 22, of the loss of cyclohexene [determined by following the loss of hydrogen pressure vs. time, in acetone, at 22 °C and starting with 40 psig H_2 and 1.2 mM (1,5-COD)Ir · $P_2W_{15}Nb_3O_{62}^{8-}$] demonstrating the excellent curve fit to the nucleation plus autocatalysis, then fast hydrogenation, three-step kinetic scheme. The resultant rate constant k_2 has been corrected by the mathematically required "scaling" and "stoichiometric" factors (22), correction factors required to take into account, respectively, the changing size of the nanocluster's surface and the use of the hydrogenation of 1400 equivalents of cyclohexene to monitor the nanocluster formation reaction. [The deviations between the observed and calculated curves late in the reaction are expected and in fact help confirm the correctness of the kinetic treatment, as explained elsewhere (22).] (Reproduced with permission from L. AM. Chem. Soc., 1997, 119, 10832. Copyright 1997, AM. Chem. Soc.)

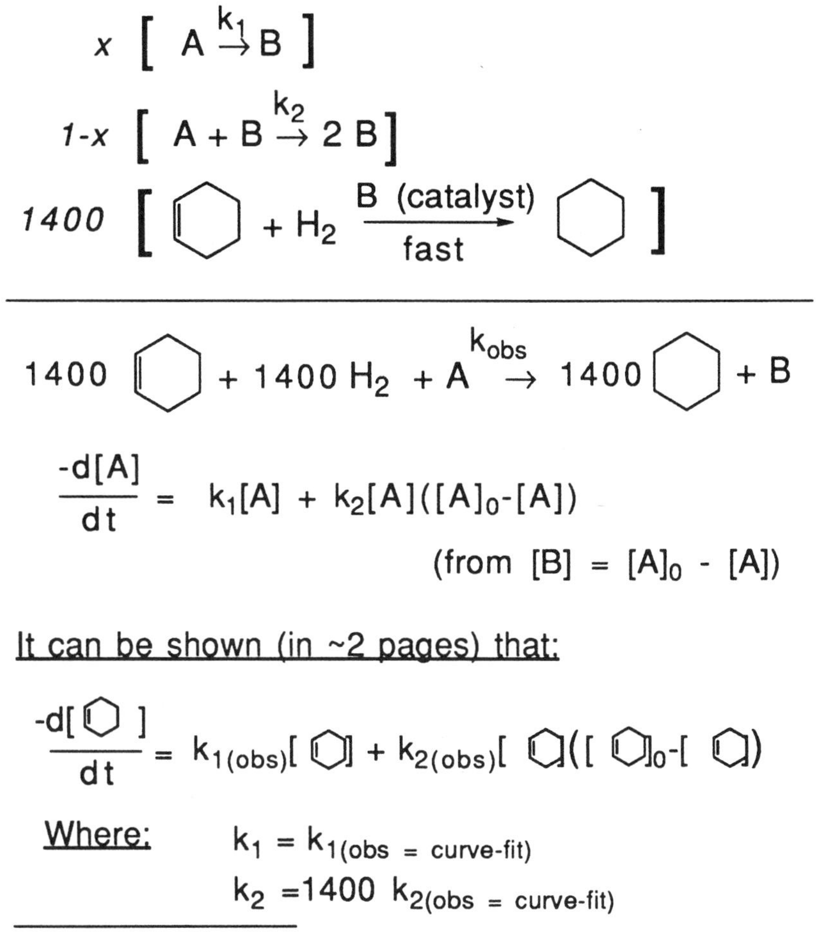

$$x \left[A \xrightarrow{k_1} B \right]$$

$$1-x \left[A + B \xrightarrow{k_2} 2B \right]$$

$$1400 \left[\bigcirc + H_2 \xrightarrow[\text{fast}]{B \text{ (catalyst)}} \bigcirc \right]$$

$$1400 \bigcirc + 1400\, H_2 + A \xrightarrow{k_{obs}} 1400 \bigcirc + B$$

$$\frac{-d[A]}{dt} = k_1[A] + k_2[A]([A]_0 - [A])$$

$$(\text{from } [B] = [A]_0 - [A])$$

It can be shown (in ~2 pages) that:

$$\frac{-d[\bigcirc]}{dt} = k_{1(obs)}[\bigcirc] + k_{2(obs)}[\bigcirc]([\bigcirc]_0 - [\bigcirc])$$

Where: $k_1 = k_{1(obs\ =\ curve\text{-}fit)}$

$k_2 = 1400\, k_{2(obs\ =\ curve\text{-}fit)}$

Fig. 12. Pseudoelementary step treatment of the kinetic data.

use the (fast) rate of ca. 1400 equivalents of olefin hydrogenation by the nanoclusters to follow their $A \to B$, $A + B \to 2B$ kinetics and mechanism of formation. Note that at first glance it seems remarkable that a two-step kinetic scheme can fit essentially exactly the data (Fig. 11) corresponding to a mechanism that must have ≥300 steps {recall the stoichiometry back in Eq. (2) involving 300 equivalents of the precursor $(Bu_4N)_5Na_3[(1,5\text{-}COD)Ir \cdot P_2W_{15}Nb_3O_{62}]$ being converted into 300 Ir(0) atoms}. However, the simple key here is that the $A + B \to 2B$ step is repeated ca. 300 times, a point made clearer in a pictorial view of the (minimal) mechanism necessary to account for the observed kinetic and other data (Fig. 13).

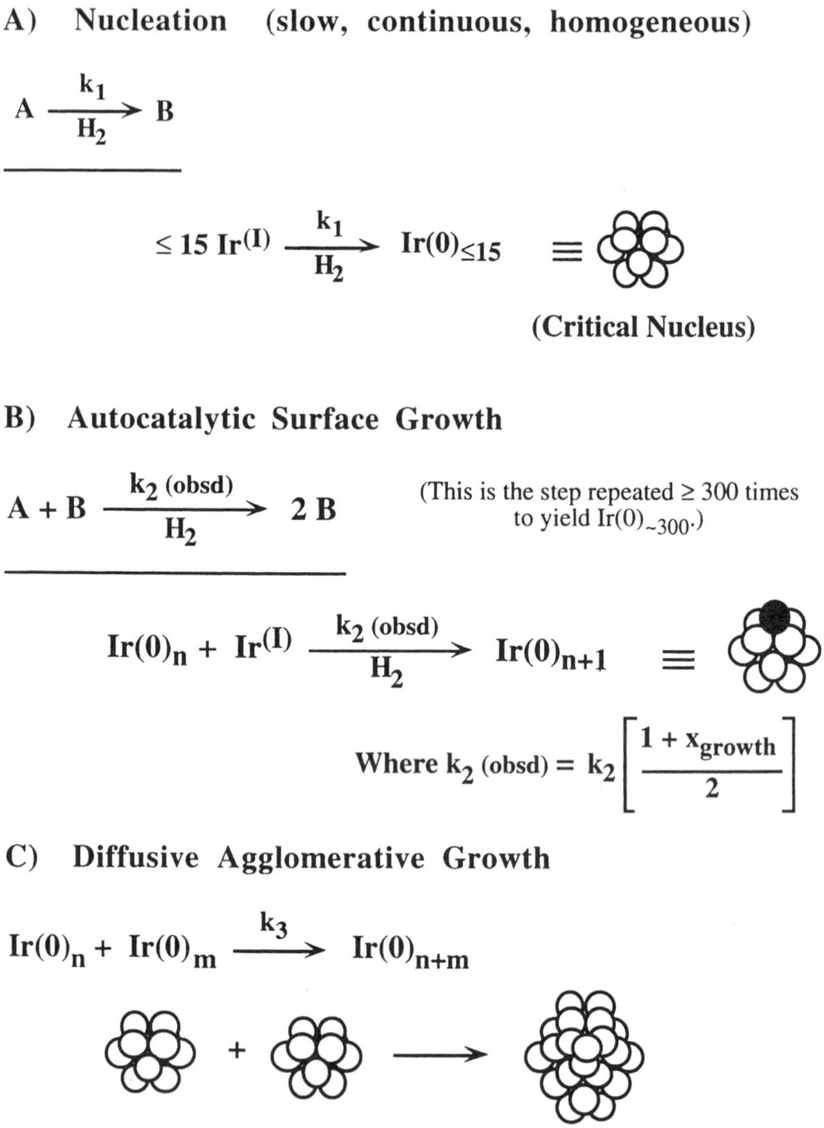

A) Nucleation (slow, continuous, homogeneous)

$$A \xrightarrow[H_2]{k_1} B$$

$$\leq 15\ Ir^{(I)} \xrightarrow[H_2]{k_1} Ir(0)_{\leq 15}\ \equiv$$

(Critical Nucleus)

B) Autocatalytic Surface Growth

$$A + B \xrightarrow[H_2]{k_2\ (obsd)} 2\ B$$

(This is the step repeated ≥ 300 times to yield $Ir(0)_{\sim 300}$.)

$$Ir(0)_n + Ir^{(I)} \xrightarrow[H_2]{k_2\ (obsd)} Ir(0)_{n+1}\ \equiv$$

Where $k_2\ (obsd) = k_2 \left[\dfrac{1 + x_{growth}}{2} \right]$

C) Diffusive Agglomerative Growth

$$Ir(0)_n + Ir(0)_m \xrightarrow{k_3} Ir(0)_{n+m}$$

Fig. 13. Pictorial view of the proposed, minimum mechanisms of formation of the Ir(0) and other transition-metal nanoclusters prepared under H_2.

We have also verified the H_2-uptake kinetic method by showing that the evolution of cyclooctane (see the stoichiometry in Fig. 2) can also be followed by GLC (22). Significantly, the GLC data give the same autocatalytic-shaped kinetic curves, are quantitatively fit by the same $A \rightarrow B$, $A + B \rightarrow 2B$ analytic kinetic expressions, and, quite pleasingly, yield the same k_1 and k_2 values within experimental error (22). Based on the resultant, reliable kinetic data available in our original mechanistic paper [and, for example, the effects of added H^+, H_2O, or $P_2W_{15}Nb_3O_{62}^{9-}$ and temperature changes therein (22)], a more intimate, but still Occam's razor (i.e., minimalistic), mechanism can be written with some confidence (Fig. 14).

3. Some Key Points from the Mechanism

- The nucleation is slow, continuous (i.e., not at all like LaMer's burst nucleation from supersaturated solution), and homogeneous (recall footnote 9).
- The growth is accomplished by fast, autocatalytic surface growth; that is, the metal nanocluster is effectively a "living metal polymer."
- The formation of near-monodisperse nanoclusters is due to the fact that the fast autocatalysis separates nucleation and growth in time, a requirement to achieve size-dispersity control in nanocluster syntheses.
- Diffusive agglomeration normally does not compete with autocatalytic surface growth (and at sufficiently high H_2 concentrations), another important reason that near-monodisperse nanoclusters are obtained.
- The occurrence of magic-number-sized nanoclusters is explained mechanistically for the first time: they are the natural consequence of kinetically controlled surface (autocatalytic) growth. Once a full-shell, magic number is formed, it is more stable and thus less reactive; the magic-number nanocluster then naturally "sits" in its local minimum for some time (recall Fig. 5) and, therefore, grows more slowly than non-full-shell, non-magic-number nanoclusters.
- The pseudoelementary step method promises to become an important tool in the study of complex reactions.
- Overall, it is possible to understand the mechanism of a ≥ 300-step reaction in near-molecular detail. The mechanism of formation of our polyoxoanion-stabilized Ir(0) nanoclusters is, to our knowledge, one of the highest multistep, nonorganic-polymer-forming reactions whose mechanism is understood in considerable detail.

4. Predictions of the Mechanism

The mechanism in Fig. 13 and 14 also contains other predictions, most of which have already been verified.

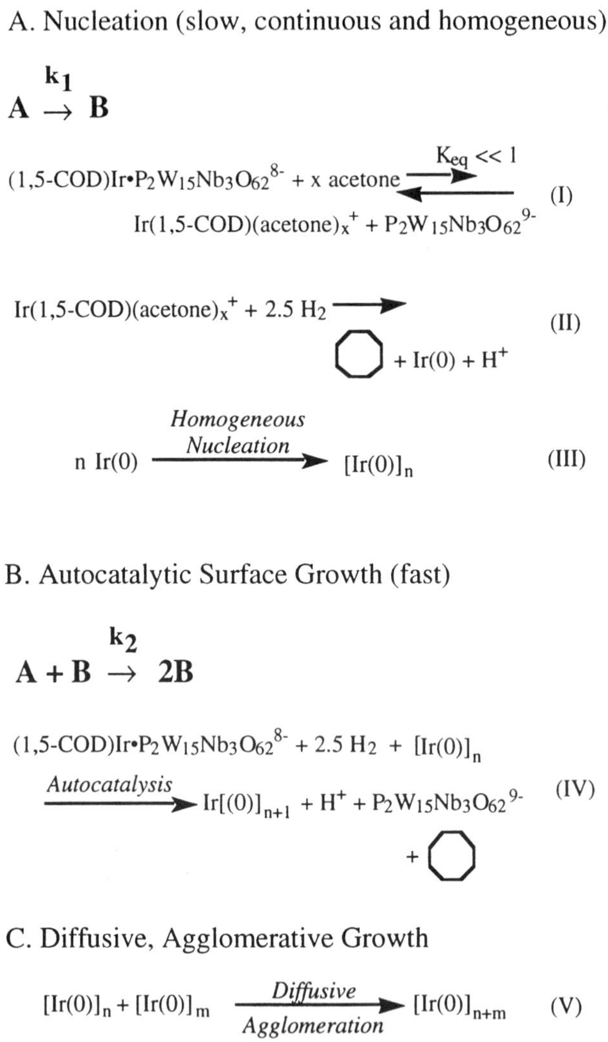

A. Nucleation (slow, continuous and homogeneous)

$$A \xrightarrow{k_1} B$$

$$(1,5\text{-COD})Ir \cdot P_2W_{15}Nb_3O_{62}{}^{8-} + x \text{ acetone} \underset{\phantom{K_{eq} \ll 1}}{\overset{K_{eq} \ll 1}{\rightleftarrows}} Ir(1,5\text{-COD})(\text{acetone})_x{}^+ + P_2W_{15}Nb_3O_{62}{}^{9-} \quad (I)$$

$$Ir(1,5\text{-COD})(\text{acetone})_x{}^+ + 2.5\ H_2 \longrightarrow \bigcirc + Ir(0) + H^+ \quad (II)$$

$$n\ Ir(0) \xrightarrow{\textit{Homogeneous Nucleation}} [Ir(0)]_n \quad (III)$$

B. Autocatalytic Surface Growth (fast)

$$A + B \xrightarrow{k_2} 2B$$

$$(1,5\text{-COD})Ir \cdot P_2W_{15}Nb_3O_{62}{}^{8-} + 2.5\ H_2 + [Ir(0)]_n$$

$$\xrightarrow{\textit{Autocatalysis}} Ir[(0)]_{n+1} + H^+ + P_2W_{15}Nb_3O_{62}{}^{9-} + \bigcirc \quad (IV)$$

C. Diffusive, Agglomerative Growth

$$[Ir(0)]_n + [Ir(0)]_m \xrightarrow{\textit{Diffusive Agglomeration}} [Ir(0)]_{n+m} \quad (V)$$

Fig. 14. More detailed version of the proposed, still minimalistic (i.e., Occam's razor), mechanism of formation of the Ir(0) and other transition-metal nanoclusters prepared under H_2.

1. At low [H_2], that is, at H_2 mass-transfer-limited conditions, such as in poorly stirred solutions, at too low H_2 pressure or at too high reactant or olefin concentration (so that the available H_2 is consumed faster than it can diffuse into solution), one should see dramatic effects where amorphous metal, produced by diffusive agglomeration replaces the normal formation of near-monodispersed size distributions (i.e., step V in Fig. 14 becomes faster than step IV). This prediction, summarized in Fig. 15, has been experimentally verified (28). It teaches that avoiding H_2 (or other reactant gas) mass-transfer-limiting conditions is crucial to obtaining near-monodisperse nanoclusters.

2. The $k_2[B]/k_1$ ratio should correlate with nanocluster size; that is, a large $k_2[B]$ (faster growth) or smaller k_1 (slower nucleation) leading to a larger $k_2[B]/k_1$ ratio predicts larger nanoclusters. This prediction has been verified (26).

3. The living-metal polymer analogy means that size control to larger nanoclusters is now easily obtained: for example, just take $Ir(0)_{\sim 150}$ nanoclusters

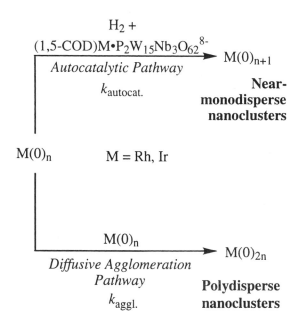

Fig. 15. Two, now well-precedented, *parallel* nanocluster growth pathways: autocatalysis (top pathway) leading to *near-monodisperse* nanoclusters, and diffusive agglomerative growth (bottom pathway) leading to *polydisperse* nanoclusters. Note that sufficient H_2 is required to yield near-monodisperse nanoclusters (i.e., so that the autocatalytic growth pathway will kinetically dominate over the diffusive agglomeration pathway), a point experimentally verified elsewhere (28).

and add ~150, ~410, and ~750 equivalents of Ir {i.e., the indicated equivalents of the nanocluster precursor $(Bu_4N)_5Na_3[(1,5\text{-COD})Ir \cdot P_2W_{15}Nb_3O_{62}]$} to obtain the first sequential series of nanoclusters centering about the four magic numbers 147, 309, 561, and 923, namely $Ir(0)_{\sim 150}$, $Ir(0)_{\sim 300}$, $Ir(0)_{\sim 560}$, and $Ir(0)_{\sim 900}$ nanoclusters. This prediction, too, has been verified elsewhere (26); as discussed therein, it is just the "seed" or "germ" growth method that has been practice since 1906 but for which little mechanistic understanding was available until now.

4. The surface-autocatalytic growth predicts that ligands that will bind to a certain face of a growing nanocluster can lead to nanocluster shape control. Elegant studies from El-Sayed's laboratories confirm this prediction: capping agents, which bind differently to the {111} or {100} faces of growing Pt nanoclusters, yield some shape control (30). Note also that El-Sayed and co-worker's studies provided independent confirmation of the surface-growth mechanism detailed in Figs. 13 and 14. The design and use of highly-face-selective capping agents to yield better shape-controlled nanocluster syntheses is an important, but unexplored, area for future research.

5. The surface-growth mechanism and its living-metal polymer nature predict how to make, for the first time and in a rational, designed, and understood way, all possible geometric isomers of bi-, tri-, and higher multimetallic nanoclusters (metallo "block co-polymers") (Fig. 16), and if good lattice matching and other constraints can be met (see Refs. 35 and 36a in Ref. 26 for cases where these constraints are *not* met). A search of the syntheses of bimetallic nanoclusters to date (see the summary of available references provided in Ref. 26) reveals that most bimetallic nanocluster syntheses to date are largely hit or miss, generally being performed without insight as to why one metal should be added first, followed by the second metal after the desired nanocluster core of the first metal is fully formed. The available heterobimetallic nanocluster literature suggests that the application of the insights in Fig. 16 to the designed synthesis of bi-, tri-, and higher multimetallic, "onion-skinned" nanoclusters ("nano-onions") of the type in Fig. 16 should prove rewarding. This remains to be experimentally verified or refuted, however.

6. Another prediction of our kinetic and mechanistic work is that the method used to follow the nanocluster growth (i.e., via the nanoclusters' catalytic activity and the pseudoelementary step method) should be more generally applicable to other metals and systems. This, too, has been verified in our study, showing that what was thought to be a "$(C_{18}H_{17})_3NMe^+[RhCl_4]^-$" benzene hydrogenation catalyst is, in fact, a Cl^- and $(C_{18}H_{17})_3NMe^+$ stabilized $Rh(0)_n$ nanocluster benzene hydrogenation catalyst (31).

7. And finally, a search of the literature [including many observations in older literature that we can now explain (22, p. 10398)] strongly suggests that the mechanism in Fig. 13 and 14—the first, new mechanism for transition-metal parti-

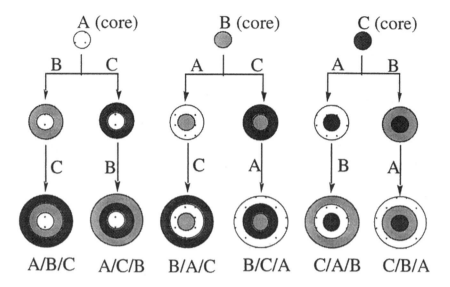

Fig. 16. Illustration of the living-metal polymer approach to the synthesis of multimetallic nanoclusters ("nano-onions") of, at least in principle, well-defined initial structure and layer thickness. The specific example shown is adapted from our earlier paper (26) and illustrates the synthesis of all possible geometric isomers of a trimetallic nano-onion. It needs to be emphasized that this scheme is idealized and intended only to illustrate the concept involved. It is already known, for example, that some bimetallic nanoclusters do not show such idealized second layers as illustrated (see Refs. 35 and 36a in Ref. 26). In addition, if a lattice or symmetry mismatch exists between the layers, thereby creating a high interfacial energy, it is more likely that imperfect (e.g., mound or island) growth will occur [e.g., as seen in molecular-beam epitaxial growth (26, Ref. 35c)]. Layer (atomic) migrations can also occur (26, Ref. 36a). However, it is also likely that closer to idealized bi- and higher multimetallic nanoclusters can be obtained by application of the autocatalytic surface-growth mechanism and the other principles illustrated in the figure and provided first in Refs. 22 and 26.

cle formation in more than 45 years[11]—is almost surely much more general, probably extending to most other systems using H_2 [and where *cis*-oxidative addition or other mechanisms of H_2 activation (32) are possible[12]], to particle formation under H_2 in certain heterogeneous catalyst syntheses, and to other nanocluster

---

[11] Note that the general mechanism of nucleation and growth steps for nanoparticle growth are, of course, well established; indeed these steps are common for first-order phase transitions. However, it is important to realize that even though LaMer's and our mechanism both contain nucleation and growth phases of the reaction, the underlying mechanistic steps are completely different: LaMer's

syntheses employing "H_2 equivalents" such as CO/H_2O, H_2CO, BH_3, N_2H_4, basic MeOH, or citrate, for example. More evidence on, and discussion of, this point is available in our original paper (22).

IV. SUMMARY AND FUTURE DIRECTIONS

This account has chronicled our efforts over the last decade to synthesize, characterize, and study the kinetics and mechanism of formation of polyoxoanion- and tetrabutylammonium-stabilized $Ir(0)_n$ nanoclusters. A number of novel findings resulted: the formation of near-monodisperse nanoclusters of well-defined sizes and with very-well-defined compositions; the finding that certain polyoxoanions can provide sizable, if not record, stabilization of transition-metal nanoclusters; the demonstration that the resultant nanoclusters have record catalytic lifetime in solution; the demonstration that the catalytic activity (and by implication the nanocluster's surface) is reproducible to $\leq \pm 15\%$, a result which disproved a previous myth claiming that this was not possible; the demonstration of a new way to monitor nanocluster growth using the pseudoelementary step concept and the catalytic activity of the nanoclusters or the GLC evolution of cyclooctane from the nanocluster precursor; and the finding of a likely universal (22) mechanism of slow, continuous nucleation followed by fast autocatalytic surface growth for transition-metal nanocluster formation under H_2 or related reductants. The mechanistic insights also demonstrate that it is possible to understand in detail the mechanism of a self-assembly reaction that has ≥ 300 steps, allow the first mechanism-based explanation for the formation of magic-number nanoclusters under

---

mechanism consists of burst nucleation from supersaturated solution followed by diffusional-based growth; our mechanism consists of slow, continuous nucleation and then autocatalytic, surface growth (which is not diffusion controlled). Restated, the terms "nucleation and growth" are really classes of reaction types, not underlying, elementary mechanistic steps (i.e., just as ligand substitution, oxidative addition, migratory insertion, and reductive elimination of organometallic chemistry are also classes of reactions). Consider, for example, the organic reaction subclasses of "S" (substitution) or "E" (elimination"); they do not imply a specific mechanism, which is why the terms S_N1, S_N2, S_H2', S_E2, S_E2', S_H2, S_Ni, S_Ni', $S_{Rn}1$, E_1, E_2, $E_{1'\text{CA}}$, $E_{1\text{-CB}}$ (and so on) were developed by Ingold and co-workers (but are now replaced by IUPAC reaction mechanisms descriptors).

This comment is not trivial; rather, it was prompted by a question received after a lecture on our work (by a solid-state materials scientist with a nonmolecular background) wondering, essentially, if all nucleation and growth mechanisms were alike. The answer is "no." This example further illustrates a main point of this chapter: the need to emphasize both the molecular and the more extended, materials, aspects in nanocluster and other areas of materials science.

[12] One would expect the two other known mechanisms of H_2 activation (heterolytic and metalloradical (32)) to be able to substitute for *cis*-oxidative addition in the A → B initiation and nucleation steps in Fig. 14. We have evidence that this is the case for at least heterolytic H_2 activation, and have reported those findings elsewhere (34).

kinetic controlled synthesis conditions, and provide seven other predictions from the mechanism, most of which have been tested and found to be correct.

Perhaps most important, however, is the demonstration that many of the key findings and insights are the result of as rigorous an adherence as the science would allow to the standards, principles, and other "rules" of smaller-molecule chemistry. Those standards, principles, and rules included using atomically-well-defined precursors for the nanocluster syntheses; obtaining full stoichiometry with complete mass and charge balances for the nanocluster evolution reaction; achieving full characterization of the nanoclusters and their average composition and associated stabilizing ligands in so far as the system would allow; and then performing detailed kinetic and mechanistic studies where warranted. While slow and frustrating at times, we can now say with confidence that the results and their often quantitative nature have been more than worth the extra effort required.

There remains much to do; in progress are additional mechanistic studies; the preparation of a variety of different nanoclusters of different metals for catalytic, magnetic, and other applications; scaling up all the nanocluster syntheses to multigram scales (23); fundamental studies of the stabilization provided by polyoxoanions in comparison to other polyanions, including even higher-charge polyoxoanions (24); attempts to better our own record of nanocluster catalytic lifetime in solution and to extend this record to high-temperature stability too, if possible, as such stability issues promise to be key hurdles to developing nanoclusters as novel "soluble heterogeneous catalysts" (4); and studies of interesting nanocluster catalytic reactions, plus the kinetics and mechanisms of the most important reactions uncovered.

It seems clear that nanoclusters will continue to provide a wealth of opportunities for fundamental and practical contributions for years to come!

ACKNOWLEDGMENTS

In this regard, this sometimes slow and tedious work would not have been possible without the generous, longer-term funding provided by the DOE, grant FG06-089ER13998, support which is gratefully acknowledged. The efforts proofreading the manuscript by Jason Widegren, Dr. Brooks Hornstein, and Professor Saim Özkâr are most appreciated as well.

REFERENCES

1. Some lead reviews: (a) P Jena, BK Rao, SN Khanna. Physics and Chemistry of Small Clusters. New York: Plenum, 1987. (b) RP Andres, RS Averback, WL Brown, LE Brus, WA Goddard III, A Kaldor, SG Louie, M Moscovits, PS Peercy, SJ Riley, RW Siegel, F Spaepen, Y Wang. J. Mater. Res. 4: 704, 1989. This is a panel report

from the United States Department of Energy, Council on Materials Science on Research Opportunities on Clusters and Cluster-Assembled Materials. (c) JM Thomas. Pure Appl. Chem. 60: 1517, 1988. (d) A Henglein. Chem. Rev. 89: 1861, 1989. (e) A superb series of papers, complete with a record of the insightful comments by the experts attending the conference, is available in Faraday Discussions 92: 1–300, 1991. (f) JS Bradley. In: G Schmid, ed. Clusters and Colloids. From Theory to Applications. New York: VCH, 1994, pp 459–544. (g) G Schmid. In: R Ugo, ed. Aspects of Homogeneous Catalysis. Dordrecht: Kluwer, 1990, Chapter 1. (h) A Fürstner, ed. Active Metals: Preparation, Characterization, and Applications. Weinheim: VCH, 1996. (i) LJ de Jongh, ed. Physics and Chemistry of Metal Cluster Compounds. Dordrecht: Kluwer, 1994. (j) G Schmid. Chem. Rev. 92: 1709, 1992. (k) DV Goia, E Matijevic. New J. Chem. 1203–1215, 1998. (l) M. José-Yacamán. Metall. Mater. Trans. A. 29A: 713, 1998. (m) G Schmid, M Bäumle, M Geerkens, - I Heim, C Osemann, T Sawitowski. Chem. Soc. Rev. 28: 179–185, 1999. (n) H Weller. Curr. Opinion Colloidal Interfacial Sci. 3: 194–199, 1998, and references therein. (o) JF Ciebien, RT Clay, BH Sohn, RE Cohen. New J. Chem. 685, 1999.
2. R Pool. Science. 248: 1186–1188, 1990. ("Clusters: Strange Morsels of Matter").
3. JD Aiken III, RG Finke. J. Mol. Cat. A: Chem. 145: 1–44, 1999. ("A Review of Modern Transition-Metal Nanoclusters: Their Synthesis, Characterization, and Applications in Catalysis.")
4. JD Aiken III, Y Lin, RG Finke. J. Mol. Cat. A: Chem. 114: 29–51, 1996. ("A Perspective on Nanocluster Catalysis: Polyoxoanion and $(n\text{-}C_4H_9)_4N^+$ Stabilized $Ir(0)_{\sim 300}$ Nanocluster 'Soluble Heterogeneous Catalysts'").
5. For a general discussion on the stability of colloids or nanoclusters, see, for example: (a) CS Hirtzel, R Rajagopalan. Colloidal Phenomena: Advanced Topics. New Jersey: Noyes, 1985, pp 27–39, 73–87. (b) RJ Hunter. Foundations of Colloid Science. Vol. 1. New York: Oxford University Press, 1987, pp 316–492. Other volumes covering colloidal stabilization and chemistry: (c) B Ranby, ed. Physical Chemistry of Colloids and Macromolecules. Boston: Blackwell Scientific, 1987. (d) JW Goodwin, ed. Colloidal Dispersions. London: The Royal Society of Chemistry, 1982. (e) J Heicklen. Colloid Formation and Growth: A Chemical Kinetics Approach. New York: Academic Press, 1976. (f) PC Hiemenz. Principles of Colloid and Surface Chemistry. New York: Marcel Dekker, 1986. (g) S Ross, ID Morrison. Colloidal Systems and Interfaces. New York: Wiley, 1988. (h) DA Fridrikhsberg. A Course in Colloid Chemistry. Moscow: Mir, 1986. (i) DH Everett. Basic Principles of Colloid Science. London: Royal Society of Chemistry, 1988.
6. (a) ME Labib. Colloids and Surfaces. 29: 293, 1988. ("The Origin of Surface Charge on Particles Suspended in Organic Liquids.") (b) JO'M Bockris, AKN Reddy. Modern Electrochemistry. Vol. II. Plenum Press, 1973.
7. JD Aiken III, RG Finke. J. Am. Chem. Soc. 121: 8803, 1999.
8. DJ Edlund, RJ Saxton, DK Lyon, RG Finke. Organometallics 7: 1692, 1988.
9. (a) K Nomiya, M Pohl, N Mizuno, DK Lyon, RG Finke. Inorg. Syn. 31: 186–201, 1997. (b) H Weiner, JD Aiken III, RG Finke. Inorg. Chem. 35: 7905–7913, 1996. (c) BJ Hornstein, RG Finke, submitted, ("The Lacunary Polyoxoanion Synthon $P_2W_{15}O_{56}^{12-}$: An Investigation of the Key Varibles in Its Synthesis Plus Multiple Control Reactions Leading to a Reliable Synthesis").

10. (a) MT Pope. Heteropoly and Isopoly Oxometalates. New York: Springer-Verlag, 1983. (b) A Müller, MT Pope, eds. Polyoxometalates: From Platonic Solids to Anti-retroviral Activity; Proceedings of the July 15–17, 1992, Meeting at the Center for Interdisciplinary Research in Bielefeld, Germany. Dordrecht, The Netherlands: Kluwer, 1992. (c) CL Hill, ed. Chem. Rev. 98: 1–390, 1998. ("Polyoxometalates.") (d) A Müller, MT Pope, eds. Polyoxometalates: Self-Assembled Beautiful Structures, Adaptable Properties, Industrial Applications; Proceedings of the October 4–6, 1999, Meeting at the Center for Interdisciplinary Research in Bielefeld, Germany. Dordrecht, The Netherlands: Kluwer, 2000 (in press).
11. MT Pope, A Müller. Angew. Chem. Int. Ed. Engl. 30: 34, 1991. ("Polyoxometalate Chemistry: An Old Field with New Dimensions in Several Disciplines.")
12. $(1,5\text{-COD})M \cdot P_2W_{15}Nb_3O_{62}^{8-}$ (M = Ir, Rh): (a) M Pohl, DK Lyon, N Mizuno, K Nomiya, RG Finke. Inorg. Chem. 34: 1413, 1995. (b) M Pohl, RG Finke. Organometallics 12: 1453–1457, 1993. (c) RG Finke, DK Lyon, K Nomiya, S Sur, N Mizuno. Inorg. Chem. 29: 1784–1787, 1990. (d) See also Ref. 9(a).
13. Y Lin, RG Finke. Inorg. Chem. 33: 4891–4910, 1994.
14. (a) M Pohl, Y Lin, TJR Weakley, K Nomiya, M Kaneko, H Weiner, RG Finke. Inorg. Chem. 34: 767, 1995. (b) T Nagata, M Pohl, H Weiner, RG Finke. Inorg. Chem. 36: 1366, 1997.
15. (a) J Kiwi, M Grätzel J. Am. Chem. Soc. 101: 7214, 1979. (b) J Wiesner, A Wokaun, H Hoffmann. Prog. Colloid Poly. Sci. 76: 271, 1988.
16. JM Lehn. Supramolecular Chemistry: Concepts and Perspective. New York: Weinheim, VCH, 1995. See p. 195 where Lehn refers to "supramolecular nanochemistry."
17. (a) A Trovarelli, RG Finke. Inorg. Chem. 32: 6034, 1993. (b) KS Suslick, JC Cook, B Rapko, MW Droege, RG Finke. Inorg. Chem. 25: 241, 1986. (c) For comparison, see the case of the mass spectrum of larger organic molecules such as polystyrene: J Yergey, D Heller, G Hansen, RJ Cotter, C Fenselau. Anal. Chem. 55: 353, 1983.
18. JU Köhler, JS Bradley. Catal. Lett. 45: 203–208, 1997.
19. LN Lewis, Chem. Rev. 93: 2693–2730, 1993. ("Chemical Catalysis by Colloids and Clusters.")
20. JS Bradley. In: G Schmid, ed. Clusters and Colloids. From Theory to Applications. New York: VCH, 1994, pp 459–544.
21. Reviews: (a) RG Finke. In: A Müller, MT Pope, eds. Polyoxometalates: From Platonic Solids to Anti-retroviral Activity; Proceedings of the July 15–17, 1992, Meetings at the Center for Interdisciplinary Research in Bielefeld, Germany. Dordrecht, The Netherlands: Kluwer, 1994. (b) RG Finke. In: A Müller, MT Pope, eds. Polyoxometalates: Self-Assembled Beautiful Structures, Adaptable Properties, Industrial Applications: Proceedings of the October 4–6, 1999, Meeting at the Center for Interdisciplinary Research in Bielefeld, Germany. Dordrecht, The Netherlands: Kluwer, 2001, in press. ("Polyoxoanions in Catalysis: From Record Catalytic Lifetime Nanocluster Catalysis to Record Catalytic Lifetime Catechol Dioxygenase Catalysis.")
22. MA Watzky, RG Finke. J. Am. Chem. Soc. 119: 10382, 1997.
23. BJ Hornstein, JD Aiken III, F Müller, RG Finke. Manuscript in preparation. ("Polyoxoanion- and Tetrabutylammonium-Stabilized Transition Metal "Soluble Analogs of Heterogeneous Catalysts": Scaled-up Synthesis and the Effect of Solvent on the Isolability and Catalytic Activity of Ir(0) Nanoclusters.")

24. S Özkâr, RG Finke. submitted ("A Comparison of the Nanocluster-Stabilizing Abilities of Commonly Employed Anions vs the $P_2W_{15}Nb_3O_{62}^{9-}$ Polyoxoanion 'Gold Standard'").
25. MT Reetz, R Breinbauer, P Wedemann, P Binger. Tetrahedron 54: 1233–1240, 1998.
26. MA Watzky, RG Finke. Chem. Mater. 9: 3083–3095, 1997.
27. Y Lin, RG Finke. J. Am. Chem. Soc. 116: 8335–8353, 1994.
28. JD Aiken III, RG Finke. Chem. Mater. 11: 1035, 1999.
29. X Peng, J Wickham, AP Alivisatos. J. Am. Chem. Soc. 120: 5343, 1998.
30. (a) TS Ahmadi, ZL Wang, TC Green, A Henglein, MA El-Sayed. Science 272: 1924–1926, 1996. (b) JM Petroski, ZL Wang, TC Green, MA El-Sayed. J. Phys. Chem. B 102: 3316, 1998.
31. KS Weddle, JD Aiken III, RG Finke. J. Am. Chem. Soc. 120: 5653, 1998.
32. JP Collman, LS Hegedus, JR Norton, RG Finke. Principles and Applications of Organotransition Metal Chemistry. Mill Valley, CA: University Science Books, 1987; see Chapter 5 on oxidative addition reactions.
33. RM Noyes, RJ Field. Acc. Chem. Res. 10: 214, 1979; and Acc. Chem. Res. 10: 273, 1979.
34. JA Widegren, JD Aiken III, S Özkâr, RG Finke. Chem. Mater. 13: 312, 2001.

3
Magic Numbers in Clusters

Nucleation and Growth Sequences, Bonding Principles, and Packing Patterns

Boon K. Teo
University of Illinois at Chicago, Chicago, Illinois

Hong Zhang
Air Force Research Laboratory (AFRL/MLPO), Wright-Patterson AFB, Ohio

I. INTRODUCTION

In clusters, atoms are often packed into polygonal (2-D) or polyhedral (in 3-D) shapes of high symmetry and high packing efficiency with characteristic numbers (1–17). For example, three, four, six, and eight atoms are often packed into triangular, tetrahedral, octahedral, and cubic arrays, respectively. We shall refer to these special characteristic numbers as *magic numbers* (18–20). The progression of these magic numbers, as a result of packing of atoms according to certain prescribed rules governed by electronic and/or steric principles, may be termed *magic sequences* (18–20). These magic numbers and magic sequences are also widely observed and/or utilized in other branches of science, technology, and engineering (21–29).

Magic numbers and sequences in cluster formation are intimately related to the *nucleation* and *growth* processes, which are governed by the often competing *bonding* and *packing* factors (30–34). While these processes may be kinetically or thermodynamically controlled, and hence sensitive to experimental conditions, the fact that bonding effects are *electronic* in origin and packing factors are *steric* in nature implies that magic sequences result from successive fillings of electronic or atomic shells, respectively.

This chapter reviews magic numbers and magic sequences in clusters as they relate to their structures dictated by bonding and packing principles. We shall give the generating functions and the first 10 magic numbers of a few representative magic sequences frequently observed, along with the underlying electronic or steric rules. More extensive listings can be found in the literature (18–20,34).

However, gas-phase clusters (35–38), as frequently detected by mass spectrometry, may or may not have well-defined structures. With increasing temperature of the nozzle, the shells of atoms tend to disappear, presumably due to the "melting" of the surface of the cluster. The "melting" temperature increases with the size of the cluster. The "melting" behavior can be analyzed by, among other techniques, laser photoabsorption. For instance, the photoabsorption spectra of sodium clusters show dramatically different spectra at different temperatures. Below 100 K, the spectrum exhibits many sharp peaks, typical of molecules with structures. However, for temperatures above 380 K, there are just two broad peaks, consistent with the *jellium* model for clusters lacking internal structures.

Clusters with magic numbers of atoms are more abundant and presumably more stable than others. Unstable clusters tend to shed extra atoms or acquire additional atoms and develop into more stable ones. As the cluster size increases, the force governing their formation may switch from bonding (electronic) to packing (steric) in nature. Still, it is often unclear how many atoms are required for the transition from molecular to bulk properties. Structurally, while large sodium clusters show evidence of icosahedral structure, the bulk sodium has the body-centered-cubic (bcc) structure. Similarly, while many high-nuclearity gold or gold-silver clusters adopt partial-icosahedral, icosahedral, or polyicosahedral structures (32), both metals have the face-centered-cubic (fcc) structure in the bulk. It is not unreasonable to assume that, like the melting behavior, the critical size and shape of such transition may vary from system to system and from property to property. It may also depend on the experimental probe used. These latter phenomena, however, are beyond the scope of this chapter.

II. FREQUENCY, TRUNCATION, AND CONJUGATION: FROM PLATONIC SOLIDS TO ARCHIMEDEAN POLYHEDRA

The majority of clusters adopt highly symmetrical polygonal or polyhedral structures. In three dimensions, a convex, regular polyhedron is a polyhedron with regular polygonal faces. Platonic polyhedra are convex, regular polyhedra with the same number of regular polygonal faces meet at each vertex. There are only five Platonic polyhedra: tetrahedron, octahedron, cube (hexahedron), icosahedron, and dodecahedron, as portrayed in Chart 1 (top row). If different kinds of

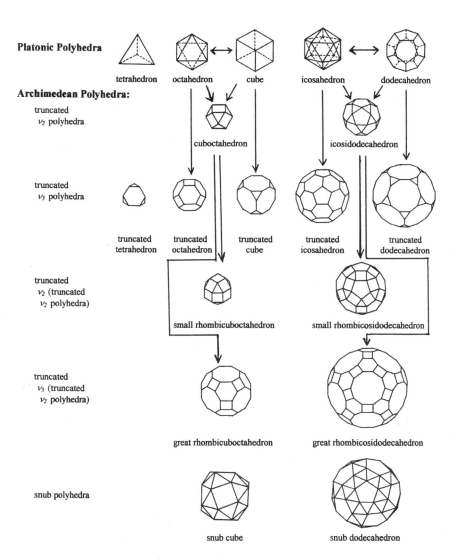

Chart 1. The 5 Platonic polyhedra and the 13 Archimedean polyhedra and their interrelations. The duality is indicated by double arrows.

polygonal faces are allowed to meet at each of the vertices, *semiregular* polyhedra are formed. In fact, there are 13 semiregular Archimedean polyhedra (also shown in Chart 1) and an infinite number of semiregular prisms and antiprisms.

Euler's theorem states that the number of edges, E, of any convex polyhedron is given by

$$E = V + F - 2 \tag{1}$$

where V and F are the numbers of vertices and faces, respectively.

One important concept for understanding the properties of various polyhedra, and the interrelationships among them, is their *duality* or *conjugation*. The dual of a polyhedron is defined as the polyhedron formed by interchanging the vertices and the faces. This relationship of polyhedra is indicated by the double arrows in Chart 1. For example, the dual of the octahedron is the cube, and the dual of the icosahedron is the dodecahedron. A tetrahedron is self-dual. According to Euler's theorem [Eq. (1)], dual polyhedra have the same number of edges.

Two types of polyhedral geometries, the deltahedra and the 3-connected polyhedra, are particularly important in cluster chemistry. A deltahedron is a polyhedron with triangular faces only. A 3-connected polyhedron is a polyhedron in which each vertex is connected to three other vertices. For example, as depicted in Chart 2, the octahedron and icosahedron are deltahedra, whereas the cube and pentagonal dodecahedron are 3-connected polyhedra. The tetrahedron, on the other hand, is both a deltahedron and a 3-connected polyhedron. The conjugation of deltahedra and the 3-connected polyhedra are indicated by double arrows in Chart 2.

A *high-frequency*, v_n, polygonal or polyhedral cluster is defined as a cluster with $n + 1$ atoms on each edge of the polygon or polyhedron, respectively. Charts 3 and 4 depict early members of the v_n triangles and the v_n centered triangles, respectively, in 2-D. Chart 5 shows early members of the v_n tetrahedra in 3-D. The nuclearities (numbers of atoms) are given in parentheses.

The *truncation* of polygonal or polyhedral figures results in "truncated polygons" or "truncated polyhedra." For example, truncation of the 5 Platonic polyhedra gives rise to 13 Archimedean polyhedra, as illustrated in Chart 1. Truncation of either a v_2 octahedron or a v_2 cube gives rise to a cuboctahedron. Further truncation of a v_2 cuboctahedron yields the small rhombicuboctahedron, and so on. The "family tree" of the 13 Archimedean polyhedra is portrayed in Chart 1.

If dissimilar arrangements of faces about the vertices of a polyhedron are allowed, a total of 92 polyhedra with regular faces can be produced. The obvious examples of this type are the pyramids. The description of these polyhedra is beyond the scope of this chapter.

Magic Numbers in Clusters

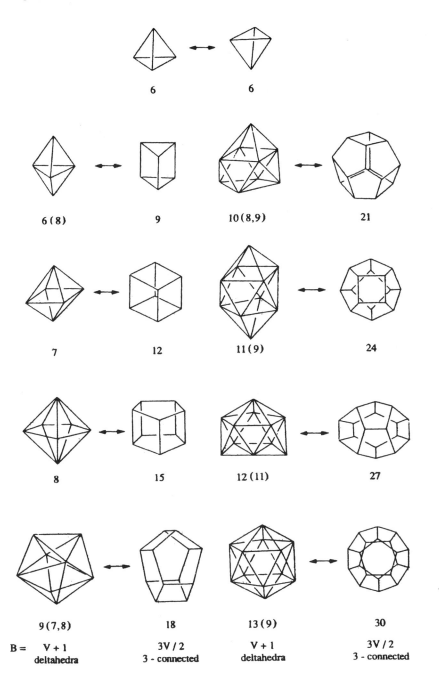

Chart 2. Deltahedra (left) and 3-connected polyhedra (right) and their duality (indicated by double arrows).

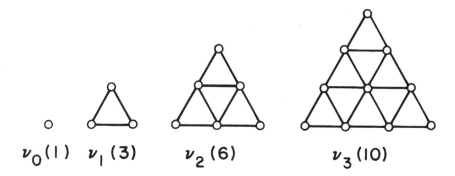

Chart 3. The v_n triangular sequence ($n = 0,1,2,3$).

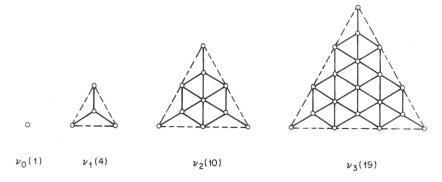

Chart 4. The v_n centered triangular sequence ($n = 0,1,2,3$).

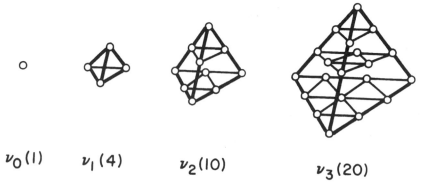

Chart 5. The v_n tetrahedral sequence ($n = 0,1,2,3$).

III. CLUSTER GROWTH: SELF-ORGANIZATION AND SELF-SIMILARITY PRINCIPLES

A. Golden Section: "The Divine Proportion"

Generating function

$G_n = \tau^n$

Magic numbers

$\ldots \tau^{-2}, \tau^{-1}, \tau^0, \tau, \tau^2, \tau^3, \ldots \{\tau = (1 + 5^{1/2})/2\}$
$\ldots 0.382, 0.618, 1, 1.618, 2.618, 4.236, 6.854, 11.090,$
$17.944, 29.034, 46.979, 76.013, \ldots$

We shall start with a magic sequence that is prevalent in nature. It is based on the positive root of the equation $x^2 = x + 1$ [which is $\tau = (1 + 5^{1/2})/2 = 1.618\ldots$]. If $x = b/a$ represents the ratio of two segments, a and b, of a straight line of total length $a + b$, the equation can be rewritten as $(b/a)^2 = b/a + 1$. The ratio $b/a = \tau = 1.618\ldots$ is known as the "Golden Proportion" or "Golden Section." It has many remarkable algebraic and geometric properties. For example, the ratio of two successive terms in the above sequence of magic numbers is τ (geometric proportion), while the sum of two consecutive terms yields the next term (arithmetic proportion). The latter can be represented by

$$\tau^n = \tau^{n-1} + \tau^{n-2}$$

It is this peculiar property of combining arithmetic and geometric proportions in the same series, producing the recurrence of similar shapes, that explains the important roles played by the Golden Section in the morphology of life and growth. Luca Pacioli called it the "Divine Proportion" (39).

B. Fibonacci Series: Gnomonic Growth

Generating function

$f_n = f_{n-1} + f_{n-2}$

Magic numbers

$1, 1, 2, 3, 5, 8, 13, 21, 34, 55, \ldots$

The Fibonacci series refers to the progression in which each term is equal to the sum of the two preceding terms. It is closely related to the Golden Section series in that the ratio of two consecutive terms approaches asymptotically the Golden Proportion of $\tau = 1.618$ very rapidly (e.g., $8/5 = 1.6$, $13/8 = 1.625, \ldots$). Like the Golden Section series, the Fibonacci series also has the property, by simple accretion via additive steps, of producing a "gnomonic" homothetic growth of similar shapes. It is this property of producing, by simple additions, a succession of numbers in geometric progression, and of similar shapes, that explains the important roles played by these two series in the morphology of life forms (39).

C. Cluster Classification and Cluster Growth

Recently, we proposed a cluster classification scheme (34,40–43) categorizing clusters into three broad classes according to their nucleation and growth pathways, as depicted in Chart 1 of Ref. 34. The primary clusters are the simplest polyhedral clusters, such as a tetrahedral or an icosahedral cluster. Two particular series of primary clusters, the deltahedra and the 3-connected polyhedra, have been discussed earlier (cf. Chart 2). These clusters, generally of low nuclearity, are of prime importance in that they may serve as "nucleation cores" or "building blocks" for the next (intermediate) stage of growth which produces the so-called secondary clusters. Two broad types of secondary clusters can be distinguished on the basis of the mechanism of the cluster growth. The first (Type I) is the layer-by-layer (LBL) or shell-by-shell (SBS) growth pathway (18) that involves the addition of successive layers of atoms onto a "nucleation core," giving rise to the "v_n polyhedral clusters." Chart 6 portrays the early members of the series v_n icosahedral clusters. The second (Type II) is the "cluster of clusters" (COC) growth pathway (42,43) that involves condensation of smaller cluster units as basic building blocks, giving rise to the "s_n supraclusters." A s_n supracluster is defined as a cluster of n smaller cluster units fused together via vertex, edge, or face sharing. Chart 7 depicts the early members of the series of the vertex-sharing polyicosahedral clusters.

D. Cluster Growth Pathways

We shall use the above-mentioned cluster classification scheme in describing the nucleation and growth of clusters, using the icosahedral structure as the basic geometry. Fig. 1 (44,45) shows the early stages of cluster growth based on the pioneering work of Hoare and Pal (46) and Briant and Burton (47). This particular

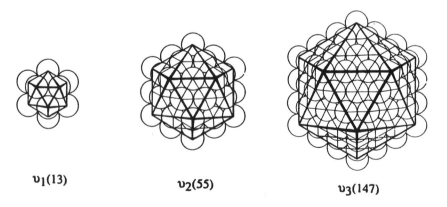

$v_1(13)$ $v_2(55)$ $v_3(147)$

Chart 6. The first three members of the v_n icosahedral (Mackay) sequence ($n = 1,2,3$).

Magic Numbers in Clusters

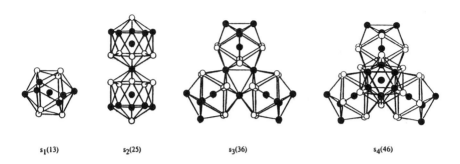

$s_1(13)$ $s_2(25)$ $s_3(36)$ $s_4(46)$

Chart 7. The early members of the vertex-sharing polyicosahedral S_n sequence (n = 1,2,3,4).

growth sequence (44) starts with one atom and adds one atom at a time. The cluster $c(n)$ grows via an atom-by-atom (ABA) mechanism, where n is the nuclearity. According to our classification scheme, these clusters are primary clusters (34). Examples of known structures can be found in the literature (44). While Briant and Burton (47) considered further growth of the 13-atom cluster to form a 33-atom pentagonal dodecahedral cluster to a 45-atom cluster to a 55-atom v_2 icosahedral cluster, we propose that the formation of the 13-atom icosahedral cluster, $c(13)$, may signify the "end" of the "early stages" of cluster growth (44). (Note that capping the 20 triangular faces of a 13-atom icosahedral cluster produces a 33-atom pentagonal dodecahedral cluster. Further capping the 12 pentagonal faces of the 33-atom pentagonal dodecahedral cluster gives rise to a 45-atom cluster.) Further growth can take on many different pathways, depending on the kinetics and thermodynamics of the system and the environment.

As discussed earlier, two distinct pathways (among others) are commonly observed. The first is the LBL or SBS growth mechanism, resulting in the cluster growth sequence based on the v_n icosahedral series of 13, 55, 147, . . . [the Mackay sequence (48,49)] As the cluster "grows" by adding successive layers or shells, it maintains the icosahedral geometry, as depicted in Chart 6. The second is the COC growth mechanism (32–34,40–43) as exemplified by the vertex-sharing polyicosahedral growth portrayed in Fig. 2. Here instead of adding one atom at a time, the cluster "grows" by adding one icosahedron at a time, giving rise to the growth sequence of 13, 25, 36, 46, . . . (50–53).

E. Self-Organization and Self-Similarity Principles

A comparison of the structures of secondary clusters, the v_n icosahedral (Chart 6) and the s_n polyicosahedral (Fig. 2) clusters, with those of primary clusters $c(n)$ (Fig. 1) reveals numerous similarities (44). Indeed, these similarities are

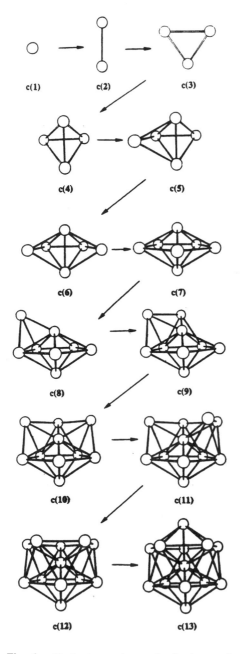

Fig. 1. Nucleation and growth of primary clusters $c(n)$ via the atom-by-atom (ABA) mechanism, where n is the nuclearity.

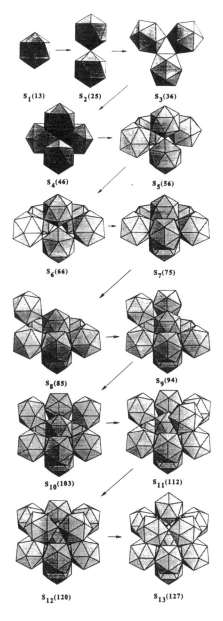

Fig. 2. Growth of secondary clusters based on the cluster-of-clusters (COC) growth mechanism as exemplified by the vertex-sharing polyicosahedral s_n sequence. Here the cluster "grows" by adding one icosahedron at a time, giving rise to the nuclearities 13, 25, 36, 46, . . . , 127 for n = 1,2,3,4, . . ., 13.

manifestations of the spontaneous self-organization and self-similarity processes in cluster growth. *Self-organization* means spontaneous assemblage of atoms to form energetically stable clusters of relatively efficient packing and high symmetry. *Self-similarity* implies that the resulting clusters look more or less alike when examined at different levels of magnification. Such underlying geometric similarity is, in fact, pattern within pattern, symmetry across scale. The resulting patterns are often referred to as *fractals* (54,55).

We shall discuss the electronic (bonding) requirements and the packing (stereochemical) patterns of various cluster geometries in Secs. IV and V, respectively.

IV. ELECTRONIC CLOSED SHELLS: BONDING PRINCIPLES

A. Topological Electron Counting (TEC) Rule: Primary Clusters

While many elegant electron counting models have been proposed in the literature (56–69), we shall make use of the topological electron counting (TEC) model developed by us (32–34,40–43). The TEC rule assumes that (a) each vertex atom contributes *three* orbitals for cluster interaction of which only the bonding contributions (B) are filled; and (b) the remaining valence orbitals (one for main-group and six for transition-metal vertex atoms) are "filled" as a result of ligand bonding or as lone pairs (cf. Chart 2 of Ref. 34). For a cluster with V_n main-group and V_m transition-metal vertices, there will therefore be $3V$ orbitals for cluster interaction (here $V = V_n + V_m$), forming B bonding and A antibonding molecular orbitals (where $A + B = 3V$). These orbitals can be visualized as hybrid orbitals based on the three valence p orbitals. The remaining valence orbitals (one s for main-group elements; five d and one s for transition metals) are used for ligand bonding or for lone pairs. (This is an oversimplification, of course; in reality, extensive mixing of the cluster valence orbitals can occur.)

If the cluster bonding orbitals are filled with B electron pairs, the total number of cluster valence electron pairs (or topological electron pairs) is (70–73)

$$T = B + V_n + 6V_m \qquad (2)$$

In fact, B is the number of electron pairs primarily responsible for "holding the cluster together." The B values for two important series of primary clusters—the deltahedral and the 3-connected—are given in Chart 2.

1. Deltahedral Clusters: Electron Deficient Clusters

Generating function $\qquad\qquad\qquad$ Magic numbers
$G_n = n$ $\qquad\qquad\qquad\qquad\qquad$ 1, 2, 3, 4, 5, 6, 7, 8, 9, 10, 11, 12, 13, . . .

Magic Numbers in Clusters

A deltahedron is a polyhedron with triangular faces only. Important examples are also shown in Chart 2 (left side of the double arrows). These are commonly called *closo* clusters. Boron hydrides of general formula $[B_nH_n]^{2-}$ are prime examples of this class of clusters. Other representative main-group and transition-metal clusters can be found in Refs. 70–73.

Skeletal electron pair (SEP) theory (58) predicts that the number of skeletal electron pairs for this class of clusters (except tetrahedron) is given by

$$B = V + 1 \quad \text{for closo-deltahedral clusters} \tag{3}$$

where V is the number of vertices. The B values for the common deltahedra are given in Chart 2. For example, an icosahedron has $V = 12$ vertices and a B value of 13, thereby satisfying the SEP theory.

It turns out that the SEP theory applies not only to closo-deltahedral clusters (excluding tetrahedron) but also to polyhedra, which can be visualized as derivable from closo-deltahedra with one, two, or three missing vertices, commonly called *nido, arachno, or hypho* clusters, respectively. Equation (3) is appropriately modified to $B = V + 2, V + 3, V + 4$ for these latter polyhedra. For example, a square pyramid ($V = 5$) can be considered as an octahedron ($V = 6$) with a missing vertex (a nido cluster); hence $B = V + 2 = 5 + 2 = 7$. This may be called the *debor* principle. Another way of stating the debor principle is that removal of a few atoms from a closo-deltahedral cluster does not alter the number of skeletal electron pairs (B).

The opposite of the debor principle is the *capping* principle (74,75), which states that capping does not alter the number of skeletal electron pairs of the parent polyhedron. For example, a tetracapped tetrahedron is a deltahedron with eight vertices, yet there are only six skeletal electron pairs ($B = 6$) within the central tetrahedron.

2. 3-Connected Clusters

Generating function | Magic numbers
$G_n = 2n$ | 2, 4, 6, 8, 10, 12, 14, 16, 18, 20, . . .

3-Connected clusters are polyhedral clusters in which each vertex is connected to three other vertices. Some examples of 3-connected polyhedra are shown in Chart 2 (right side of the double arrows). Polyhedral C_nH_n belongs to this class of clusters. Other examples of main-group or metal clusters can be found in Refs. 70–73. Since each vertex is connected to three other vertices via three cluster orbitals, these clusters (with a total of V vertices) can be considered as electron precise in that one-half of the $3V$ orbitals will be bonding:

$$B = \frac{3V}{2} \quad \text{for 3-connected polyhedral clusters} \tag{4}$$

The other half, of course, will be antibonding ($A = 3V/2$). According to Euler's theorem [Eq. (1)], for 3-connected polyhedra the number of edges, E, is also given by $3V/2$. Hence, the B value for a 3-connected polyhedral cluster is equal to the number of edges ($B = E$). The B values for other important 3-connected polyhedral clusters are given in Chart 2.

B. Shell Model: Close-Packed v_n Polytopal Metal Clusters (Type I Secondary Clusters)

The TEC rule can be generalized to include closed-packed polyhedral transition-metal clusters. According to the generalized TEC rule, a v_n polytopal (polygonal or polyhedral) *metal cluster* is required to have a total of T_n cluster valence (topological) electron pairs (34,70,71):

$$T_n = 6S_n + B_n \tag{5}$$

where S_n is the number of "surface" atoms and B_n is the number of shell electron pairs, which is the number of the skeletal electron pairs of the v_n polytopal metal cluster. Here the interior atoms are considered merely as electron contributors.

For close-packed high-nuclearity metal clusters, we may invoke the Shell model (18,34), which states that

$$T_n = 6G_n + K \tag{6}$$

Here G_n is the total number of metal atoms (nuclearity), and K is related to the B value of the center of the cluster (which may be an atom, an edge, center of a face, or a polyhedral hole). The K values of commonly observed polygonal or polyhedral are tabulated in Table 2 of Ref. 34.

One can use Eq. (6) to predict the electron counts of v_n polygonal or polyhedral close-packed metal clusters. We shall consider three examples here: (a) v_n triangular clusters (Chart 3), (b) v_n tetrahedral clusters (Chart 5), and (c) v_n icosahedral clusters (Chart 6).

1. 2-D Metal Clusters

An examination of Chart 3 reveals that a v_n triangle repeats itself in steps of 3. In other words, removal of the peripheral atoms of a v_n triangle produces a v_{n-3} triangle. Hence we have the following three situations: (1) for v_0, v_3, v_6, . . . , v_{n3} triangles centering at an atom, $K = 7$; (2) for v_1, v_4, v_7, . . . , v_{3n+1} triangles centering at a triangle, $K = 6$; and (3) for v_2, v_5, v_8, . . . , v_{3n+2} triangles centering at a triangle, $K = 6$. Only the first two structures are known in metal clusters.

For the v_1 triangular metal cluster (cf. Chart 3, and Table 2 of Ref. 18), $K = 6$, $T_1 = 6G_1 + K = 6 \times 3 + 6 = 24$, $N_1 = 2T_1 = 48$, which agrees with $N_{obs} = 3 \times$

$8 + 12 \times 2 = 48$ in $Os_3(CO)_{12}$ (76,77). For the ν_2 triangular metal cluster centering at a triangle structure (cf. Chart 3 and Table 1), $K = 6$, $T_2 = 6G_2 + K = 6 \times 3 + 6 = 42$, $N_2 = 2T_2 = 84$, which agrees with $N_{obs} = 3 \times 11 + 3 \times 8 + 12 \times 2 + 3 = 84$ in $[Cu_3Fe_3(CO)_{12}]^{3-}$ (78).

2. 3-D Metal Clusters

As portrayed in Chart 5, a ν_n tetrahedron repeats itself in steps of 4. We have the following four situations: (1) since $\nu_1, \nu_5, \nu_9, \ldots$ tetrahedra all center at a tetrahedral hole, $K = 6$; (2) since $\nu_2, \nu_6, \nu_{10}, \ldots$ tetrahedra all center at an octahedral hole, $K = 7$; (3) since $\nu_3, \nu_7, \nu_{11}, \ldots$ tetrahedra all center at a tetrahedral hole, $K = 6$; (4) since $\nu_4, \nu_8, \nu_{12}, \ldots$ tetrahedra all center at an atom, $K = 7$. For instance, for the ν_1 tetrahedral metal cluster (cf. Chart 5), $K = 6$, $T_1 = 6G_1 + K = 6 \times 4 + 6 = 30$, $N_1 = 2T_1 = 60$, which agrees with $N_{obs} = 4 \times 9 + 12 \times 2 = 60$ in $Co_4(CO)_{12}$ (79). For the ν_2 tetrahedral metal cluster centered on an octahedron hole (cf. Chart 5), $K = 7$, $T_2 = 6G_2 + K = 6 \times 10 + 7 = 67$, $N_2 = 2T_2 = 134$, which agrees with $N_{obs} = 10 \times 8 + 4 + 24 \times 2 + 2 = 134$ in $[Os_{10}C(CO)_{24}]^{2-}$ (80).

The electron counts of ν_n icosahedral clusters [Mackay sequence (48,49)] can easily be calculated via the Shell model. Here, the number of cluster valence electron pairs, T_n, of a ν_n polyhedral cluster, is given by Eq. (6). The Mackay sequence has nuclearities $\nu_1(13), \nu_2(55), \nu_3(147), \nu_4(309), \nu_5(561), \ldots$. The early members are portrayed in Chart 6. As we shall see, for this growth sequence, the cluster grows by adding $S_n = 10n^2 + 2$ atoms per shell and the cluster maintains icosahedral symmetry. The nuclearities (in parentheses) are given by $G_n = \frac{1}{3}(2n + 1)(5n^2 + 5n + 3)$. For example, $\nu_1(13), \nu_2(55)$, and $\nu_3(147)$ depicted in Chart 6 are predicted by the Shell model [Eq. (6)] to have 85, 337, and 889 electron pairs. (Here, $K = 7$ since the clusters are centered at an atom.) Only the $\nu_1(13)$ icosahedron is known in metal cluster chemistry (although the Mackay sequence has been observed in the adiabatic expansion of inert gases such as Xe, to be discussed later).

C. C^2 Model: Cluster of Clusters (Type II Secondary Clusters)

In order to understand the electronic requirements of Type II secondary clusters, we developed a simple "cluster of clusters" (C^2) model. According to the C^2 model (42,43), for a secondary cluster, which can be regarded as a cluster formed by n polyhedral clusters fused together, the B value is given by the sum of the B_j values ($j = 1$ to n) of the n individual cluster units (building blocks) minus the sum of the B_k values ($k = 1$ to s) of the s shared element(s) such as vertice(s), edge(s), or face(s). The B values for the commonly observed cluster units and shared moieties can be found in Chart 5 of Ref. 34. Some interesting examples of vertex-, edge-, and face-sharing supraclusters in 2-D, 3-D, or mixed 2-D and 3-D cluster systems can be found in Refs. 42 and 43. The total number of electron pairs (T) can be calculated from Eq. (2), and the total electron count is given by $N = 2T$. Two examples will be discussed.

1. Polyoctahedral Clusters

There are three ways of joining two octahedra: sharing a vertex, an edge, or a face (see Chart 9 of Ref. 34). Since each octahedron) (O_h) contributes a B value of 7 (cf. Chart 2), two octahedra sharing a vertex, an edge, or a face will give rise to the predicted B values of $B = 2 \times 7 (O_h) - 3$ (atom) $= 11, 2 \times 7 (O_h) - 5$ (edge) $= 9$, and $2 \times 7 (O_h) - 6$ (triangle) $= 8$, respectively. The latter two structures are exemplified by clusters $[Ru_{10}C_2(CO)_{24}]^{2-}$ (81) and $[Rh_9(CO)_{19}]^{3-}$ (82) with $N_{obs} = 10 \times 8 + 2 \times 4 + 24 \times 2 + 2 = 138$, and $9 \times 9 + 19 \times 2 + 3 = 122$, which agree reasonably well with the predicted electron counts of $N = 2T = 2(6V_m + B) = 2 \times (6 \times 10 + 9) = 2 \times 69 = 138, 2 \times (6 \times 9 + 8) = 124$, respectively.

2. Vertex-Sharing Polyicosahedral Clusters

The electronic requirements of the vertex-sharing polyicosahedral clusters (Fig. 2), as exemplified by an interesting series of Au-Ag nanoclusters, have been theoretically defined by us on the basis of the C^2 model (34,42,43). In fact, the agreement between the experimentally observed electron counts and those predicted by the C^2 model provides strong electronic evidence for the cluster of clusters concept for this series of supraclusters (42). As described in Ref. 42, the number of skeletal electron pairs for an $s_n(N)$ supracluster with n vertex-sharing icosahedral units is

$$B_n = 4n + 18 \tag{7}$$

and the total number of cluster valence electron pairs is

$$T_n = 58n + 54 \tag{8}$$

In terms of the total number of cluster valence electrons,

$$N_n = 116n + 108 \tag{9}$$

[For $n = 2$, Eqs. (7)–(9) become $B_n = 4n + 15$, $T_n = 58n + 45$, and $N_n = 116n + 90$.] Equations (7)–(9) have been successfully used to rationalize and/or predict the electron counts of a wide variety of vertex-sharing polyicosahedral supraclusters (42). It should also be pointed out that the $4n$ term in Eq. (7) may be interpreted as the "polyoctet rule" in that each icosahedron in polyicosahedral supracluster contributes eight electrons (i.e., four electron pairs) to the B value (32,42):

$$B_p = 4n \tag{10}$$

If we assume that each coinage metal atom contributes one s electron (as a pseudoalkali metal) to, and that each halide ligand withdraws one electron from, the cluster, then B_p can be calculated by

$$B_p = \frac{M - X - Q}{2} \tag{11}$$

where M, X, and Q refer to the numbers of metal atoms, halide ligands, and the overall charge, respectively.

The above equations are useful in the systematization and rationalization of a series of Au-Ag clusters synthesized and structurally characterized by us. The metal frameworks of these clusters are based on vertex-sharing polyicosahedra, as portrayed in Fig. 2. For example, the 25-metal-atom cluster $[(p - \text{Tol}_3\text{P})_{10}\text{Au}_{13}\text{Ag}_{12}\text{Br}_8]^+$ (83) has $B_p = (25 - 8 - 1)/2 = 8$ electron pairs as predicted for a bi-icosahedral supracluster ($4n = 4 \times 2 = 8$ electron pairs). Similarly, and the 25-metal-atom dicationic cluster $[(p - \text{Tol}_3\text{P})_{10}\text{Au}_{13}\text{Ag}_{12}\text{Cl}_7]^{2+}$ (84), $B_p = (25 - 7 - 2)/2 = 8$ electron pairs, also as expected.

For the 38-metal-atom cluster $[(p - \text{Tol}_3\text{P})_{12}\text{Au}_{18}\text{Ag}_{20}\text{Cl}_{14}]$ (85,86) with three icosahedra sharing three vertices, $B_p = (38 - 14)/2 = 12$ electron pairs, which agrees with the triicosahedral model ($4n = 4 \times 3 = 12$ electron pairs). Similarly, the 37-metal-atom cluster $[(p - \text{Tol}_3\text{P})_{12}\text{Au}_{18}\text{Ag}_{19}\text{Br}_{11}]^{2+}$ (87) has $B_p = (37 - 11 - 2)/2 = 12$ pairs of electrons, once again, as expected.

For the 25-atom cluster formed by two icosahedra sharing one vertex, the C^2 model predicts $B = 2 \times 13$ (icosahedron) $- 1 \times 3$ (sharing one vertex) $= 23$ skeletal electron pairs, $T = 6V_m + B = 6 \times (25 - 2) + 23 = 161$ total electron pairs or a total valence electron count of $N = 2 \times 161 = 322$. One example is the $[(p - \text{Tol}_3\text{P})_{10}\text{Au}_{13}\text{Ag}_{12}\text{Br}_2(\mu - \text{Br})_2(\mu_3 - \text{Br})_4]^+$ cluster (83) for which the N_{obs} of $(10 \times 2 + 25 \times 11 + 2 \times 1 + 2 \times 3 + 4 \times 5 - 1) = 322$ valence electrons is in accordance with the calculated value. More examples can be found in the literature (32–34).

D. Jellium Model: Jelliumic Clusters

Generating function Magic numbers
1s, 1p, 1d, 2s, 1f, 2p, 1g, 2d, 3s, 1h, . . . 2, 8, 18, 20, 34, 40, 58, 68, 70, 92

The *jellium* model (35,88) treats the cluster as a smooth jelly of positively charged ions to which electrons are attracted. In terms of a free-electron picture, the valence electrons interact with a smooth one-particle effective potential and an electron-electron interaction potential. A spherically symmetric potential well is employed, along with parameters derived from the bulk. Solving the Schroedinger equation yielded discrete energy levels characterized by the angular momentum quantum number L in the order 1s, 1p, 1d, 2s, 1f, 2p, 1g, 2d, 3s, 1h, ...

For clusters dominated by the electronic effects, the magic numbers correspond to the numbers of valence electrons that completely fill one of the electronic shells. Clusters with closed electronic shells are exceptionally stable and more or less spherical. As shown in Fig. 1 of Ref. 35 the jellium model predicts successive filling of the closed-shell electronic configurations $1s$, $1p$, $1d$, $2s$, $1f$, $2p$, $1g$, ($2d$, $3s$, $1h$), . . . , giving rise to total numbers of electrons of 2, 8, 18, 20, 34, 40, 58, 68, 70, 92, ... On the other hand, if they do not have enough electrons, they attain the *electronic subshells* and adopt either a prolate (elongated) or an oblate (compressed) structure, as illustrated in Fig. 1 of Ref. 88.

As an example, in the gas phase, the magic numbers for the cationic Cu_n^+, Ag_n^+, and Au_n^+ bare clusters (89) occur at n = 3, 9, 21, 35, 41, 59, 93, . . . , whereas that for anionic Cu_n^-, Ag_n^-, and Au_n^- bare clusters occur at n = 1, 7, 19, 33, 39, 57, 91, ... These observations are consistent with the jellium model if the coinage metals are considered as pseudo alkali metals with only one electron in the valence shell.

The jellium model is applicable only to clusters up to a few thousand atoms, as we shall discuss next, using gas-phase sodium clusters as an example.

a. Small Sodium Clusters. Among the first experiments on sodium clusters are the one performed by Knight and co-workers at the University of California, Berkeley in 1984 (35). They obtained the mass spectra for a molecular beam of sodium clusters embedded in argon gas after ionization by UV light. Each peak represents the number of atoms detected in a fixed time window. The observed peaks are followed by abrupt decrease for n = 2, 8, 20, 40, 58, 92, . . . since the latter sequence is associated with the electronic shell structures of Na_n clusters characterized by large energy gaps between energy levels (and hence extra stabilities).

b. Large Sodium Clusters. In 1990 at the Max Planck Institute in Germany, Martin et al. (37) measured abundance spectra for Na_n clusters from a few atoms to about 22,000. They confirmed that the jellium model can predict magic numbers for Na_n (n = 2, 8, 20, 40, 58, 92, 138, 198, 263, 341, 443, 557, . . .) up to about n = 1500. There, however, the sequence weakens and a new one appears corresponding not to the filling of electronic shells but rather to the packing of shells of atoms (as observed for Xe_n or Ar_n clusters to be discussed later).

The experimental procedure for the large sodium clusters involves adiabatic expansion followed by photoionization and analysis by a time-of-flight mass spectrometer. Increasing the wavelength of the ionizing light produces an increase in the average cluster size produced and is performed in order to span the range from about N = 300 to 22,000. An evident feature is that the period of oscillation changes abruptly in the region 1400–2000. The peaks below 1400 correspond to the filling of electronic shells (jellium model), while the peaks above 2000 correspond to the addition of successive layers of atoms on icosahedra or cuboctahedra (shell-by-shell growth). It is interesting to point out that the transition from electronic (jellium model) to atomic packing occurs at, approximately, v_8 icosahedron or cuboctahedron, both of which have 2057 atoms.

It is not unreasonable to assume that, for small clusters, the atoms are very mobile, and as each atom is added the shape changes to accommodate it into a sphere-like configuration, as in liquid droplets. Hence, the electronic effect prevails. As the size increases, changes in shape become increasingly more difficult, and a new growth pattern emerges. Each new atom added condenses on the surface and remains on the outer shell of the cluster; further growth takes place by the accumulation of shells of atoms around a rigid core, giving rise to the LBL or SBS growth discussed earlier. For large Na clusters, the magic numbers observed indicate v_n icosahedral or v_n cuboctahedral structures consistent with sphere packing. We discuss this layer-by-layer (or shell-by-shell) growth next.

V. ATOMIC CLOSED SHELLS: PACKING PATTERNS

A. Two Dimensional v_n Polygonal Clusters

Two dimensional (2-D) polygonal clusters of frequency n (18) can be thought of as made up of v_n triangles. In fact, the triangle is the basic building block in 2-D. Chart 3 shows early members of the v_n triangle series. Note that each polygon may be divided in two different ways. In the first the polygon is subdivided from one corner, and the total number of triangles is $F = p - 2$. In the second decomposition, the polygon is subdivided from the center, and the total numbers of triangles is $F = p$. Thus, Chart 4 depicts the centered triangular numbers.

General Formulas

There are simple general formulas for the numbers of points (atoms) in these figures. Let G_n denote the total number of points in a v_n polygon, S_n the number on the boundary (or perimeter), and $I_n = G_n - S_n$ the number in the interior. The general formulas for a p-sided polygon (which can be decomposed into F triangles) are

$$G_n = \frac{1}{2}Fn^2 + \frac{1}{2}pn + 1 \tag{12}$$

$$S_n = pn \tag{13}$$

$$I_n = \frac{1}{2}Fn^2 - \frac{1}{2}pn + 1 \tag{14}$$

The two different decompositions of a polygon (decomposed either from a corner or from the center) have F values differing by 2. It follows from Eqs. (12)–(14) that

$$G_n \text{ (centered polygon)} = G_n \text{ (polygon)} + n^2 \tag{15}$$

S_n (centered polygon) = S_n (polygon)   (16)

I_n (centered polygon) = I_n (polygon) + n^2 (17)

Listed below are a few examples.

1. Triangular Numbers (Chart 3)

Generating function

$G_n = \frac{1}{2}n^2 + \frac{3}{2}n + 1$

$S_n = 3n$

Magic numbers

G_n = 1, 3, 6, 10, 15, 21, 28, 36, 45, 55, 66, ...

S_n = 1, 3, 6, 9, 12, 15, 18, 21, ...

The triangular numbers, t_n, can be rewritten as

$$t_n = \frac{1}{2}n^2 + \frac{3}{2}n + 1$$

whose coefficients can easily be determined via the general formulas (12)–(14). Thus, for a triangle (Charts 3), of course, $p = 3$, $F = 1$, and Eqs. (12)–(14) become

$$G_n = \frac{1}{2}n^2 + \frac{3}{2}n + 1 = t_n$$

and $S_n = 3n$ and $I_n = \frac{1}{2}(n-1)(n-2)$. Note that Eq.(15) is satisfied.

2. Centered Triangle (Chart 4)

Generating function

$G_n = \frac{3}{2}n^2 + \frac{3}{2}n + 1$

$S_n = 3n$

Magic numbers

G_n = 1, 4, 10, 19, 31, 46, 64, 85, 109, 136, 166, ...

S_n = 1, 3, 6, 9, 12, 15, 18, 21, ...

B. Three-Dimensional v_n Polyhedral Clusters

Three dimensional polyhedra of frequency n (18) can be thought of as made up of v_n tetrahedra (Chart 5). In fact, the v_n tetrahedron is the basic building block of any given v_n polyhedron.

General Formulas

Let G_n denote the total number of points in a v_n polyhedron, S_n the number of points on the surface or outermost shell, and I_n the number of interior points. Then

$$G_n = I_n + S_n \quad (18)$$

Formulas for 3-D polyhedra depend on three parameters: (a) C, the number of v_n tetrahedral cells into which the polyhedron is divided; (b) F_s, the number of triangular faces on the surface; and (c) V_i, the number of vertices in the interior. For example, an icosahedron may be decomposed into 20 tetrahedra by placing a

Magic Numbers in Clusters

vertex at the center and joining it to the 12 boundary vertices. Each of the 20 tetrahedra has one of the 20 triangular faces as its base and the central vertex as its apex. Thus we have $C = 20$, $F_s = 20$, and $V_i = 1$.

The general formulas for 3-D polyhedra are:

$$G_n = \alpha n^3 + \tfrac{1}{2}\beta n^2 + \gamma n + 1, \quad n \geq 0 \tag{19}$$

where

$$\alpha = \tfrac{C}{6} \tag{20a}$$
$$\beta = \tfrac{1}{2}F_s \tag{20b}$$
$$\gamma = \tfrac{F_s}{4} + V_i + 1 - \tfrac{C}{6} \tag{20c}$$
$$S_n = \beta n^2 + 2, \quad n \geq 1 \tag{21}$$
$$I_n = G_n - S_n = \alpha n^3 - \tfrac{1}{2}\beta n^2 + \gamma n - 1, \quad n \geq 1 \tag{22}$$

By definition, $S_0 = 1$ and $I_0 = 0$. [These equations are unexpectedly simple. Special cases of Eq. (21) for bodies that can be embedded in the fcc lattice were discovered earlier by Buckminster Fuller (90) and proved by Coxeter (91).]

There is no direct analog of Eqs. (15)–(17) in three dimensions, the following formulas for centered polyhedra are nonetheless useful. Suppose a polyhedron is divided into C tetrahedral cells from a central vertex, so that $V_I = 1$ and $F_s = C$. Then Eqs. (19), (21), (22) become

$$G_n = \tfrac{1}{12}(2n + 1)(Cn^2 + Cn + 12), \quad n \geq 0 \tag{23}$$
$$S_n = \tfrac{1}{2}Cn^2 + 2, \quad n \geq 1 \tag{24}$$
$$I_n = \tfrac{1}{12}(2n - 1)(Cn^2 - Cn + 12), \quad n \geq 1 \tag{25}$$

In this case we also have the identity

$$I_n = G_{n-1} \tag{26}$$

We shall discuss a few examples of polyhedra of frequency n.

1. Tetrahedron (Chart 5)

Generating functions

$G_n = \tfrac{1}{6}n^3 + n^2 + \tfrac{11}{6}n + 1, \quad n \geq 0$

$S_n = 2n^2 + 2, \quad n \geq 1$

Magic numbers

$G_n = 1, 4, 10, 20, 35, 56, 84, 120, 165, 220, 286, \ldots$

$S_n = 1, 4, 10, 20, 34, 52, 74, 100, 130, 164, 202, \ldots$

Chart 5 shows regular tetrahedra of frequency 0, 1, 2, 3, ... In this case we have $C = 1$, $F_s = 4$, $V_i = 0$, so from Eqs. (19)–(22) we find that there are a total of

$$G_n = \tfrac{1}{6}n^3 + n^2 + \tfrac{11}{6}n + 1, \quad n \geq 0$$

There are

$$S_n = 2n^2 + 2, \quad n \geq 1$$

atoms on the surface and

$$I_n = \tfrac{1}{6}n^3 + n^2 - \tfrac{11}{6}n - 1, \quad n \geq 1$$

interior atoms.

The numbers of atoms in a centered tetrahedron are given by.

$$G_n = \tfrac{1}{3}(2n + 1)(n^2 + n + 3), \quad n \geq 0$$
$$S_n = 2n^2 + 2, \quad n \geq 1$$
$$I_n = \tfrac{1}{3}(2n - 1)(n^2 - n + 3), \quad n \geq 1$$

2. Octahedron (see Chart XVI of Ref. 34)

Generating functions

$$G_n = \tfrac{2}{3}n^3 + 2n^2 + \tfrac{7}{3}n + 1, \quad n \geq 0$$

$$S_n = 4n^2 + 2, \quad n \geq 1$$

Magic numbers

G_n = 1, 6, 19, 44, 85, 146, 231, 344, 489, 670, 891, . . .
S_n = 1, 6, 18, 38, 66, 102, 146, 198, 258, 326, 402, . . .

A ν_n octahedron may be decomposed into four tetrahedra (slicing it, for example, along the equatorial plane and by a vertical plane through the north and south poles). Thus $C = 4$, $F_s = 8$, and $V_i = 0$, and from Eqs. (19)–(22) we obtain the formulas for G_n, S_n, and I_n shown above.

3. Icosahedron (Chart 6)

Generating functions

$$G_n = \tfrac{10}{3}n^3 + 5n^2 + \tfrac{11}{3}n + 1, \quad n \geq 0$$

$$S_n = 10n^2 + 2, \quad n \geq 1$$

Magic numbers

G_n = 1, 13, 55, 147, 309, 561, 923, 1415, 2057, 2869, 3871, . . .
S_n = 1, 12, 42, 92, 162, 252, 362, 492, 642, 812, 1002, . . .

An icosahedron (18) may be decomposed into 20 tetrahedra. Thus, $C = F_s = 20$ and $V_i = 1$, and we obtain the formulas shown above.

Listings of generating functions and magic numbers and sequences of a wide variety of high-frequency polyhedral clusters, along with the way the polyhedra are decomposed into ν_n tetrahedra, can be found in the literature (18,34). We shall discuss ν_n icosahedra as an example. In 1981, Echt, Sattler, and Recknagel (92), in West Germany, were among the first to demonstrate experimentally that magic numbers exist in gas-phase clusters. The Xe atoms aggregate in clusters when undergoing adiabatic expansion plus cooling. The clusters produced in

Magic Numbers in Clusters 77

this way are then ionized and analyzed by a time-of-flight mass spectrometer (92–94). The magic numbers observed (Fig. 3) are n = 13, 19, (23), 25, 55, 71, (81), 87, (101), 147. The numbers in parentheses are magic numbers observed experimentally but do not fit the v_n icosahedral sequence. They relate to other growth pathways or packing patterns (e.g., the peak at 19 can be explained by a six-atom pentagonal cap of an icosahedron).

The majority of the peaks in Fig. 3 can be explained in terms of a high-frequency icosahedral structure series (Mackay sequence). Here we have $S_n = 10n^2 + 2$, $I_n = G_{n-1}$, and $G_n = I_n + S_n = 1, 13, 55, 147, 309, 561, \ldots$; in fact, all the magic numbers predicted are experimentally observed. The icosahedral structure was confirmed by electron diffraction experiments for a broad size range (95).

However, the behavior of Xe clusters does not extend to other noble-gas clusters. In fact, studies on Ar revealed a different set of magic numbers (95,96). It has been shown that the growth of Ar clusters adopts the form of polyicosahedra (cluster-of-clusters growth by either joining together or interpenetrating

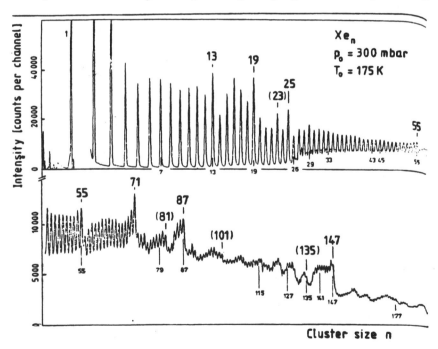

Fig. 3. Mass spectrum of Xe clusters. Observed magic numbers are consistent with the v_n icosahedral (Mackay) sequence. Peaks with numbers in parentheses indicate other less pronounced growth pathways. (Reproduced with permission from O. Echt, K. Sattler, and E. Recknagel, Phys. Rev. Lett., 1981, 47, 1121. Copyright 1981, American Physical Society.) (From Ref. 92.)

icosahedral units) instead of layer-by-layer growth (Mackay sequence) observed for Xe. The reason for this difference in behavior is not fully understood.

It should be emphasized that hard spheres cannot really be packed into a v_n icosahedral structure such that neighboring spheres on successive shells are touching. As a result, strains build up and at some stage it becomes more favorable for it to transform to a v_n cuboctahedral structure in which all atoms are of the same size, thereby relieving the strain energy. The cuboctahedron is, in fact, the basic unit of the fcc structure of bulk Xe. Note that the cuboctahedral structure has the same magic number sequence as the icosahedral structure (18,34).

C. Concepts of Peeling an Onion and Slicing Cheese for Atom Counting

In order to understand the packing of spheres (or atoms) in a v_n polyhedron, Teo and Sloane (18) introduced the concepts of "peeling an onion" and "slicing a cheese." Peeling an onion refers to the removal of the S_n surface atoms (the "skin") to reveal the I_n interior atoms (the "core") of a v_n polyhedral cluster. There are, in fact, three classes of v_n polyhedral clusters. In the first class, exemplified by the icosahedron (cf. Chart 6), cuboctahedron, and twinned cubotahedron, the removal of the surface atoms produces a v_{n-1} polyhedron of the same shape. In this case the whole onion can be built up from layers of the same shape. The second class is exemplified by a v_n octahedron which has the property that removal of the outer shell produces a v_{n-2}, not a v_{n-1}, octahedron. The onion is now composed of layers of the same shape, but the size of the layers decreases by 2 each time, viz., step size of $s = 2$. Similarly, for a v_n tetrahedron the interior is a v_{n-4} tetrahedron, in steps (s) of 4 (cf. Chart 5). A third class comprises those in which the interior polyhedron has a shape different from the original. For example, removing the outer shell from a truncated v_3 octahedron does not produce a similar polyhedron of a smaller size. The core of a truncated v_3 octahedron (cf. Chart 8a), in fact, is a v_1 octahedron.

The concept of slicing the cheese involves removing the atoms layer by layer, i.e., in planar layers, each layer being parallel to the next, as illustrated in Chart 8 for the truncated v_3 octahedron along the threefold (Chart 8b) and fourfold (Chart 8c) symmetry axes. Here we assume close-packing arrangements between adjacent layers of atoms. The most common types of close packing are icosahedral (icp), face-centered cubic (fcc), hexagonal close packing (hcp), cubic, and body-centered cubic (bcc), as we discuss in Sec. VI.

D. Concept of Tiling: Atomic Subshells

Recent experimental evidence suggests that while closed atomic shells are extraordinarily stable, there may exist, in some cases, subshells that lie between successive closed shells (97–101). The number of subshells depends on the number

Magic Numbers in Clusters

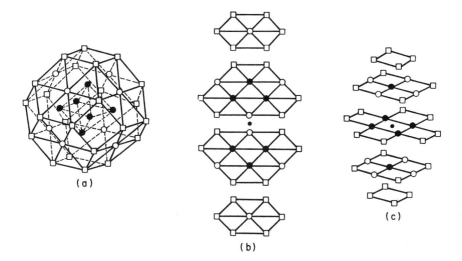

Chart 8. The "peeling onion" concept: the removal of the surface atoms of a truncated v_3 octahedral cluster gives rise to a simple octahedron.

of faces of a v_n polyhedral cluster that need to be covered (or tiled) to produce a v_{n+1} polyhedron. We may refer to the stepwise formation of the atomic subshells as a stepwise tiling process. For example, it is only necessary to cover four of the eight triangular faces of a v_n octahedron in order to produce a v_{n+1} octahedron, as illustrated in Fig. 4. Hence, it is expected to have three subshells, or four steps, which correspond to the tiling of the four adjacent triangular faces, as indicated in Fig. 4.

$$V_n \xrightarrow{\Delta} V_n^{1\Delta} \xrightarrow{\Delta} V_n^{2\Delta} \xrightarrow{\Delta} V_n^{3\Delta} \xrightarrow{\Delta} V_{n+1}$$

Here, tiling four triangles of a v_n octahedron produces a v_{n+1} octahedron: $v_n^{4\Delta} = v_{n+1}$. The tiling of the *atomic subshells* in the atomic packing of clusters may be likened to the filling of *electronic subshells* in the electronic jellium model, both via the Aufbau principle. As with the jellium model, the shape of the cluster may change as the subshells are filled or, more appropriately, tiled, but for distinctly different reasons.

Atomic subshell structures were first observed in aluminum clusters containing 250 to 400 atoms in 1991. The structures were discussed in terms of the jellium model with the assumption that each Al atom contributes three quasi-free electrons (97). Later, the size range was extended to 1400 atoms (98,99), but the periodicity observed indicated that the shells contained too few electrons expected from the jellium model. Different explanations were suggested. The most convincing one seems to be the one put forth by the Max Planck Institute researchers.

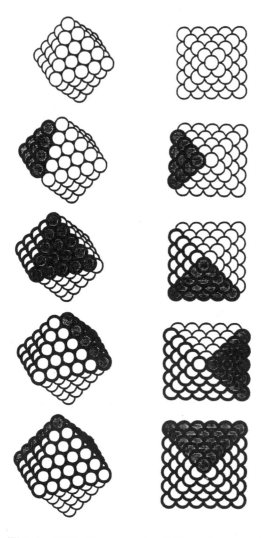

Fig. 4. "Tiling" concept: the addition of atoms to four triangular faces of a v_n octahedral cluster produces a v_{n+1} octahedral cluster. The coverage of one triangular face is referred to as a subshell; hence, there are three subshells between successive full shells. (From Ref. 101.) (Reproduced with permission from T. P. Martin, U. Zimmerman, N. Malinowski, U. Naher, S. Frank, F. Tast and K. Wirth, Surf. Rev. Lett., 1996, 3, 281. Copyright 1996, World Scientific Publications.)

Magic Numbers in Clusters

They suggested that the Al atoms organize themselves in octahedral geometry (100, 101). As mentioned earlier, for octahedral geometry one needs to cover only four triangular faces to obtain the next (larger) octahedron, resulting in three subshells between the consecutive full shells, as was indeed observed (Fig. 4).

VI. CLOSE-PACKED CIRCULAR AND SPHERICAL CLUSTERS

Since the majority of the known high-nuclearity metal clusters are more or less close-packed and either circular in two dimensions or spherical in three dimensions in nature, Sloane and Teo (19,20) explored the size and shape of possible circular and spherical clusters derivable from the common close-packed lattices: in 2-D, the square and the hexagonal lattices and in 3-D, the simple cubic (sc), fcc, bcc, and hcp lattices. For 2-D lattices, the choice of the origin may be a lattice point, the midpoint of an edge, or the center of a square or triangle. Circles of increasing radius are then drawn to pass through lattice points. For 3-D lattices, the origin may, in addition, be at the center of a tetrahedral or an octahedral hole (a hole is a point within the lattice most distant from the adjacent lattice points), and, instead of drawing circles, we draw spheres passing through the lattice points. We are interested in the number and arrangement of lattice points or atoms in each shell. Given these information, the size and shape of the clusters can easily be determined.

The atoms can be classified according to their distances from the origin. Let S_n denote the number of atoms at distance $(n)^{1/2}$ from the origin. These S_n atoms form a "shell" of radius $(n)^{1/2}$. Then the set of all atoms inside or on this shell forms a *close-packed circular* (2-D) or *spherical* (3-D) *cluster*, centered at the origin, containing a total of

$$G_n = \sum_{n=1}^{n} S_m \qquad (27)$$

atoms (the *nuclearity* of the cluster).

The numbers S_n and G_n may be found analytically from the theta series of the packing. Extensive tables of the numbers S_n and G_n including the first 80 or so layers of each type of cluster, can be found in the literature (Tables 1–27 in Ref. 19), along with the coordinates of the atoms.

For a particular close packing, the theta series (with respect to a particular choice of origin) is

$$\Sigma q^{N(x)} = \Sigma S_n q^n \qquad (28)$$

Equation (28) is a power series of the variable q, and $N(x) = x \cdot x$ is the *norm* of the vector x, which is its squared length. The coefficient of each term q^n gives the number of atoms, S_n, on the nth shell. The theta series can be expressed in terms of the Jacobi series (102–104). The theta series can be found in Refs. 19 and 20, along with examples in cluster chemistry.

A. Two-Dimensional Circular Clusters

1. Square Lattice with Respect to Lattice Point (Chart 9)

Generating function

$\theta_3(q)^2 = 1 + 4q + 4q^2 + 4q^4 + \cdots$

Magic numbers

$G_n = 1, 5, 9, 13, 21, 25, 29, 37, 45, 49, 57, \ldots$
$S_n = 1, 4, 4, 4, 8, 4, 4, 8, 8, 4, 8, \ldots$

Note that S_n is given by the coefficient of q^n in the theta series.

2. Square Lattice with Respect to Center of Square (Chart 10)

Generating function

$\theta_2(q)^2 = 4q^{1/2} + 8q^{5/2} + 4q^{9/2} + \cdots$

Magic numbers

$G_n = 4, 12, 16, 24, 32, 44, 52, 60, 68, \ldots$
$S_n = 4, 8, 4, 8, 4, 12, 8, 8, 8, \ldots$

Note that G_n and S_n of a square lattice with respect to a square hole (center of square) are twice the corresponding values for a square lattice with respect to an edge.

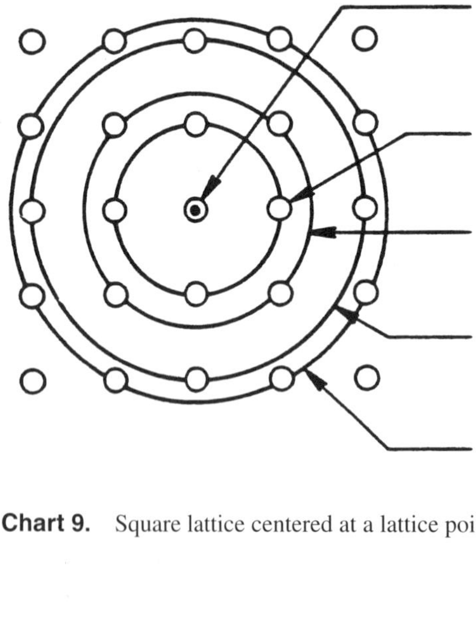

R_n	S_n	T_n
$\sqrt{0}$	1	1
$\sqrt{1}$	4	5
$\sqrt{2}$	4	9
$\sqrt{4}$	4	13
$\sqrt{5}$	8	21

Chart 9. Square lattice centered at a lattice point (see text).

Magic Numbers in Clusters

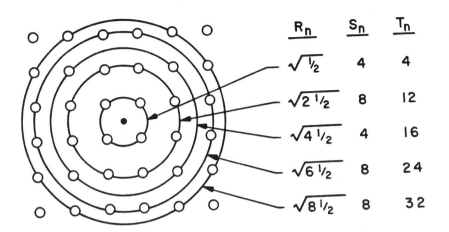

	R_n	S_n	T_n
	$\sqrt{1/2}$	4	4
	$\sqrt{2\,1/2}$	8	12
	$\sqrt{4\,1/2}$	4	16
	$\sqrt{6\,1/2}$	8	24
	$\sqrt{8\,1/2}$	8	32

Chart 10. Square lattice centered at a square hole (see text).

B. Three-Dimensional Spherical Clusters

1. Simple Cubic Lattice with Respect to Lattice Point (Chart 11)

Generating function
$\theta_3(q)^3 = 1 + 6q + 12q^2 + 8q^3 + \cdots$

Magic numbers
$G_n = 1, 7, 19, 27, 3, 57, 81, 81, 93, 123, 147, \ldots$
$S_n = 1, 6, 12, 8, 6, 24, 24, 0, 12, 30, 24, \ldots$

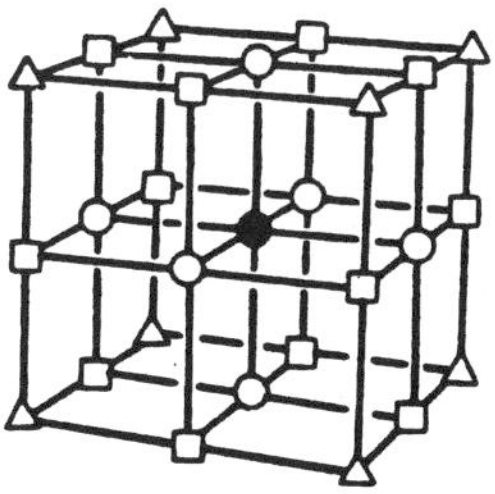

	R_n	S_n	T_n	coordinates
●	$\sqrt{0}$	1	1	000
○	$\sqrt{1}$	6	7	100
□	$\sqrt{2}$	12	19	110
△	$\sqrt{3}$	8	27	111

Chart 11. Simple cubic lattice centered at a lattice point (see text).

VII. CONCLUSION

It is known that clusters with magic numbers of atoms are more abundant than others, presumably because they are particularly stable. In this chapter, we presented representative examples of sequences of magic numbers frequently observed in clusters. It is believed that the formation of these magic sequences are intimately related to the nucleation and growth processes that are dictated by the often competing bonding and packing factors. While these processes may be kinetically or thermodynamically controlled, and hence sensitive to experimental conditions, the facts that the bonding effects are electronic in origin and the packing factors are steric in nature often give rise to polyhedra clusters of high symmetry and packing efficiency. Shapes such as the 5 Platonic, the 13 Archimedean, and other, related, more-or-less spherical, polyhedral are frequently found. The progressions of these clusters, via either layer-by-layer or cluster-of-clusters growth pathways, form sequences of magic numbers as a result of filling the electronic or atomic shells or both. In many cases, these progressions correspond to the succession of high-frequency polyhedral clusters of increasing size while maintaining overall shape and symmetry. Such a global symmetry is the result of the self-organization and self-similarity principles that permeate all levels of different stages of the cluster growth processes. These magic numbers and sequences are also commonly observed and/or utilized in other branches of science and technology as well as in many engineering research and applications.

ACKNOWLEDGMENTS

We are grateful to the National Science Foundation for the financial support of this work. We thank Enrico DeVita, Chris Keh, and Dominic Papandria (UIC) for their help in the preparation of this chapter.

REFERENCES

1. WN Lipscomb. Boron Hydrides. New York: W.A. Benjamin, 1963.
2. EL Muetterties, ed. Boron Hydride Chemistry. New York: Academic Press, 1975.
3. EL Muetterties, WH Knoth. Polyhedral Boranes. New York: Marcel Dekker, 1968.
4. RW Rudolph. Acc. Chem. Res. 9:466, 1976.
5. RN Grimes. Carboranes. New York: Academic Press, 1970.
6. RN Grimes, ed. Metal Interactions with Boron Clusters. New York: Plenum Press, 1982.
7. ME O'Neill, K Wade. In: G Wilkinson, FGA Stone, EW Abel, eds. Comprehensive Organometallic Chemistry, Vol. 1. Oxford: Pergamon Press, 1982.

8. T Onak. In: G. Wilkinson, FGA Stone, EW Abel, eds. Comprehensive Organometallic Chemistry, Vol. 1. Oxford: Pergamon Press, 1982.
9. RN Grimes. In: G. Wilkinson, FGA Stone, EW Abel, eds. Comprehensive Organometallic Chemistry, Vol. 1. Oxford: Pergamon Press, 1982.
10. JD Corbett. Prog. Inorg. Chem. 21:129, 1976.
11. RJ Gillespie. Chem. Soc. Rev. 8:315, 1979.
12. P Chini. Gazz. Chim. Ital. 109:225, 1979.
13. P Chini. J. Organomet. Chem. 200:37, 1980.
14. P Chini, G Longoni, V Albano. G. Adv. Organomet. Chem. 14:285, 1976.
15. BFG Johnson, ed. Transition Metal Clusters. Chichester: Wiley-Interscience, 1980.
16. RE Benfield, BFG Johnson. Top. Stereochem. 12:253, 1981.
17. DMP Mingos. In: G Wilkinson, FGA Stone, EW Abel, eds. Comprehensive Organometallic Chemistry, Vol. 1. Oxford: Pergamon Press, 1982.
18. BK Teo, NJA Sloane. Inorg. Chem. 24:4545, 1985.
19. NJA Sloane, BK Teo. J. Chem. Phys. 83:6520, 1985.
20. BK Teo, NJA Sloane. Inorg. Chem. 25:2315, 1986.
21. JP Borel, J Buttet. Small Particles and Inorganic Clusters. Amsterdam: North-Holland, 1981.
22. J Wronka, DP Ridge. J. Am. Chem. Soc. 106:67, 1984.
23. MF Jarrold, JE Bower, JS Kraus. J. Chem. Phys. 86:3876, 1987.
24. S Leutwyler, A Hermann, L Woeste, E Schumacher. Chem. Phys. 48:253, 1980.
25. DE Powers, SG Hansen, ME Geusic, AC Puiu, JB Hopkins, TG Dietz, MA Duncan, PRR Langridge-Smith, RE Smalley. J. Phys. Chem. 86:2556, 1982.
26. JL Gole, JH English, UE Bondybey. J. Phys. Chem. 86:2560, 1982.
27. TM Barlak, JR Wyatt, RJ Colton, JJ deCorpo, JE Campana. J. Am. Chem. Soc. 104:1212, 1982.
28. DA Lichtin, RB Bernstein, V Vaida. J. Am. Chem. Soc. 104:1831, 1982.
29. TF Magnera, DE David, R Tian, D Stulik, J Michl. J. Am. Chem. Soc. 106:5040, 1984.
30. AW Castleman Jr, RG Keese. Science 241:36, 1997.
31. S Bjornholm. Contemp. Phys. 31:309, 1990.
32. H Zhang, BK Teo. Inorg. Chim. Acta. 265:213, 1997.
33. H Zhang, BK Teo. Coor. Chem. Rev. 143:611, 1995.
34. BK Teo, H Zhang. Polyhedron. 9:1985, 1990.
35. W Knight, K Clemenger, W deHeer, W Saunders, M Chou, ML Cohen. Phys. Rev. Lett. 52:2141, 1984.
36. WD Knight, W deHeer, WA Saunders. Z Phys D—Atoms, Molecules and Clusters 3:109–114, 1986.
37. TP Martin, T Bergmann, H Golich, T Lange. Chem. Phys. Lett. 172:209, 1990.
38. M Brack. Sci Am 50, 1997.
39. M Ghyka. The Geometry of Art and Life. New York: Dover, 1977.
40. BK Teo, H Zhang. J. Cluster Sci. 1:223–228, 1990.
41. BK Teo, H Zhang. J. Cluster Sci. 1:155–187, 1990.
42. BK Teo, H Zhang. Inorg. Chem. 27:414–417, 1988.
43. BK Teo, H Zhang. Inorg. Chim. Acta. 144:173–176, 1988.
44. BK Teo, H Zhang. Proc. Natl. Acad. Sci. 88:5067–5071, 1991.

45. BK Teo, H Zhang. J. Cluster Sci. 6:203, 1995.
46. MR Hoare, P Pal. Adv. Phys. 20:161–196, 1971.
47. CL Briant, J Burton. J. Phys. Status. Solidi. 85:393, 1978.
48. AC Mackay. Acta. Crystallogr. 15:916, 1962.
49. B Bagley, Nature (London) 208:674, 1965.
50. CE Briant, BRC Theobald, JW White, LK Bell, DMP Mingos, AJ Welch. J. Chem. Soc. Chem. Commun. 201–202, 1981.
51. BK Teo, H Zhang, X Shi. Inorg. Chem. 29:2083–2091, 1990.
52. BK Teo, MC Hong, H Zhang, DB Huang. Angew. Chem. Int. Ed. Engl. 26:897–900, 1987.
53. BK Teo, H Zhang, X Shi. J. Am. Chem. Soc. 112:8552, 1990.
54. BB Mandelbrot. The Fractal Geometry of Nature. New York: Freeman, 1983.
55. HO Peitgen, P Richter. The Beauty of Fractals. Heidelberg: Springer, 1986.
56. R Hoffmann, WN Lipscomb. J. Chem. Phys. 36:2179, 1962.
57. RE Williams. Inorg. Chem. 10:210, 1971.
58. KJ Wade. Chem. Soc. Chem. Commun. 792, 1971.
59. KJ Wade. Adv. Inorg. Chem. Radio. Chem. 18:1, 1976.
60. DMP Mingos. Acc. Chem. Res. 17:311, 1984.
61. A Stone. Inorg. Chem. 20:563, 1981.
62. JW Lauher. J. Am. Chem. Soc. 100:5305, 1978.
63. JW Lauher. J. Am. Chem. Soc. 101:2604, 1979.
64. JW Lauher. J. Organomet. Chem. 213:25, 1981.
65. RB King. Inorg. Chim. Acta. 116:99, 1986.
66. RB King. Inorg. Chim. Acta. 116:109, 1986.
67. RB King. Inorg. Chim. Acta. 116:119, 1986.
68. RB King. Inorg. Chim. Acta. 116:125, 1986.
69. RB King, DH Rouvray. J. Am. Chem. Soc. 99:7834, 1977.
70. BK Teo. Inorg. Chem. 24:1627, 1985.
71. BK Teo. Inorg. Chem. 24:4209, 1985.
72. BK Teo. Inorg. Chem. 23:1251, 1984.
73. BK Teo, G Longoni, FRK Chung. Inorg. Chem. 23:1257, 1984.
74. MI Forsyth, DMP Mingos. J. Chem. Soc. (Dalton Trans) 610, 1977.
75. DMP Mingos. Nature 236:99, 1972.
76. ER Corey, LF Dahl. Inorg. Chem. 1:521, 1962.
77. MR Churchill, BG DeBoer. Inorg. Chem. 16:878, 1977.
78. G Doyle, KA Eriksen, D Van Engan. J. Am. Chem. Soc. 107:7914, 1985.
79. CH Wei, LF Dahl. J. Am. Chem. Soc. 88:1821, 1966.
80. PF Jackson, BFG Johnson, J Lewis, WJH Nelson, M McPartlin. J. Chem. Soc. (Dalton):2099, 1982.
81. C-MT Hayward, JR Shapley, MR Churchill, C Bueno, AL Rheingold. J. Am. Chem. Soc. 104:7347, 1982.
82. S Martinengo, A Fumagalli, R Bonfichi, G Ciani, A Sironi. J. Chem. Soc. Chem. Commun. 825, 1982.
83. BK Teo, H Zhang, X Shi. Inorg. Chem. 29:2083, 1990.
84. BK Teo, H Zhang. Inorg. Chem. 30:3115, 1991.
85. BK Teo, H Zhang, X Shi. J. Am. Chem. Soc. 112:8552, 1990.

86. BK Teo, M Hong, H Zhang, D Huang, X Shi. J. Chem. Soc., Chem. Commun. 204, 1988.
87. BK Teo, MC Hong, H Zhang, DB Huang. Angew. Chem., Int. Ed. Engl. 26:897, 1987.
88. K Clemenger. Phys. Rev. B32:1359, 1985.
89. I Katakuse, T Ichihara, Y Fujita, T Matsuo, T Sakurai, H Matsuda. Int. J. Mass. Spec. Ion. Proc. 74:33, 1986.
90. RW Marks. The Dynamic World of Buckminster Fuller. Carbondale, Ill: Southern Illinois University Press, 1960.
91. HSM Coxeter. In: RS Cohen et al., eds. For Dirk Struik. Dordrecht, Holland: Reidel, 1974, pp. 25.
92. O Echt, K Sattler, E Recknagel. Phys. Rev. Lett. 47:1121, 1981.
93. J Farges, MF deFeraudy, B Raoult, G Torchet. J. Phys. (Paris) Colloq. 38:C2–47, 1977.
94. J Farges, MF deFeraudy, B Raoult, G Torchet. Surf. Sci. 106:95, 1981.
95. J Farges, MF deFeraudy, B Raoult, G Torchet. J. Chem. Phys. 84:3491, 1986.
96. J Farges, MF deFeraudy, B Raoult, G Torchet. J. Chem. Phys. 78:5067, 1983.
97. JL Persson, RL Whetten, HP Cheng, RS Berry. Chem. Phys. Lett. 186:215, 1991.
98. J Lerme, M Pellarin, JL Vialle, B Baguenard, M Broyer. Phys. Rev. Lett. 68 (18):2818, 1991.
99. M Pellarin, B. Baguenard, M Broyer, J Lerme, JL Vialle, A Perez. J. Chem. Phys. 98:944, 1993.
100. TP Martin, U Naher, H Schaeber. Chem. Phys. Lett. 199:470, 1992.
101. TP Martin, U Zimmerman, N Malinowski, U Naher, S Frank, F Tast, K Wirth. Surf. Rev. Lett. 3:281–286, 1996.
102. R Bellman. A Brief Introduction to Theta-Functions. New York: Holt, Rinehart, and Winston, 1961.
103. ET Whittaker, GN Watson. A Course of Modern Analysis, 4th ed. Cambridge: Cambridge University, 1963.
104. C Tannery, J Molk. Eléments de la théorie des fonctions elliptiques, 2nd ed., Chelsea: New York, 1972, Vols. 1–4.

4
Modeling Metal Nanoparticle Optical Properties

K. Lance Kelly, Traci R. Jensen, Anne A. Lazarides, and George C. Schatz
Northwestern University, Evanston, Illinois

I. INTRODUCTION

The interaction of light with noble-metal nanoparticles has been a significant stimulus for scientific research (1). Colloidal nanoparticles are responsible for the brilliant reds (gold particles) and yellows (silver particles) in stained glass windows, and it was one of the great triumphs of classical physics when, in 1908, G. Mie presented a complete solution to Maxwell's equations for a sphere (2), which provided an accurate description of the extinction spectra of spheres of arbitrary size. Mie's solution remains of great interest to this day, but the modern generation of metal nanoparticle experiments has provided new challenges to theory that have spawned new directions for research. In this chapter we describe recent advances in theoretical research in this area, emphasizing especially the linear optical properties (extinction, absorption, scattering) of isolated noble-metal particles of arbitrary shape and sizes in the 10–200 nm range.

One of the reasons why Mie's theory has remained important for so long is that the only routine method for preparing metal nanoparticles with somewhat controllable properties has been colloid chemistry. Under the best of circumstances, such as for 10–15 nm Au particles, typical wet-chemistry methods for colloid synthesis make unaggregated spherical particles with ± 1-nm dispersion (3). Obviously this is a good situation for applying Mie theory. However, most colloids involve particles that have significant dispersion in particle size and shape. In addition, some fraction of the particles is aggregated (4). These problems make it difficult to use Mie theory in a rigorous fashion, but it is still used for qualitative information.

Recently there has been growing interest in characterizing the optical properties of metal nanoparticles that are made by lithographic methods, such as nanosphere lithography (5), e-beam lithography (6), and by other methods (7,8) that produce well-defined shapes and sizes without aggregation. This has stimulated new interest in the electrodynamics of isolated nanoparticles (9–12), but this time specific attention has been directed to nonspherical nanoparticles with structures that are known from AFM or STM measurements. Other complicating factors include the presence of a substrate that supports the particles, a solvent layer on top of the particles, and particles that are sometimes close enough together that their electromagnetic coupling changes the spectra. Although extinction, absorption, and scattering are still the primary optical properties of interest, other spectroscopic techniques are also being brought to bear on these particles, including surface-enhanced Raman spectroscopy (SERS) (5,13–16) and a variety of nonlinear scattering measurements [hyper-Rayleigh (17), hyper-Raman (18), SHG (19), etc.]. These techniques are sensitive to the electromagnetic fields at or near the particle surfaces (whereas extinction is more sensitive to fields in the interior of the particle), thus providing new challenges to the development of accurate methods.

In this chapter we give an overview of recent theoretical work that is aimed at describing the optical properties of noble-metal nanoparticles. This includes a fairly brief introduction to Mie theory, as well as theories that have been developed to treat particles other than spheres, particularly the Discrete Dipole Approximation. We then consider the application of these theories to a variety of problems of recent interest for gold and/or silver particles. These include studies of the particle size and shape dependence of extinction spectra, the treatment of substrate and solvent effects for nonspherical particles, the treatment of particle interactions in dimers, and the calculation of surface electromagnetic fields. We will not consider particle interaction effects for aggregates larger than dimers, but we should note several recent papers on this topic (20–22) that use methods related to those discussed here.

II. MIE THEORY AND LONG-WAVELENGTH APPROXIMATIONS

A. Mie Theory

In this section we want to solve Maxwell's equations for light of frequency ω, represented as a plane wave, scattering from a spherical particle. Except at the particle surface, this requires that we be able to describe light propagating through a linear, isotropic, homogeneous medium. In this case, Maxwell's equations may be reduced to solving the wave equations (2)

$$\nabla^2 \mathbf{E} + k^2 \mathbf{E} = 0$$
$$\nabla^2 \mathbf{H} + k^2 \mathbf{H} = 0 \qquad (1)$$

for the electric field $\mathbf{E}$ and the magnetic field $\mathbf{H}$. Here the constant k is related to ω by (in SI units) $k^2 = \omega^2 \epsilon \mu$, where ϵ is the permittivity and μ is the permeability. Often we will replace ϵ by $\epsilon_0 \varepsilon$, where ε is a dimensionless quantity known as the dielectric function, and ϵ_0 is the permittivity of free space. One can similarly treat the permeability, but in this chapter we will always assume that $\mu = \mu_0$, where μ_0 is the permeability of free space.

Equation (1) must be solved subject to the divergence-free constraints $\nabla \cdot \mathbf{E} = 0$ and $\nabla \cdot \mathbf{H} = 0$. In addition, $\mathbf{E}$ and $\mathbf{H}$ are related to each other via

$$\nabla \times \mathbf{E} = i\omega\mu\mathbf{H}$$
$$\nabla \times \mathbf{H} = -i\omega\epsilon\mathbf{E} \qquad (2)$$

To solve the vector wave equations, we express $\mathbf{E}$ and $\mathbf{H}$ as linear combinations of the vector functions $\mathbf{M}$ and $\mathbf{N}$. These functions satisfy the wave equation and are related to each other via $\mathbf{N} = k^{-1}\{\nabla \times \mathbf{M}\}$. In addition, $\mathbf{M}$ may be expressed in terms of the scalar function ψ by using $\mathbf{M} = \nabla \times \mathbf{r}\psi$, where $\mathbf{r}$ is the coordinate vector.

The scalar function ψ may be determined by solving the equation

$$\nabla^2 \psi + k^2 \psi = 0 \qquad (3)$$

subject to the boundary condition at the sphere surface that the tangential components of $\mathbf{E}$ and $\mathbf{H}$ derived from ψ should be continuous. The general solution of this equation may be written in terms of the two functions

$$\psi_{emn} = \cos m\varphi \, P_n^m(\cos\theta) \, z_n(kr)$$
$$\psi_{omn} = \sin m\varphi \, P_n^m(\cos\theta) \, z_n(kr) \qquad (4)$$

where $m = 0, \pm1, \pm2, \pm3, \ldots, \pm n$ and $n = 0, 1, 2, \ldots$. The functions P_n^m are the associated Legendre polynomial, familiar from wavefunctions for the hydrogen atom, and z_n is a spherical Bessel function. The Bessel functions may be chosen in four different ways; i.e., $z_n = j_n(kr), y_n(kr), h_n^+(kr)$, and $h_n^-(kr)$, depending on what boundary conditions are imposed. The asymptotic ($r \to \infty$) limits of these four functions are $\sin(kr - n\pi/2)$, $\cos(kr - n\pi/2)$, $e^{i(kr - n\pi/2)}$, and $e^{-i(kr - n\pi/2)}$, respectively. Another important issue concerns the behavior of the Bessel functions for $r = 0$. All the spherical Bessel functions diverge at $r = 0$ except $j_n(kr)$. This means that only $j_n(kr)$ may be used to represent a finite field at the coordinate origin.

The functions ψ_{emn} and ψ_{omn} can be used to construct magnetic and electric fields, using the definitions $\mathbf{M}_{emn} = \nabla \times \mathbf{r}\psi_{emn}$ and $\mathbf{N}_{emn} = k^{-1}\{\nabla \times \mathbf{M}_{emn}\}$ and similarly for the *omn* subscripts. We omit the details here. To use these fields to define complete solutions to light scattering off a spherical particle, we need to express the plane-wave field in terms of $\mathbf{M}_{emn}$, $\mathbf{N}_{emn}$, and their *omn* counterparts. If we consider polarized light, where $\mathbf{k}$ is along the z axis and the x axis defines the polarization direction, then the resulting "partial wave" expansion is

$$\mathbf{E}_i = E_0 e^{ikr\cos\theta} \mathbf{e}_\mathbf{x} = E_0 \sum_{n=1}^{\infty} i^n \frac{2n+1}{n(n+1)} (\mathbf{M}_{o1n}^{(1)} - i\mathbf{N}_{e1n}^{(1)}) \tag{5}$$

where $\mathbf{e}_\mathbf{x}$ is a unit vector along the polarization direction, and the sum expands the plane wave into spherical-wave solutions. The superscript (**1**) in the vector spherical harmonics stands for the $j_n(kr)$ solution. The scattered field $\mathbf{E}_s$ outside the spherical surface is likewise written as

$$\mathbf{E}_s = E_0 \sum_{n=1}^{\infty} i^n \frac{2n+1}{n(n+1)} (ia_n \mathbf{N}_{e1n}^{(3)} - b_n \mathbf{M}_{o1n}^{(3)}) \tag{6}$$

where the superscript (**3**) stands for the outgoing wave spherical Bessel function h_n^+. The complete solution is now written as the sum of the incident plus scattered waves solution for geometries outside the sphere. A form similar to Eq. (6) applies inside the sphere, except that the superscript (**1**) is used so that the field will be finite at the origin. By matching tangential components of the inside and outside fields at the sphere surface and applying similar boundary conditions to $\mathbf{H}$, one can express the unknown coefficients a_n and b_n in terms of known quantities. The resulting expressions are

$$a_n = \frac{\mu m^2 j_n(x)[xj_n(x)]' - \mu_1 j_n(x)[mxj_n(mx)]'}{\mu m^2 j_n(mx)[xh_n^{(1)}(x)]' - \mu_1 h_n^{(1)}(x)[mxj_n(mx)]'} \tag{7}$$

$$b_n = \frac{\mu_1 j_n(x)[xj_n(x)]' - \mu j_n(x)[mxj_n(mx)]'}{\mu_1 j_n(mx)[xh_n^{(1)}(x)]' - \mu h_n^{(1)}(x)[mxj_n(mx)]'} \tag{8}$$

where $x = ka$, $m = k_1/k$ is the ratio of indices of refraction (i.e., the ratio of the square roots of the dielectric functions inside and out), and μ and μ_1 are the magnetic permeabilities (which in this chapter are assumed to be the same). The subscript 1, wherever it appears, always refers to properties inside the sphere.

With these coefficients determined, one can evaluate the complete electric field $\mathbf{E}_i + \mathbf{E}_s$ outside the sphere. If the magnetic field is similarly determined and the resulting outgoing radiative flux is calculated in the $r \to \infty$ limit, then one can determine the scattering and extinction cross sections. The scattering cross section C_{sca} is the ratio of outgoing radial flux to the incoming flux associated with the plane-wave solution, while the extinction cross section C_{ext} is proportional to the ratio of the scattered flux in the forward direction to incoming flux. In Mie theory, C_{sca} and C_{ext} only depend on the coefficients a_n and b_n:

$$C_{sca} = \frac{2\pi}{k^2} \sum_{n=1}^{\infty} (2n+1)(|a_n|^2 + |b_n|^2) \qquad (9)$$

$$C_{ext} = \frac{2\pi}{k^2} \sum_{n=1}^{\infty} (2n+1)\text{Re}(a_n + b_n) \qquad (10)$$

These expressions for the cross sections are in a convenient form for numerical computations, as methods for evaluating the Bessel functions needed to determine the coefficients a_n and b_n are available for a wide range of orders and arguments. Also, decomposition of the cross section into electric (a_n) and magnetic (b_n) contributions from different partial waves is often useful for numerical simplifications and physical insight. For small particles (small x) one finds that $a_n \propto x^{(2n+1)}$ and $b_n \propto x^{(2n+3)}$, so the dominant term is the dipole term involving a_1, and higher electrical terms contribute equally with magnetic terms of the previous order. The leading order contribution to a_1 is

$$a_1 = -\frac{2ix^3}{3} \frac{\varepsilon - 1}{\varepsilon + 2} \qquad (11)$$

where we have used the conversion $\varepsilon = m^2$ between dielectric function and index of refraction (and for particles not in vacuum, ε refers to the ratio of inside to outside dielectric functions). This expression for a_1 connects us to the long-wavelength approximation, which is considered in the next section; it also shows us that the cross sections will have a peak whenever the real part of the denominator vanishes (Re $\{\varepsilon + 2\} = 0$). This corresponds to excitation of the electric dipole plasmon resonance. This only occurs in metals and similar materials that have negative real dielectric functions that increase from $-\infty$ at zero frequency to 0 at the bulk plasmon frequency.

To complete this section we present some representative results of Mie theory calculations. In an earlier paper (9) we did this for Ag spheres, so here we consider another important metal, Au, using dielectric functions from Palik (23). The results are presented in Fig. 1A and 1B. Figure 1a shows extinction spectra for Au particles having radii from 10 to 200 nm. This shows the well-known dipole plasmon resonance maximum, which for small particles occurs at about 520 nm (corresponding to the wavelength where Re $\{\varepsilon + 2\} = 0$). As the particle gets larger, the dipole resonance red-shifts and broadens for reasons that will be explained in the next section. In addition, for a radius of 200 nm we see the appearance of three peaks in the extinction profile. For this size, the dipole plasmon resonance has shifted so far to the red (to beyond 1000 nm) that a quadrupole plasmon resonance is now visible at 620 nm and an octopole resonance at 520 nm.

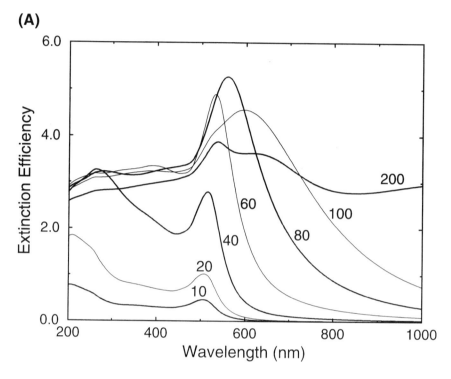

Fig. 1. (A) Mie theory results for the extinction spectra (extinction efficiency vs. wavelength) of gold spheres having radii from 10 to 200 nm. The extinction efficiency is the ratio of the extinction cross section to the area of the sphere. (B) Mie theory results for 30- and 100-nm Au spheres, showing the decomposition of the extinction cross section into contributions from scattering and absorption.

Figure 1B shows the decomposition of the extinction for 30- and 100-nm radius particles into contributions from absorption and scattering. For the 30-nm case, the scattering contribution is small, and all three curves have a similar dependence on wavelength. This situation changes dramatically for the 100-nm particle. Here we see that scattering is more important than absorption for wavelengths longer than 500 nm. The scattering contribution is also considerably more red-shifted than absorption, so we see that most of the red-shifting and broadening of the extinction as particle size is increased is due to the growth of scattering.

B. Long-Wavelength Approximations

If the particles are very small compared to the wavelength of light (typically <1%), then it is possible to replace the electrodynamics treatment just described by electrostatics. By this we mean that one takes the limit of zero frequency in

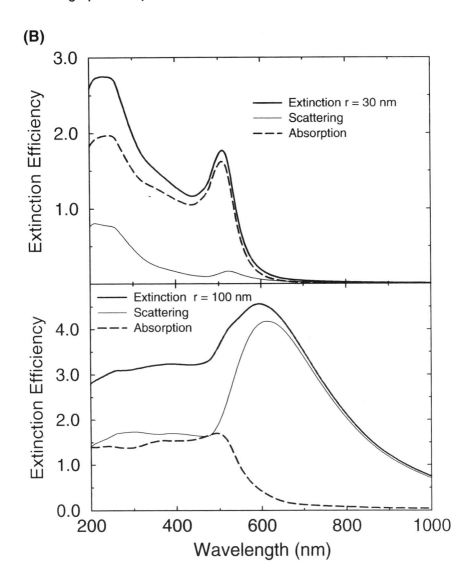

solving Maxwell's equations [thereby allowing us to solve Laplace's equation for the electric potential instead of Eq. (1)], but the correct frequency-dependent dielectric function is used (sometimes this is called the quasistatic limit). The virtue of this approximation is that it is possible to solve Laplace's equation analytically for several important particle shapes, including spheres. The details of this approach are described by Zeman and Schatz (24,25). In its lowest-order treatment, one finds that the induced polarization **P** in a sphere resulting from the imposition of a field **E** is

$$\mathbf{P} = \alpha \mathbf{E} \tag{12}$$

where the polarizability α is

$$\alpha = a^3 \frac{\varepsilon - 1}{\varepsilon + 2} \tag{13}$$

where a is the sphere radius. This expression shows that α is proportional to the long-wavelength expression for the Mie a_1 parameter [Eq. (11)]. As a result this theory predicts that the plasmon resonance occurs when Re $\{\varepsilon + 2\} = 0$, independent of particle size.

The real power of the long-wavelength limit occurs when one includes perturbative corrections for electrodynamic effects through order $1/\lambda^3$. This allows us to extend the electrostatic treatment to particles that are up to 10% of the wavelength of light in any dimension. We will call this corrected electrostatic treatment the modified long-wavelength approximation (MLWA) (26).

The electrodynamic corrections can be thought of as including for the leading terms in Eq. (12) that arise when left- and right-hand sides are expanded in powers of $1/\lambda$, and then the result is averaged over the particle volume. The lowest-order term is, of course, the electrostatic solution. For spherical particles with polarization **P**, the next higher-order corrections involve rewriting Eq. (12) as (27)

$$\mathbf{P} = \alpha[\mathbf{E} + \mathbf{E}_{rad}] \tag{14}$$

where the radiative correction field $\mathbf{E}_{rad}$ is

$$\mathbf{E}_{rad} = \frac{2}{3}ik^3 \mathbf{P} + \frac{k^2}{a}\mathbf{P} \tag{15}$$

The first term in this expression describes *radiative damping*. It arises from spontaneous emission of radiation by the induced dipole. This emission grows rapidly with particle size, eventually reducing the size of the induced dipole and increasing the plasmon linewidth. The second term comes from depolarization of the radiation across the particle surface due to the finite ratio of particle size to wavelength. This *dynamic depolarization* term causes red-shifting of the plasmon resonance as the particle size is increased. An equivalent theory for other particle shapes has been described by Zeman and Schatz (24,25).

The net effect of both of these terms is to produce a modified polarization in which the polarization **P** of the particle is given by Eq. (12) multiplied by the following correction factor:

$$F = (1 - \frac{2}{3}ik^3 \alpha - \frac{k^2}{a}\alpha)^{-1} \tag{16}$$

Modeling Optical Properties

Note that the radiative damping contribution to the correction factor is proportional to the product of the polarizability (proportional to particle volume) times k^3 [$= (2\pi/\lambda)^3$]. The dynamic depolarization term is proportional to α/a (proportional to particle area) times k^2. Clearly both terms will be of order unity when the particle radius becomes comparable to $\lambda/2\pi$, which for $\lambda = 600$ nm implies $a = 100$ nm.

Figure 2a shows the MLWA predictions of extinction for Au spheres. This figure is intended to be analogous to Fig. 1a, and indeed we see many similarities for small particle sizes. However, the larger particle sizes show considerable variation, with the MLWA plasmon considerably more red-shifted than the exact result. Figure 2b shows MLWA results analogous to Fig. 1b. This treatment shows that the smaller particle is again dominated by absorption, while the larger one is mostly scattering. In fact a major difference between the exact and MLWA results is the absence of significant absorption in the MLWA results for the larger size sphere. This indicates that the large absorption seen in Fig. 1b for wavelengths less than 500 nm is due to the contribution of higher multipoles.

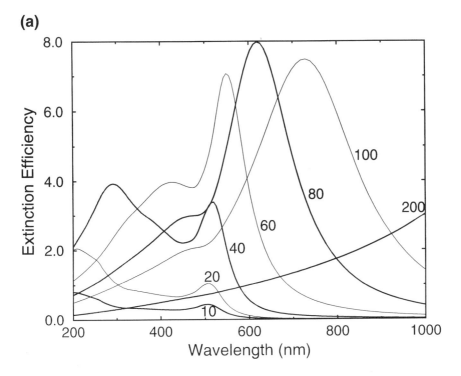

Fig. 2a. MLWA extinction spectra for gold spheres analogous to those in Fig. 1a.

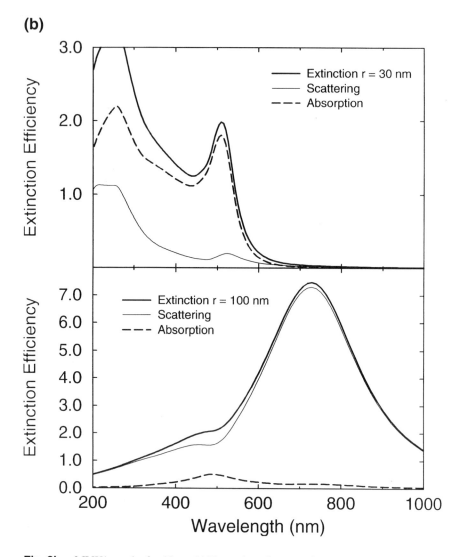

Fig. 2b. MLWA results for 30- and 100-nm Au spheres analogous to Fig. 1b.

III. TREATING NONSPHERICAL PARTICLES

A. General Discussion

As soon as one considers particles other than spheres, it is usually not possible to obtain analytical solutions to Maxwell equations. Because of this, there has been a great deal of effort put into developing numerical methods. In fact, since the classical electrodynamics problem is also relevant to studying radar, the literature on numerical methods is enormous. However, there are some important challenges and simplifications associated with metal nanoparticles that limit the applicability of methods used to study problems with much longer wavelengths and length scales. [Here we note that Maxwell's equations are not invariant to scaling both the wavelength and size by the same amount, as dielectric functions vary with frequency but not size (except for very small particles).]

Among the numerically exact methods that have been used to describe nonspherical metal nanoparticles are

1. Discrete dipole approximation (DDA) (9–12,26,28,29)
2. Multiple multipole methods (MMP) (30)
3. Finite-difference time-domain (FDTD) methods (31)
4. T-matrix methods (32)

DDA methods divide the particle into a large number of polarizable cubes. The induced dipole polarizations in these cubes are determined self-consistently (see Sec. III.C), and then properties such as the extinction cross section are determined in terms of the induced polarizations. The MMP method divides the particle into domains with shapes that allow for series expansion solutions of Maxwell's equations. The coefficients in these expansions are then determined by matching boundary conditions at the domain interfaces. These boundary conditions are matched by using least squares, so the resulting solution is exact within each domain but approximate at the boundaries. FDTD methods solve Maxwell's equations as a function of time (rather than at fixed frequency as with the other methods) using a 3D spatial grid for expressing the spatial derivatives. There is also a finite element version of the same theory. T-matrix methods express the fields inside and outside the object as expansions in vector spherical harmonics that are similar to Eqs. (5) and (6). The coefficients in this expansion are determined by matching boundary conditions at the particle surfaces.

B. MLWA Treatment of Spheroids

In addition to the numerically exact methods for solving Maxwell's equations, it is possible to develop approximate electrodynamic methods that are useful for gaining insight and for generating qualitative information about scattering and absorption. One of the most useful methods in this regard is the long-wavelength generalization of the MLWA theory that was described in Sec. II.B. Various versions of the theory for spheroids have been described by Adrian (33), Kerker (34), Gersten and Nitzan (35), and Zeman and Schatz (25). The simplest version of the theory involves solving the Laplace equation by using separation of variables in spheroidal coordinates. A spheroid is a particle whose surface is governed by

$$\frac{x^2 + y^2}{a^2} + \frac{z^2}{c^2} = 1 \qquad (17)$$

where $c > a$ for a prolate spheroid and $c < a$ for an oblate spheroid. In the long-wavelength limit, one finds that Eq. (12) still governs the induced polarization, but the polarizability (along the z axis) is given by

$$\alpha_\| = c^3 \frac{1 + \chi_\|}{3} \frac{\xi_0^2 - 1}{\xi_0^2} \frac{\varepsilon - 1}{\varepsilon + \chi_\|} \tag{18}$$

Here $\xi_0 = (1 - a^2/c^2)^{-1/2}$, and $\chi_\|$ is the shape-dependent parameter

$$\chi_\| = -1 + \frac{1}{Q_1(\xi_0)(\xi_0^2 - 1)} \tag{19}$$

where

$$Q_1 = \frac{\xi_0}{2} \ln\left[\frac{\xi_0 + 1}{\xi_0 - 1}\right] - 1 \tag{20}$$

Equation (18) shows that the plasmon resonance condition is Re $\{\varepsilon + \chi_\|\} = 0$, so the value of $\chi_\|$ determines the resonance wavelength. Simple substitution into Eq. (19) reveals that for a prolate object $\chi_\|$ is greater than 2, while for an oblate argument it is less than 2. Since ε for a free-electron metal gets more negative as wavelength is increased, if $\chi_\| > 2$, the plasmon resonance will be red-shifted compared to the corresponding sphere resonance. Of course this same resonance will be blue shifted for oblate spheroids that are excited along the symmetry axes.

If the initial field is perpendicular to the symmetry axis, then an expression similar to Eq. (18) still applies, but $\chi_\|$ is replaced by $\chi_\perp$ and the resonance condition is now Re $\{\varepsilon + \chi_\perp\} = 0$. The expression for $\chi_\perp$ is

$$\chi_\perp = -1 - 2\left(\zeta_0^2 - \frac{\zeta_0(\zeta_0^2 + 1)}{2} \cos^{-1}\left[\frac{\zeta_0^2 - 1}{\zeta_0^2 + 1}\right]\right)^{-1} \tag{21}$$

where $\zeta_0 = (a^2/c^2 - 1)^{-1/2}$. Direct substitution shows that $\chi_\perp$ is less than 2 for a prolate spheroid and greater than 2 for an oblate spheroid. This means that for perpendicular polarization, oblate spheroids have red-shifted plasmon resonances relative to a sphere. For silver and gold, the resonances that are blue-shifted relative to that of a sphere are broadened by interband transitions, so it is the red-shifted resonances that are more noticeable. Our emphasis in the next few sections will be on red-shifted reonances associated with oblate spheroids.

Zeman and Schatz (25) have developed electrodynamic expansions (MLWA) that correct Eq. (18), and similar expressions for oblate spheroids, through order λ^3. Figure 3 presents extinction spectra for 2:1 and 5:1 oblate gold spheroids (with polarization perpendicular to the symmetry axis) based on MLWA theory. In these figures the dimensions of the spheroids have been chosen such that the volume is the same as a 30-nm-radius sphere. Comparisons with a more accurate theory (DDA), described in the next section, are presented to

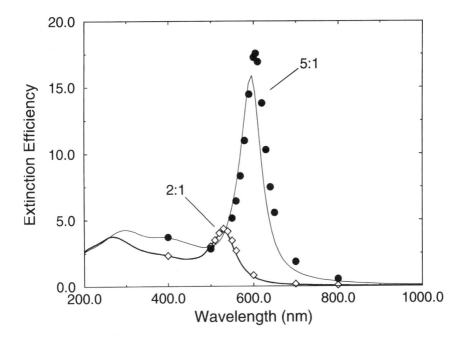

Fig. 3. MLWA (solid curves) and DDA (circles and diamonds) extinction spectra for 2:1 and 5:1 gold spheroids. These spheroids have a volume equivalent to that for a 30-nm-radius sphere, and the semimajor/semiminor axes are 37.8 nm/18.9 nm for the 2:1 spheroid and 51.3 nm/10.3 nm for the 5:1 spheroid. The extinction efficiency plotted is defined as the cross section divided by the area of a sphere with the same volume as the spheroid. The DDA calculations are based on grids of 19 × 38 × 38 for the 2:1 and 10 × 51 × 51 for the 5:1.

demonstrate that the MLWA result is accurate. Figure 3 shows the expected red shift as the ratio a/c (and hence $\chi_\perp$) increases.

C. Discrete Dipole Approximation (DDA)

In the DDA approach, one represents the object of interest as a cubic lattice of N polarizable points. There is no restriction as to which of the cubic lattice sites is occupied, which means that DDA can represent an object or multiple objects of arbitrary shape. We take the ith element to have a polarizability α_i (and no higher multipole polarizabilities) and with its center at a position denoted $\mathbf{r}_i$. The polarization induced in each element as a result of interaction with a local electric field $\mathbf{E}_{\text{loc}}$ is (omitting the frequency factors $e^{i\omega t}$)

$$\mathbf{P}_i = \alpha_i \cdot \mathbf{E}_{\text{loc}}(\mathbf{r}_i) \tag{22}$$

E_{loc}, for isolated particles, is the sum of an incident field and a contribution from all other dipoles in the same particle:

$$\mathbf{E}_{loc}(\mathbf{r}_i) = \mathbf{E}_{loc,i} = \mathbf{E}_{inc,i} + \mathbf{E}_{self,i}$$
$$= \mathbf{E}_0 \exp(i\mathbf{k} \cdot \mathbf{r}_i) - \sum_{j \neq i} \mathbf{A}_{ij} \cdot \mathbf{P}_j \qquad (23)$$

$\mathbf{E}_0$ and $\mathbf{k}$ are the amplitude and wave vector of the incident wave, respectively, and the interaction matrix $\mathbf{A}$ has the form

$$\mathbf{A}_{ij} \cdot \mathbf{P}_j = \frac{\exp(ikr_{ij})}{r_{ij}^3} \{k^2 \mathbf{r}_{ij} \times (\mathbf{r}_{ij} \times \mathbf{P}_j) + \frac{1 - ikr_{ij}}{r_{ij}^2} \times [r_{ij}^2 \mathbf{P}_j - 3\mathbf{r}_{ij}(\mathbf{r}_{ij} \cdot \mathbf{P}_j)]\} \qquad (j \neq i) \qquad (24)$$

where $k = \omega/c$. Note that the metal dielectric constant enters the calculation through the polarizabilities α_i. The explicit formula for α_i was developed by Draine and Goodman (29) such that the dipole lattice for an infinite solid exactly reproduces the continuum dielectric response of that solid to electromagnetic radiation.

Substituting Eqs. (23) and (24) into Eq. (22) and rearranging terms in the equation, we generate an equation of the form

$$\mathbf{A}' \cdot \mathbf{P} = \mathbf{E} \qquad (25)$$

where $\mathbf{A}'$ is a matrix built out of the matrix $\mathbf{A}$ from Eq. (24). For a system with a total of N elements, $\mathbf{E}$ and $\mathbf{P}$ in Eq. (25) are $3N$-dimensional vectors and $\mathbf{A}'$ is a $3N \times 3N$ matrix. By solving these $3N$ complex linear equations, the polarization vector $\mathbf{P}$ is obtained, and with this the extinction cross sections and other optical properties may be calculated. In practice there are significant advantages associated with performing the sum over dipole fields in Eq. (23), using fast Fourier transform methods, and solving Eq. (25) by complex conjugate gradient techniques. This is the implementation developed in the work of Draine and Flatau (29), and it is what we have used in the present studies. Further details of the method are described in Ref. 26.

Tests of the DDA method are provided by Figs. 3 and 4. Figure 3, which refers to 2:1 and 5:1 Au spheroids, shows results which match the MLWA calculations quite well. These results were based on grids of $19 \times 38 \times 38$ for the 2:1 calculations, and $10 \times 51 \times 51$ for 5:1. Calculations with larger grids give essentially the same results, indicating that the results are converged. Figure 4 presents analogous Ag results for particles whose sizes (see caption) were taken to match recent measurements (11). These particles are generated by a technique known as nanosphere lithography (5), which produces a periodic particle array (PPA) of silver particles whose shapes are very close to uniform for all the particles being

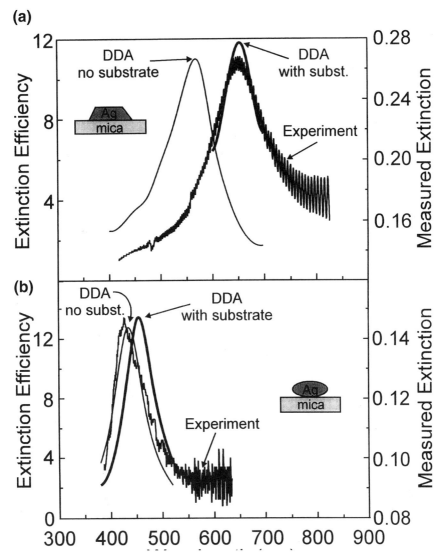

Fig. 4. DDA extinction spectra for silver PPA particles. (a) refers to PPA I of Jensen et al. (11), while (b) refers to PPA IV. Particle I is modeled as a truncated tetrahedron with base dimension of the equilateral triangle equal to 93 nm and a height of 28 nm. Particle IV refers to an annealed particle represented as an oblate spheroid having a major axis (diameter) of 90 nm and a minor axis (height) of 42 nm. Insert shows the particle shapes for particles I and IV. The results from experiment (11) are indicated in each figure. DDA results are presented for an isolated particle (no substrate) and for a particle on a mica substrate. The substrate is modeled as a cylinder of height 30 nm and diameter 153 nm for PPA I, and of height 45 nm and diameter 120 nm for PPA IV. Total number of dipoles in these calculations is 63,126 for PPA I and 18,855 for PPA IV.

irradiated. In these measurements, several sizes and shapes were considered; Figure 4 considers only two. They are denoted PPA I and PPA IV (following the notation in Ref. 11), and we note that PPA I is modeled as a truncated tetrahedron, while PPA IV has been annealed so that its shape is approximately spheroidal. Note that PPA I has a plasmon resonance that is far to the red of that for PPA IV. This reflects the fact that PPA I is much flatter (heights of 28 nm vs. 42 nm, while the widths are approximately the same). The comparison with experiment for the theoretical results labeled "no substrate" is quite good for PPA IV, but is not good at all for PPA I. The difference between the two results has been studied (11) and is primarily due to the much stronger interaction of particle I with the mica substrate that is used to support the particles. This point is apparent from the results labeled "with substrate" in Fig. 4, and it will be further considered in the next section.

IV. SUBSTRATE AND SOLVENT EFFECTS

A. Mie Theory for Core-Shell Particles

The theory developed for a single solid sphere in Eqs. (1)–(10) can be modified to treat a spherical core having one or more shells of arbitrary thickness. This is done by applying electromagnetic boundary conditions at both the inner and outer shell surfaces, and it leads to equations analogous to Eqs. (9) and (10) for the extinction spectrum, but with expressions for the coefficients a_n and b_n that are more complex than in Eqs. (7) and (8). The results have been described for the case of a single shell by Bohren and Huffman (2). Here we present results from a program which they provide wherein we have studied several model systems of relevance to the PPA particles described in the previous section.

Figure 5 shows the results of calculations with this theory for the specific case of a 30-nm-radius Ag sphere, with shells having dielectric functions of mica (Fig. 5a), Au (Fig. 5b), and Pt (Fig. 5c). In the results for Au and Pt, we have included for the influence of the finite thickness of the metal shell on the dielectric functions using a surface scattering correction that has been discussed previously for Au (25,36). The Pt corrections involve estimates of the plasmon frequency (2.22×10^{15} s^{-1}) and width (2.45×10^{14} s^{-1}) based on fitting the results in Ref. 37 to a Drude expression.

In all cases the zero coverage result shows a plasmon resonance at 380 nm. With a dielectric function corresponding to mica, the plasmon shifts to the red as coverage increases, with even a 1-nm shell giving a detectable shift (few nm). Ultimately in the limit of an infinite shell, one expects the resonance to shift to what would be obtained for a particle in a dielectric continuum. For small particles, this would be determined by Re $\varepsilon + 2\varepsilon_{med} = 0$, where ε_{med} is the dielectric function of the surrounding medium. In the present case this leads to a resonance at 490 nm.

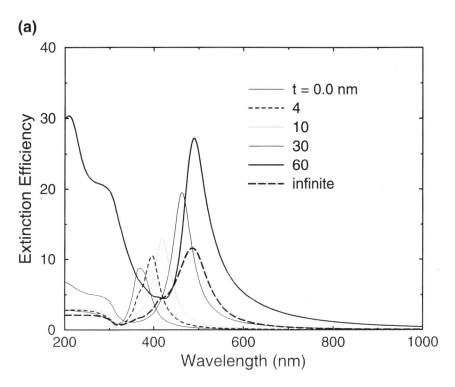

Fig. 5a. Mie core-shell results for the following systems: (a) Shell having the dielectric function of mica and a thickness of 0, 4, 10, 30, and 60 nm surrounding a Ag core whose radius is 30 nm. Also shown is the bulk mica (infinite thickness) result.

What we see from Fig. 5a is that this result is achieved for a shell of 60 nm; i.e., the shell thickness is equal to the particle diameter. This result is consistent with rapid decay of the induced dipole field about the particle. Note that for the 60-nm shell, there is substantial scattering near a wavelength of 200 nm that is not present for the smaller figures. This is Rayleigh scattering associated with the entire core-shell system, and it has a $(\lambda)^{-4}$ dependence on wavefunction λ. This effect makes a thick shell different from a dielectric continuum no matter how thick is the shell, which means that the apparent convergence in Fig. 5a between the 60-nm spectrum and that for infinite thickness is an oversimplified conclusion. However, the Rayleigh scattering contribution can be subtracted out, and what remains in the thick shell result differs from the infinite shell limit by an interference term that oscillates about the infinite result.

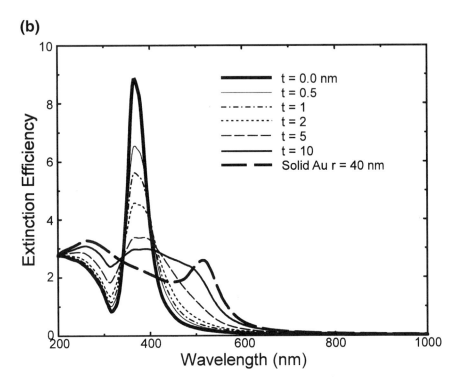

Fig. 5b. (b) Au shell having thickness of 0, 0.5, 1, 2, 5, and 10 nm surrounding a Ag core having a radius of 30 nm. Also shown is the result for a pure Au core of radius 40 nm.

Figure 5b shows that a gold shell gives behavior quite different from a mica shell. Here we see that, although the plasmon resonance is broadened for shells up to 5 nm, the resonance is not red-shifted. Then for thicker shells, a gold plasmon resonance at 520 nm gradually builds in. Figure 5c shows still different behavior for Pt shells. Here the Pt resonance frequency lies to the blue of that for Ag, so thick films show blue- rather than red-shifting. However, there is essentially no shift for thin films, similar to what we see for Au.

B. DDA Theory of Substrate and Solvent Effects

The application of DDA theory to a core-shell particle involves only one complication in the theory presented previously, namely one needs to use different finite element polarizabilities for the lattice dipoles associated with the shell and the core. These are straightforwardly derived from the appropriate dielectric functions. It is, in principle, possible to use different grid dimensions for the core

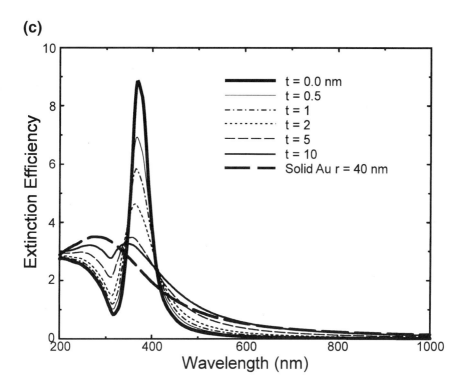

Fig. 5c. (c) Pt shell having thickness of 0, 0.5, 1, 2, 5, and 10 nm surrounding a Ag core whose radius is 30 nm. Also shown is the result for a pure Pt core of radius 40 nm.

and shell; however, doing this would make it impossible to use a Fourier representation to evaluate dipole sums, so this is not recommended.

Figure 6 presents a test of the DDA method for a core-shell system that matches the parameters considered in Fig. 5a. The figure shows good, though not perfect, agreement, indicating that DDA is nearly correct. We have tested this result for convergence with respect to the grid size, and find that it is converged. This means that the application of the bulk lattice dispersion relationships for the various components becomes less accurate when the system is heterogeneous like a core-shell particle.

One of the great virtues of the DDA is that it can treat complicated composites with essentially the same effort as a homogeneous material having the same dimensions. Figure 4 shows two examples of this for the particles labeled PPA I and PPA IV in Sec. III. The experimental preparation using nanosphere lithography (5) typically deposits these particles on a substrate of mica. To describe the effect of this substrate on the optical response, we have done DDA calculations

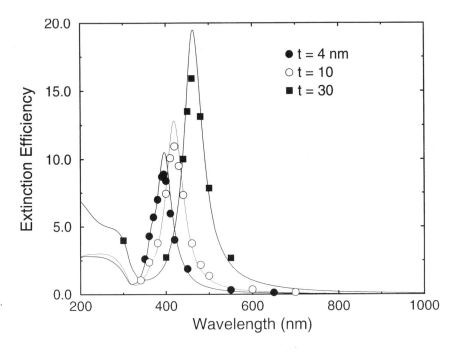

Fig. 6. DDA core-shell results (circles and square) for silver core with radius 30 nm and mica shell with thickness 4, 10, and 30 nm. The solid curves show corresponding results from Mie theory.

for an assembly that consists of the PPA particle on a cylinder of mica whose height equals the particle height and whose width is twice the particle width. Based on the results in Fig. 5a, one would expect that this should give a result that is close to being converged with respect to cylinder size, and indeed it does. What Fig. 4 shows is that the plasmon resonance for the system of particle + substrate is greatly red-shifted compared to that for the isolated particle, indicating that the substrate effect is important. For PPA I, the DDA plasmon peak is in close match with measurements once the substrate is included. The situation for PPA IV is not quite so good, but here we note that the spheroid model of particle shape is not as quantitative a representation of what is prepared by the experiment as it is for the truncated tetrahedron (11).

Figure 7 shows another application of DDA theory, this time to the case where the part of the PPA IV particle that is above the mica substrate is surrounded by a solvent, whose index of refraction is denoted n_{med} ($n_{med}^2 = \varepsilon_{med}$). This plot of peak plasmon wavelength versus n_{med} includes recent experimental results (11) that demonstrate an approximately linear dependence on n_{med} over the range considered. To model these results, we show the results of three different DDA calculations. The first (filled circles) considered the particle (no slab) to be

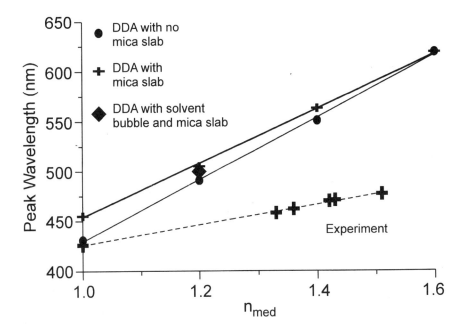

Fig. 7. DDA calculations for PPA IV, showing the wavelength of peak extinction as a function of the index of refraction n_{med} of the surrounding medium. Results from experiment (11) are also included. The DDA calculations include (a) DDA calculations with no slab, with the solvent treated as an infinite continuum (filled circles); (b) DDA with a mica slab, with the solvent treated as a continuum (pluses); and (c) DDA with a mica slab, with the solvent treated as a small droplet that coats the surface (filled diamond).

embedded in a homogeneous dielectric material whose index of refraction is that of the solvent. Here the DDA calculation is by applying the original DDA method (i.e., DDA for a particle in vacuum) with two modifications. One is that the dielectric function ε is replaced by $\varepsilon/\varepsilon_{med}$. The second is that the wavevector k is replaced by $n_{med}k$, where n_{med} is the index of refraction of the medium. The second DDA result (pluses) is the same as the first except that a mica slab is explicitly included in the calculation. The third DDA result (diamonds) includes the solvent as explicit dipoles forming a nanodrop on top of the PPA particle. The substrate is also included.

The DDA results in Fig. 7 show good qualitative agreement with experiment, but the slope of the curves is always overestimated by theory. Note the consistency between the various DDA results, with the explicit solvent (nanodrop) result being very close to the continuum solvent result. Also, the PPA particle with an explicit slab gives the same result for $n_{med} = 1.6$ (which is where the slab and solvent dielectric functions are the same) as the continuum solvent result with no substrate.

V. NEAR-FIELD PROPERTIES

For many spectroscopic properties, one needs the electric fields that come from solving Maxwell's equations evaluated close to or at the particle surfaces. This would be appropriate for applications to surface-enhanced Raman spectroscopy (SERS) where molecules on the surface interact with the incident field at the particle surface. This field induces an oscillating dipole in the molecule, and the radiation from this dipole, which also interacts with the surface, results in Raman scattering. Surface fields are also important for second-harmonic generation (SHG), as the gradient in the electron density near the particle surface is responsible for the nonlinear polarization.

A. Mie Theory Results

Evaluation of fields near the particle surfaces in Mie theory requires evaluation of $\mathbf{E} = \mathbf{E}_i + \mathbf{E}_s$ from Eqs. (5) and (6). Figure 8a shows the results of this evaluation for a 30-nm radius Ag sphere, where we have plotted contours of $|\mathbf{E}|^2$ versus the coordinates x and z, where z is the propagation direction and x is the polarization direction. In Fig. 8b we show the analogous results for a 60-nm sphere. The wavelength in both figures is 366 nm.

To understand Fig. 8a, we note that close to the sphere (but still outside), one expects that the field will look like a static dipole field. Such fields have a $1/r^3$ dependence on distance and a $\sin\theta \cos\varphi$ dependence on angles for the present choice of propagation and polarization directions. This means that in the x–z plane, the contours should have the angular dependence of a p-orbital (peaking at $\theta = 90°$ and $\varphi = 0°$ or $180°$), and they should drop off rapidly as one goes away from the particle surface. This is exactly what Fig. 8a shows, for locations that are out to about twice the radius from the particle center. Beyond that the field shows oscillatory contours that reflect the plane-wave behavior of the incident field. Figure 8b shows a more complex situation that arises when the particle is large enough to show significant quadrupolar excitation. In this case the field involves a superposition of dipole and quadrupole behavior such that the peak surface field now occurs at points where $\theta = 135°$ and $\varphi = 0°$ or $180°$. This illustrates the complexity that can arise in the evaluation of surface fields, and this complexity gets worse for nonspherical particles.

Figure 9 shows several electrodynamic properties for the 30- (top) and 60- (bottom) nm Ag spheres as a function of wavelength. Included for each size are extinction spectra, the surface average electric field $\langle|\mathbf{E}|^2\rangle$, and the peak field $(|\mathbf{E}|^2)_{max}$. This figure shows that the average and peak field strengths occur at wavelengths that are red-shifted compared to where the peaks in the extinction spectrum occur. This arises because extinction measures the volume polarization of the particle, and thus is less sensitive to the electrodynamic (finite wavelength) corrections than

Modeling Optical Properties

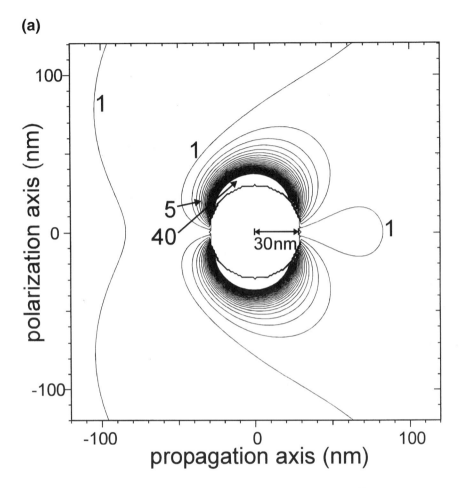

Fig. 8a. Mie theory electric fields (plotted as contours of constant ($|\mathbf{E}|^2$) outside a silver sphere with radius (a) 30 nm and (b) 60 nm. The wavelength in both figures is 366 nm.

is the surface field. The average field strength for the 30-nm sphere is enhanced by a factor of over 30 at its peak value compared to what is the field in the absence of the sphere, and the peak value is enhanced by a factor of nearly 100. For the 60-nm sphere, there are both dipole and quadrupole maxima in all the curves, with the dipole maximum dominant in extinction and in the average field. For the peak field the quadrupole resonance peak is almost as high as the dipole maximum.

The red shift in the peak field compared to the extinction maximum should be observable by comparing SERS excitation spectra with extinction spectra.

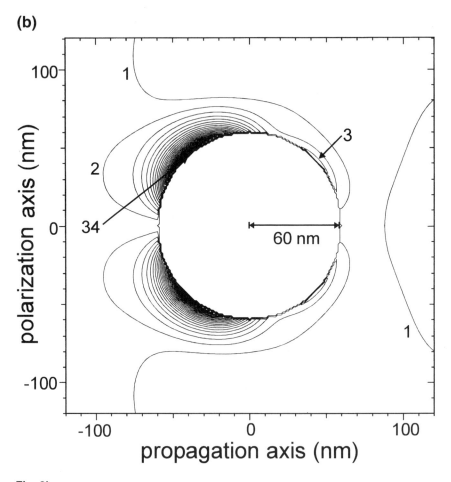

Fig. 8b.

This is because SERS is sensitive to local fields at the incident and Stokes-shifted wavelengths. However, quantitative confirmation of this result has not been reported.

B. DDA Results

To determine surface fields near particle surfaces for particles other than spheres requires a numerical method for solving Maxwell's equations. One such method is the DDA approach (9,26), discussed earlier. This approach is easy to apply, requiring the evaluation of Eq. (20) at the surfaces of the particle, where the particle surface may be approximated as the exposed surfaces of the cubical elements.

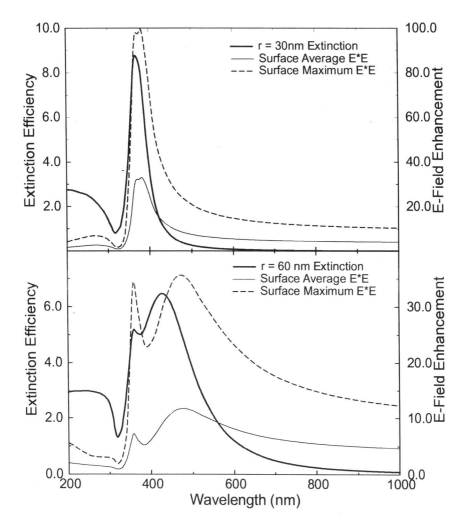

Fig. 9. Extinction efficiency (thick curve, left axis) peak [$(|\mathbf{E}|^2)_{max}$ dashed curve, right axis] and average ($\langle|\mathbf{E}|^2\rangle$ thin curve, right axis) electric field enhancement as a function of wavelength for 30- (top panel) and 60- (bottom panel) nm silver spheres.

Reference 9 presented just such an application for several particle shapes. Where the particles are spherical, these results are essentially the same as in Fig. 8 for distances at least two or more grid points away from the surface. Closer than that, the graininess of the DDA solution becomes evident, and the method becomes inaccurate.

VI. DIMERS AND AGGREGATES

To conclude this chapter, we mention the electrodynamics of interacting particles. This is actually a rich topic in its own right, particularly when considering nanoparticle aggregates, and we refer the reader to several papers which cover this problem in detail (9,20–22). Here we consider just the case of two interacting particles.

Aggregation leads to electrodynamic interactions between the particles due to mutually induced dipoles, and typically this produces red shifts in plasmon features. This result can be derived by representing each particle as a single induced dipole, yielding a result that is even quantitatively correct provided that the particles are farther than one radius in separation. For closer separations it becomes important to include higher multipoles in the interactions between particles, and if enough multipoles are included, the result is, in principle, exact. For spherical particles, a coupled multipole theory and computer code has been developed by Mackowski (38). Figure 10 gives an example of an application of this code, showing the wavelength of the plasmon maximum (only the dipole plasmon is significant) associated with two 30-nm radius gold spheres aligned along the polarization direction of the incident light as a function of the interparticle separation distance. Coupled dipole and DDA results for this system are also presented, as is the result for a single sphere (same as two spheres with infinite separation). Note that the dipole polarizabilities in the coupled dipole calculations are extracted from Mie theory [i.e., not using the electrostatic limit in Eq. (11)]. The figure shows the expected red shift relative to the 510-nm peak that occurs for infinite separation (in vacuum), with a 2-nm separation giving a shift of about 60 nm. The DDA and coupled multipole results match each other closely. The coupled dipole curve matches the more exact results for larger separations (>20 nm), but it underestimates the shift for smaller separations.

In Ref. 9 we presented similar results for coupled silver particles, showing the complete DDA extinction spectrum as a function of particle size. There it was shown that when the particles touch there is a dramatic red shift in the plasmon resonance, even relative to that for a 2-nm separation. Multiple resonances also occur. If the dimer is further "squashed together" to produce a peanut-shaped object, the resulting particle behaves as a prolate spheroid (9), with the plasmon resonance shifting blue relative to the zero separation limit as the ratio of length to width of the spheroid decreases toward one.

As a final application, we consider a dimer composed of two silver PPA particles. Figure 11 shows the results of DDA calculations for two particles that lie on the same plane with their tips pointed together and the polarization direction along the symmetry axis on which the particles are aligned. Each particle has a base dimension of 120 nm and a height of 46 nm, with a single-particle plasmon resonance peak at 574 nm. The results are presented as wavelength shift versus

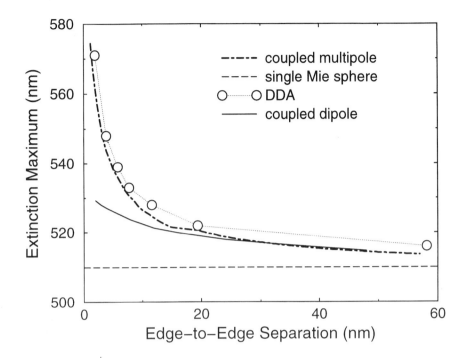

Fig. 10. Wavelength of dipole plasmon maximum as a function of the distance between the surfaces of two 30-nm-radius gold spheres, showing results of coupled multipole (dash-dot), coupled dipole (solid), and DDA (circles) calculations. The single-particle Mie theory maximum (same as two particles with infinite separation) at 510 nm is also shown (dashed). The polarization vector in this calculation is taken along the axis that connects the two particles.

the tip-to-tip distance. Note that the shift is small (<20 nm) until the tip separation drops below 50 nm. This means that for the separation distance that is appropriate for the PPA fabrication (114 nm), the shift is almost negligible. This result is important because it indicates that PPA particles can be considered to be isolated in their response to an applied field. For smaller separations, the shift becomes substantial.

ACKNOWLEDGMENTS

This research was supported by ARO Grant DAAG55-97-1-0133. We thank R. P. Van Duyne for useful comments.

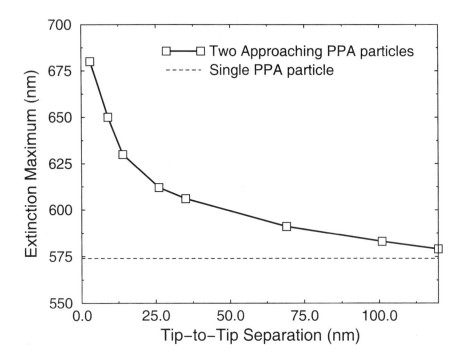

Fig. 11. Extinction spectra for two coupled PPA particles, each represented as a truncated tetrahedron, as obtained from DDA calculations, showing the evolution of the spectra as the interparticle separation is varied. Each particle has a base dimension of 120 nm and a height of 46 nm, and the two particles are aligned on a common symmetry axis (chosen to be the polarization direction) such that they point toward each other. The single-particle result (same as two particles with infinite separation) is also indicated. A total of 10,234 dipoles was used. No substrate or solvent effect was included in this calculation.

REFERENCES

1. U Kreibig, M Vollmer. Optical Properties of Metal Clusters. Springer Series in Materials Science 25. Berlin: Springer, 1995; U. Kreibig, In: Handbook of Optical Properties, Vol. II: Optics of Small Particles, Interfaces and Surfaces, Ed: RE Hummel, P. Wissmann. Boca Raton: CRC Press, 1997, pp. 145–190.
2. CF Bohren, DR Huffman, Absorption and Scattering of Light by Small Particles, New York: Wiley-Interscience, 1983.
3. CA Mirkin, RL Letsinger, RC Mucic, JJ Storhoff. Nature 382: 607–609, 1996; R Elghanian, JJ Storhoff, RC Mucic, RL Letsinger, CA Mirkin. Science 277: 1078–1080, 1997.

4. JM Petroski, ZL Wang, TC Green, MA El-Sayed, J. Phys. Chem. B 102: 3316–3320, 1998; YY Yu, SS Chang, CL Lee, CRC Wang. J. Phys. Chem. B 101: 6661–6664, 1997; SE Roark, KL Rowlen. Chem. Phys. Lett. 212: 50–56, 1993.
5. RP Van Duyne, JC Hulteen, DA Treichel. J. Chem. Phys. 99: 2101–2115, 1993; JC Hulteen, RP Van Duyne. J. Vac. Sci. Tech. A13: 1553–1558, 1995.
6. M Kahl, E Voges, W Hill. Spectrosc. Eur. 10: 8, 10, 12–13, 1998.
7. GL Hornyak, CJ Patrissi, CR Martin, JC Valmalette, J Dutta, H Hofmann. Nanostruct. Mater. 9: 575–578, 1997; SJ Oldenburg, RD Averitt, SL Westcott, NJ Halas. Chem. Phys. Lett. 288: 243–247, 1998.
8. PF Liao, MB Stern. Opt. Lett. 7: 483–485, 1982; CP Collier, RJ Saykally, JJ Shiang, SE Henrichs, JR Heath. Science 277: 1978–1981, 1997.
9. T Jensen, L Kelly, A Lazarides, GC Schatz. J. Cluster Sci. 10: 295–317, 1999.
10. TR Jensen, GC Schatz, RP Van Duyne. J. Phys. Chem. B103: 2394–2401, 1999.
11. TR Jensen, ML Duval, KL Kelly, AA Lazarides, GC Schatz, RP Van Duyne. J. Phys. Chem. 103: 9846–53, 1999.
12. N Felidj, J Aubard, G Levi. J. Chem. Phys. 111: 1195–1200, 1999.
13. H Metiu, P Das. Annu. Rev. Phys. Chem. 35: 507–536, 1984; H Metiu. Prog. Surf. Sci. 17: 153–320, 1984.
14. GC Schatz. Acc. Chem. Res. 17: 370–376, 1984.
15. M Kerker. Acc. Chem. Res. 17: 271–277, 1984.
16. M Moskovits. Rev. Mod. Phys., 57: 783–826, 1985.
17. JI Dadap, J. Shan, KB Eisenthal, TF Heinz. Phys. Rev. Lett. 83: 4045–4048, 1999.
18. WH Yang, J Hulteen, GC Schatz, RP Van Duyne. J. Chem. Phys. 104: 4313–4323, 1996.
19. CK Chen, ARB de Castro, YR Shen. Phys. Rev. Lett. 46: 145–148, 1981; CT Boyd, T Rasing, JRR Leite, YR Shen. Phys. Rev. B30: 519–526, 1984.
20. AA Lazarides, GC Schatz. J. Phys. Chem. 104: 460–467, 2000.
21. AA Lazarides, GC Schatz. J. Chem. Phys. 112: 2987–2993, 2000.
22. AA Lazarides, GC Schatz. To be published.
23. DW Lynch, WR Hunter. In: ED Palik, ed. Handbook of Optical Constants of Solids, New York: Academic Press, 1985, pp. 350–356.
24. EJ Zeman, GC Schatz. In: B Pullman, J Jortner, B Gerber, A. Nitzan, eds. Dynamics of Surfaces, Proceedings of the 17th Jerusalem Symposium; Dordrecht: Reidel, 1984, pp. 413–424.
25. EJ Zeman, GC Schatz. J. Phys. Chem. 91: 634–642, 1987.
26. WH Yang, GC Schatz, RP Van Duyne. J. Chem. Phys. 103: 869–875, 1995.
27. M Meier, A Wokaun. Opt. Lett. 8: 581–583, 1983; A Wokaun, JP Gordon, PF Liao. Phys. Rev. Lett. 48: 957–960, 1982.
28. EM Purcell, CR Pennypacker. Astrophys. J. 186: 705–714, 1973.
29. BT Draine, PJ Flatau. J. Opt. Soc. Am. A. 11: 1491–1499, 1994; BT Draine, JJ Goodman. Astrophys. J. 405: 685–697, 1993; Program DDSCAT, by BT Draine, P.J. Flatau. University of California, San Diego, Scripps Institute of Oceanography, La Jolla, CA.
30. L Novotny, DW Pohl, B Hecht. Opt. Lett. 20: 970–972, 1995; L Novotny, RX Bian, XS Xie. Phys. Rev. Lett. 79: 645–648, 1997.

31. RX Bian, RC Dunn, XS Xie, PT Leung. Phys. Rev. Lett. 75: 4772–4775, 1995; A Taflove. Computational Electrodynamics: The Finite-Difference Time-Domain Method, Boston. Artech House, 1995, 599 pp.
32. PW Barber, SC Hill. Light Scattering by Particles: Computational Methods, Singapore: World Scientific, 1990; PW Barber, RK Chang, H Massoudi. Phys. Rev. B. 27: 7251–7261, 1983.
33. F Adrian. Chem. Phys. Lett. 78: 45–49, 1981.
34. DS Wang, M Kerker. Phys. Rev. B 24: 1777–1790, 1981.
35. J Gersten, A Nitzan. J. Chem. Phys. 73: 3023–3037, 1980.
36. RD Averitt, SL Westcott, NJ Halas. J. Opt. Soc. Am. B 16: 1824–1832, 1999.
37. VL Rideout, SH Wemple, J. Opt. Soc. Am. 56: 749–751, 1966.
38. DW Mackowski, MI Mishenko. J. Opt. Soc. Am. A 13: 2266–2278, 1996.

5
Electrochemical Template Synthesis of Nanoscopic Metal Particles

Colby A. Foss, Jr.
Georgetown University, Washington, D.C.

I. INTRODUCTION

The very earliest studies of nanoscopic metal particles proceeded without the benefit of technology (e.g., electron microscopes) that could ascertain the actual geometry of the particles (1). The success of the Mie (2) and Maxwell-Garnett (3) theories in explaining the optical spectra of metal nanoparticles in glass or liquid matrices is due largely to the fact that the assumed geometry, namely spherical, was essentially correct for the systems under examination at the time. Spherical particles are indeed a ubiquitous outcome of many synthesis regimes, whether they involve the gas, liquid, or solid phase. However, other geometries, such as rods, platelets, or polyhedra, can occur, especially in solutions containing surfactants or other species that adsorb on metal surfaces and influence the thermodynamics or kinetics of particle growth (4). The mechanism for the formation of nonspherical particles is not well understood and is certainly a rich topic for fundamental investigation.

If the chemistry of nonspherical metal nanoparticle formation is still in its infancy, the demand for such species certainly is not. As mentioned in Chapter 4 the surface enhancement of Raman scattering and other optical processes depends strongly on particle size and shape. Also, new theories of the optical properties of nonspherical metal particles beg for experimental results from well-defined real systems. The topic of this chapter is *template synthesis*, a brute-force method that relies not on elegant chemical principles (which have yet to be uncovered), but on the geometry and dimensions of the pores of a host material to direct the growth of nanoparticles (Fig. 1).

(a)

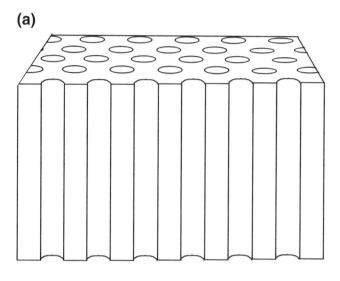

(b)

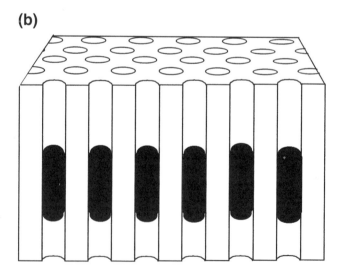

Fig. 1. General scheme for template synthesis: the porous structure of a host material (a) directs the growth of nanoparticles prepared within it (b).

The attractiveness of template synthesis lies in its conceptual simplicity and typically low cost; the only instruments required are an electrochemical potentiostat and a metal-sputtering device. Depending on the template materials employed, the subsequent optical and microscopic characterization of the metal nanoparticles can also be very straightforward. We will summarize not only the synthetic strategies used in our laboratory and in other groups, but a few of the linear and nonlinear optical studies done on such particles as well.

II. SYNTHESIS METHODS

A. Porous Host Materials

1. Anodic Aluminum Oxide (AAO)

The origin of template-based synthesis of metal nanoparticles can be traced to early studies on the colorization of anodized aluminum (5). Goad and Moskovits appear to be the first workers to recognize that the optical spectra of certain colorized anodic alumina arise from the plasmon resonance of nanoscopic metal particles deposited within the pores of the oxide (6). Andersson et al. proposed the application of such materials as selective as selective solar absorbers (7); they also realized that the cylindrical geometry of the pores of the anodic alumina likely was directing the growth of nonspherical metal particles (7). While the electrochemistry and structure of anodic alumina have been reviewed exhaustively by others (8), we will provide a brief overview here.

Fig. 2A shows a schematic of the aluminum oxide film that forms when aluminum metal is anodized in acidic media. In a simple two-electrode cell where the cathode is lead or stainless steel and the anode is aluminum metal, the application of a positive voltage (anode versus cathode) in the presence of cold acid (for example, 6% w/v oxalic or sulfuric at 0°C) results in an oxide layer which contains a parallel array of nominally cylindrical pores. The oxide formation half-reaction is

$$2Al_{(s)} + 3H_2O_{(l)} = Al_2O_{3(s)} + 6H^+_{(aq)} + 6e^- \qquad (1)$$

The growth of the oxide layer is such that it advances into the aluminum phase, with simultaneous formation and dissolution of the oxide occurring at the base of the pore. During pore formation, the aluminum anode material is never directly exposed to solution, but is always coated with the so-called barrier layer.

It is well known that the diameter of the pores increases with the anodization voltage (9,10). This feature has been exploited in the application of porous anodic films as size-selective microfiltration membranes (10) and, of course, in the control of the diameter of particles deposited within the pores (vide infra). The dependence of pore size on anodization voltage is also an incredibly fortuitous

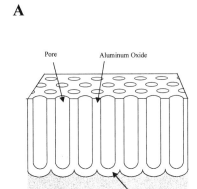
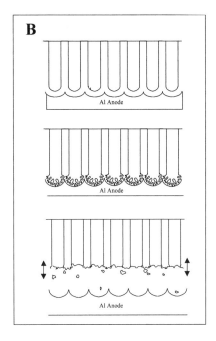

Fig. 2. (A) Schematic of porous anodic aluminum oxide (AAO) layer that forms on aluminum during certain conditions of anodization. (B) The voltage reduction/film separation process. (Top) Larger primary pore growth stage. (Middle) Voltage reduction and infiltration of barrier layer by branched pore network. (Bottom) Dissolution of branched network region of AAO film and separation from aluminum anode.

occurrence, since it allows for the ultimate separation of the porous films from the aluminum substrate, the destruction of the barrier layer, and the open communication of the pores from one face of the film to the other.

Figure 2B shows a schematic of the film removal process. After the porous oxide film has grown to the desired thickness (2B, top), the anodization voltage is decreased in 5% increments. With each reduction increment, the current decreases to a low level and then begins to recover as new pores are formed, whose smaller size corresponds to the lower voltage. After the voltage reduction/current recovery process is repeated several times, what was once the barrier layer of the high-voltage pore structure is now a rootlike network of smaller pores (2B, middle). The oxide-coated anode is removed from the cell and immersed in a higher-concentration acid solution (e.g., 25% w/w sulfuric acid at room temperature). Since the rootlike network has a higher exposed surface area than the primary large pores, dissolution occurs there first; the destruction of the network then results in the separation of the porous oxide film from the aluminum substrate (2B, bottom). Both faces of the porous film tend to be rough. The roughness of the former barrier layer face is due to the remnants of the rootlike pore network. The solution side can rough because of exposure to the anodizing solution, especially when long anodization times (and thick films) are desired. Nonetheless, the voltage reduction procedure allows for film separation with open pores. Also, surface roughness is not necessarily undesirable (vide infra).

2. Track-Etch Membranes

Another convenient template material is the track-etch membrane (11). Track-etch membranes are prepared by exposing a thin polymer film such as polycarbonate or polyester to high-energy nuclear fission fragments (Fig. 3A), which create tracks in the material. Subsequent exposure to base solution increases the diameter of the tracks to the desired value (Fig. 3B). Track-etch membranes with pore diameters as small as 10 nm are commercially available.

In the context of metal nanoparticle template synthesis, the advantages of track-etch membranes over porous alumina films lie in their flexibility (alumina films are brittle) and, in some applications, their smooth surfaces. One drawback is their low pore density, typically 10^9 pores/cm^2 (12), as opposed to 10^{11} pores/cm^2 in the case of anodic aluminas (12). Thus, the yield of particles per unit area of material is low in the case of track-etch membranes. Another potential problem lies in the fact that the pores created by fission fragment tracks are not always parallel to

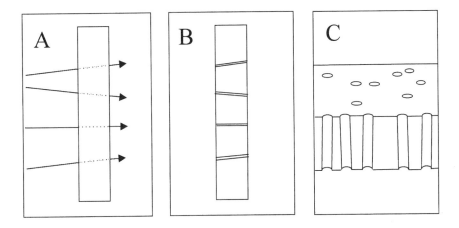

Fig. 3. Track-etch process for nanoporous polymer film formation. (A) Polymer film exposed to high-energy fission fragments. (B) Base solution increases diameter of damage tracks. (C) Resulting random placement of pores.

each other (or even perpendicular to the membrane surface), and the pore positions are random (Fig. 3C). This property is not important if the template-synthesized nanoparticles are removed from the membrane before their ultimate application. However, orientational and positional disorder can complicate interpretation of physical properties if study focuses on the metal nanoparticle/template host composite.

B. Metal Deposition

1. Deposition in Anodic Aluminas with Intact Barrier Layers

The first studies of metal nanoparticles prepared in porous anodic alumina films involved the deposition of metals within the pores while the barrier layer was intact. For example, Goad and Moskovits prepared 12-μm-thick alumina films by anodizing pure aluminum in 15% (w/w) sulfuric acid at constant current (14 mA/cm^2) at 21°C (6). Various metals were then deposited into the 15-nm-diameter pores by an AC deposition technique.

The primary advantage of this method is ease of material handling; since the anodic alumina film remains on the aluminum metal substrate, very thin oxide layers can be used without fear of their breaking. However, a drawback of electrodeposition in anodic aluminas whose barrier layer is intact is that pore filling can be uneven (13) and the resulting metal structures can be granular and disordered (6,7,14). However, some groups have recently prepared continuous metal nanowires with this technique (15), so it should be considered a viable option for nanoparticle preparation.

2. Deposition in Open Pore Templates

Electrodeposition into template films whose pores are entirely open is somewhat more involved, but allows for continuous, monolithic structures, and, in some cases, good linear correlation between the nanoparticle length and the integrated deposition current (16). The general scheme, which applies to either track-etched polymer membranes or porous anodic alumina films which have been detached from their aluminum substrate via the voltage reduction method, is shown in Fig. 4.

The first step involves vacuum or plasma deposition of a metal (usually silver) onto one face of the template film to render it conductive (4A). The coated film is then placed in an electrochemical deposition cell. The silver-coated alumina is the working electrode, and a counter- and reference electrode are immersed in a metal-plating solution. The first metal electrodeposited is usually silver, or some other metal that can be etched away easily. In other words, it merely forms a foundation for the deposition of the metal of interest (4B). The first plating solution is removed, and the cell is rinsed and refilled with the plating solution

Template Synthesis of Metal Particles

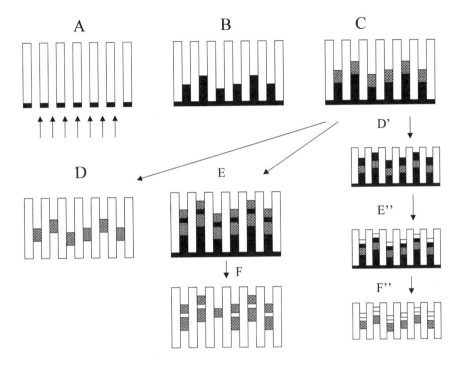

Fig. 4. General scheme for electrochemical template synthesis (open-pore host method). (A) Silver sputtering provides conductive layer for subsequent electrodeposition. (B) Electrodeposition of silver foundation. (C) Electrodeposition of metal nanoparticle phase. (D) Composite structure after silver foundation is etched away with nitric acid. (D′) Deposition of second silver layer. (E) Second gold layer deposition. (F) Paired gold nanoparticles after silver etched away. (E″) Conversion of silver to semiconducting silver iodide via electrochemical oxidation in presence of potassium iodide solution. (F″) Paired gold/silver iodide nanoparticles after silver etch.

corresponding to the metal of interest. The second metal is deposited in the desired amount (4C). The silver foundation is etched away with nitric acid, leaving the nanoparticles composed of the second metal in the pores of the template (4D). In our laboratories, the silver foundation is electrodeposited from a silver thiocyanate/ferrocyanide solution, and the second metal (gold) is deposited from a commercially available Au(I) cyanide solution (16–18).

The advantage of this method is the simplicity of the DC electrodeposition. Furthermore, the amount of metal deposited can be controlled since it is easy to monitor the total number of coulombs passed during the process. The principal drawbacks are the requirement of a plasma or vacuum deposition device and freestanding films which must be mechanically robust enough to tolerate manipulation (basically this means that the films must be thick, >10 μm for track-etch polymers and >30 μm for anodic aluminas). Also, when anodic aluminas are used, the foundation layer (4B) must be thick enough to position the final particles deep in the film where the pore structure is more well defined; both the solution face and former barrier layer face of detached anodic aluminas tend to be rough and the pore structure less well defined. As it turns out, the rougher barrier side forms a better anchor for the initial metal deposition (17).

If removing the template synthesized particles from their host is desired, this is easily accomplished by dissolving the host film. Track-etched polycarbonate films dissolved readily in methylene chloride, and anodic aluminas are destroyed in base (e.g., 0.10 M NaOH). Gold nanoparticles prepared via this method are reasonably uniform in diameter [less than 10% relative standard deviation (18)], but can suffer from significant polydispersity in length [up to ca. 20% in relative standard deviation (18)]. Figure 5 shows transmission electron microscope (TEM) images of gold nanorods formed in porous alumina via the above method.

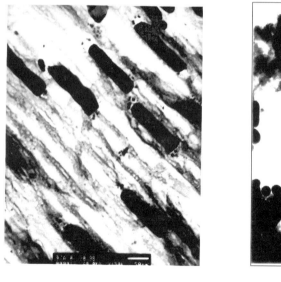

A   B

Fig. 5. Transmission electron microscope (TEM) image of template synthesized Au nanoparticles prepared according to scheme in Fig. 4. (A) Particles in AAO host. (B) Particles recovered after host dissolved in base. Scale bar corresponds to 50 nm.

Template Synthesis of Metal Particles

3. A Hybrid Conducting Substrate Method

When the alignment of metal nanoparticles within their free-standing template host is not a primary objective, a method developed by van der Zande et al. seems most promising for the fast preparation of uniform particles (19). Rather than preparing free-standing porous aluminas via the time-consuming voltage reduction method, the van der Zande method begins with the vacuum deposition of a thin layer (ca. 75 nm) of platinum on a titanium-coated silicon wafer (the titanium is added to improve adhesion of Pt on Si). A layer of aluminum (ca. 1 μm) is then vacuum-deposited on the platinum layer (Fig. 6A). The Al layer is then anodized in acid in a manner similar to that discussed in Sec. A.1 (6B). The key feature of this method is that the anodization of Al on a Pt substrate leads to a porous alumina layer whose pores are open all the way down to the Pt layer. During the anodization, the moment all of the Al is consumed and the conductive Pt layer is exposed is easily detected via the sudden increase in the evolution of hydrogen at the cathode.

A small layer of copper is then deposited as a foundation, not to avoid ill-defined pore structure (which does not occur in this method) but to allow for eventual release of the desired metal nanoparticles from the substrate (6C). The

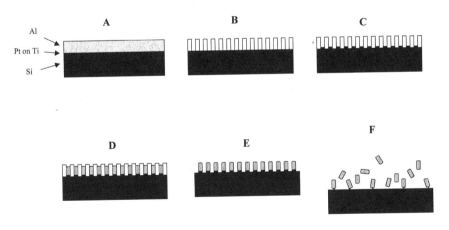

Fig. 6. Schematic of hybrid conducting substrate method (adapted with permission from van der Zande, B.M.I.; Gohmner, M.R.; Fokkink, L.G.J.; Schonenberger, C. Langmuir 2000, 16, 451–458. Copyright 2000 American Chemical Society). (A) Silicon wafer coated with titanium and platinum (the Ti improves adhesion of Pt on Si) and aluminum layers. (B) Al layer anodized to form porous AAO. (C) Cu foundation layer electrodeposited on Pt at pore base. (D) Electrodeposition of Au to form nanoparticles. (E) AAO etching in hydroxide/poly(vinylphyrrolidone) solution. (F) Acid peroxide etch of Cu base to release Au nanoparticles into solution.

copper plating solution is removed, excess copper ions reduced, and then gold plating solution is added to the cell. Gold is deposited onto the copper foundation in an amount corresponding to the desired particle length (6D). The alumina host is then dissolved in hydroxide solution (6E), which also contains poly(vinylpyrrolidone) (PVP), which adsorbs to gold surfaces and facilitates the eventual dispersion of particles in aqueous solution when then system is exposed to an acidic peroxide copper etching solution (6F).

Monodispersity of gold nanoparticles prepared with this method is about the same as that of particles prepared in the thick free-standing voltage-reduced aluminas (18,19). However, this method is experimentally much more convenient, since all electrodeposition, etching, and particle release steps can be done in the same cell.

III. OPTICAL PROPERTIES

A. Linear Optical Properties of Metal Nanoparticle/ Template Host Composite Films

One attractive feature of template hosts such as porous anodic aluminas is the parallel alignment of their pores. This allows for the straightforward interpretation of the plasmon resonance spectra of nonspherical metal particles, which are prepared within such pores. Aluminum oxide is particularly convenient for visible and near-infrared spectral analyses, since it is transparent for most of the 300-nm to 2.0-μm range. Fig. 7A shows the plasmon resonance spectra of a series of gold nanorod/porous alumina composite films measured with the light incident normal to the film surface (20). Since the rods are also aligned perpendicular to the film surfaces, the electric field of the light is incident only along the diameter, or short axis, of the rods. Thus, only one plasmon resonance band appears in each spectrum (the transverse resonance).

As the aspect ratio (length/diameter) of the particles increase, there is a slight blue shift in the plasmon resonance maximum. This effect can be explained on the basis of Eqs. (4) and (5) in Chapter 1, and the fact that the screening parameter κ for the case of the electric field incident along the short axis of a rod or ellipsoid of revolution decreases as the particle aspect ratio increases (20).

Fig. 7B shows a series of spectra at different incidence angles (θ) of a single composite film system containing aligned gold nanorods. The incident light is p-polarized; that is, the electric field is aligned parallel to the plane of incidence. At $\theta = 0$ the spectrum shows only a single plasmon band, as in the case of Fig. 7A. However, as θ increases, a second spectral band grows in, corresponding to the long-axis resonance. Converse to the transverse resonance case in Fig. 7A, the wavelength maximum of the longitudinal resonance increases with increasing particle aspect ratio (18,20).

Template Synthesis of Metal Particles

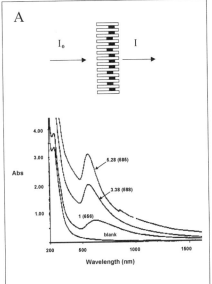
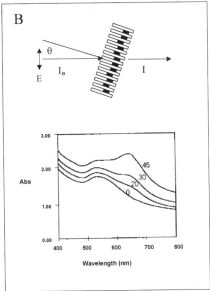

Fig. 7. UV/visible spectra of oriented Au nanoparticle composites. (A) Normal incidence spectra of 60-nm-diameter Au rods in AAO (reprinted in part with permission from Foss, C.A., Jr.; Hornyak, G.L.; Stockert, J.A.; Martin, C.R. The Journal of Physical Chemistry 1994, 98, 2963–2971. Copyright 1994 American Chemical Society). Particle aspect ratios (length/diameter) shown next to curves. Numbers in parentheses are plasmon resonance λ_{max} value in nanometers. Note blue shift with increasing aspect ratio. (B) Polarization spectra of Au nanorods (diameter = 30 ± 4 nm, length = 54 ± 7 nm) in host AAO at different incident angles θ measured relative to AAO surface normal (reprinted in part with permission from Sandrock, M.L.; Pibel, C.D.; Geiger, F.M.; Foss, C.A., Jr. The Journal of Physical Chemistry B 1999, 103, 2668–2673. Copyright 1999 American Chemical Society). Incident light is *p*-polarized (electric field parallel to plane of incidence; see schematic above spectra).

B. Linear Optical Properties of Template Synthesized Particles Liberated from Their Hosts

If the deposition of metal into the pores of the host template results in nanoparticles which are continuous, these particles can be removed from the host with conservation of their geometry. For example, van der Zande et al. released rodlike gold nanoparticles into solution by destroying the host oxide (21). The spectra of the resulting aqueous dispersions are very similar to those obtained from solutions of gold nanorods prepared via electrolysis in bulk solution (see Chapter 7). Because the nanorods are randomly oriented in solution, two plasmon resonance bands are observed: a strong one at long wavelengths due to the long-axis resonance, and a weaker one at shorter wavelengths that originates from the transverse resonance.

It is also possible to orient metal nanoparticles in a new host medium after they have been liberated from their template host. For example, we have impregnated Au nanorod/porous alumina composites with polyethylene (PE), etched away the alumina with hydroxide, and then friction-oriented the fibrous Au nanoparticle/PE (18). Figure 8 shows the polarization spectra of two Au/PE composite films: one with low-aspect-ratio nanorods (8A), and one containing higher-aspect-ratio particles (8B). As one would expect, the degree of linear dichroism increases with particle aspect ratio. Judging from the low intensity of the transverse plasmon resonance band for the case of the electric field polarized in the gross orientation direction ($\theta = 90°$ in Fig. 8), the extent of orientation is quite

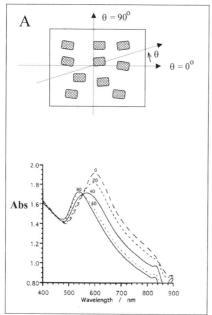
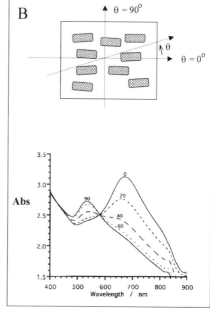

Fig. 8. Polarization UV/visible spectra of Au nanorod/polyethylene (PE) composite films (adapted from Ref. 18). Schematics above spectra show optical setup. Incidence vector is perpendicular to plane of the page, and the polarization ϕ is measured relative to gross orientation axis of PE fibers. Numbers next to spectral curve indicate ϕ values. (A) Low-aspect-ratio particles (diameter = 32 ± 1 nm, length = 40 ± 5 nm). (B) High-aspect-ratio particles (diameter = 32 ± 1 nm, length = 67 ± 7 nm).

high. Unfortunately, the physical properties of the PE matrix preclude reliable TEM evaluation of the orientation.

Orientation of template synthesized metal nanoparticles is also possible via (a) stretch orientation of a polymer host (22) and (b) application of an external electric field (23). In the stretch orientation method, gold nanorods prepared in porous alumina are freed from their host and stabilized in aqueous suspension via surface derivatization with poly(vinylpyrrolidone). The aqueous mixture is added to an ethylene glycol solution of poly(vinylalcohol) (PVA). A thin (40–90 μm) PVA film is cast from the glycol solution and stretched under gentle heating. Polarization spectral analyses indicate that the degree of alignment is virtually complete when the PVA are stretched four to six times their original length (23).

The external field method involves placing an aqueous suspension of liberated nanoparticles between two PT-coated stainless steel electrode plates, ca. 1 mm apart. The optical probe beam is directed between the plates and polarized either parallel or perpendicular to their surfaces. Application of an AC electric field between the plates induces orientation of the gold nanorods, and the orientation is evident in the polarization spectra. Significantly, the extent of the observed field-induced orientation agrees well with theoretical treatments which take into account the disruptive effects of thermal motion in the solution (24).

While template host materials such as porous anodic alumina can provide orientational order, the order is limited to the case of alignment perpendicular to the film surfaces. Stretch and friction orientation allow for alignment in the plane of the composite film. The electric-field-based method aligns particles perpendicular to the electrode plates enclosing a thin layer of solution, thus essentially recreating the geometry present in the original nanoparticle/host template system. While future experiments may be expected to extend the method to thicker samples, a fundamental drawback of the electric-field-based orientation is that thermal motion in the solution precludes complete alignment (23).

The friction and stretch orientation methods seem capable of yielding more complete alignment. However, in the context of controlling nanoscale structure, the electric-field method may have the added feature of inducing formation of nanoparticle aggregates. These "strings" of metal nanoparticles are postulated to arise from electrostatic interaction of electric-field-induced dipoles (25).

C. Complex Nanoparticle Structures

While the electrochemical template synthesis method is perhaps most noted for its application to the production of nanorods or nanowire structures, it also offers a means for preparing paired nanoparticles (26–28) and layered nanostructures (29–31). The preparation of layered structures is straightforward and involves the electrodeposition of different metals in a prescribed order. The thickness of each layer, and thus the overall final pattern of the nanoparticle structure, can be controlled electrochemically. Examples of multilayered metal particles prepared with the template synthesis include the Pt/Au structures synthesized by Mallouk and co-workers (31).

We have recently shown that gold nanoparticle pair structure preparation is only slightly more complicated. Beginning with the scheme in Fig. 4, instead of etching the silver foundation away after step 4C, a second layer of silver is deposited onto the existing gold layer (4D'). The silver plating solution is removed, the cell is rinsed, and gold plating solution is reintroduced. A second gold layer is then deposited onto the second silver layer (4E). Subsequent etching away of all of the silver (the foundation plus the second silver layer) with nitric acid leaves two gold nanoparticles in each pore (4F). Using this modified template synthesis method, we have prepared gold sphere pairs, rod pairs, and sphere-rod pairs. Paired particles such as these provide an added dimension to structural complexity and open up new avenues for structure-optical property studies.

A key feature of this method is that the interparticle spacing can be varied via the thickness of the second silver deposit. For the paired sphere structures, we have observed a subtle dichroism in the plasmon resonance band when the spheres are close together (i.e., the center-to-center distance corresponds to less than about two particle diameters). When the spheres are farther apart, no dichroism is observed, since the particles behave as independent spheres (27).

Perhaps more significant is the observation of second-harmonic generation (SHG) from noncentrosymmetric Au rod pairs (26). It is well known that at least for molecular systems, the lack of inversion symmetry is a requirement for second-order nonlinear optical (NLO) behavior. Figure 9 shows the SHG counts as a function of incidence angle (under p-polarization) for two aluminum oxide films, each containing a comparable total amount of gold, but one containing noncentrosymmetric rod pairs and one containing centrosymmetric rods. The noncentrosymmetric pairs are composed of a smaller rod [diameter (d_1) = 26 ± 3 nm, length (l_1) = 27 ± 3 nm] and a larger rod (d_2 = 26 ± 3, l_2 = 37 ± 6 nm). The centrosymmetric rods have an average diameter equal to 30 ± 4 nm and an average length of 54 ± 7 nm (26). As the incidence angle increases, an increasing component of the incident electric field is parallel to the pore axis (which is also the asymmetry axis). Thus, the SHG counts increase with incidence angle for the noncentrosymmetric pair structures. The SHG counts are low for the centrosymmetric rod system and, within error, do not increase with θ.

Using template synthesized gold nanoparticles, we were able to demonstrate conclusively that SHG activity in certain limits depends on nanoparticle symmetry. Furthermore, because particle aspect ratio and size are easily controlled, we are able to probe the effects of these parameters on SHG efficiency. For example, a key finding is that symmetry (or lack thereof) is key to SHG intensity in small metal nanoparticle structures, whose primary response to an

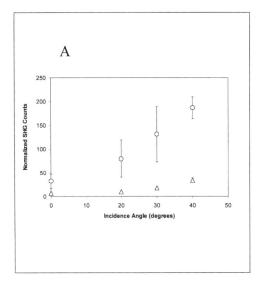 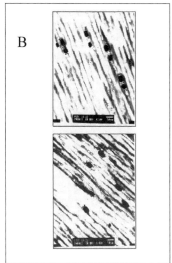

Fig. 9. (A) Second-harmonic generation (SHG) of Au nanoparticles in AAO as a function of incidence angle under p-polarization (adapted from Ref. 32). Open circles: SHG counts for noncentrosymmetric Au nanoparticle pairs. Large-pair member dimensions: length = 37 ± 6 nm, diameter = 26 ± 3 nm. Small-pair member dimensions: length = 27 ± 5 nm, diameter = 26 ± 3 nm. Edge-edge separation = 22 ± 8 nm. Open triangles: SHG counts for centrosymmetric Au rods (length = 54 ± 7 nm, diameter = 30 ± 4 nm). All SHG counts normalized to SHG output from a potassium dihydrogen phosphate slurry in decahydronaphthalene (see Ref. 26). (B) TEM images of noncentrosymmetric Au nanoparticle pair (top) and centrosymmetric rods (bottom) in AAO host oxide. Scale bar = 50 nm in both images.

external electromagnetic wave is electric dipole induction. As the particle dimensions increase, contributions from magnetic dipole and electric quadrupole modes become more pronounced (32). Since these modes can give rise to SHG even in spherical or other centrosymmetric particles, it is not surprising that larger noncentrosymmetric pairs ($d_1 = d_2 = 33 ± 4$ nm, $l_1 = 100 ± 10$ nm, $l_2 = 31 ± 4$ nm) show essentially the same SHG counts as centrosymmetric rods of similar overall dimensions ($d = 33 ± 4$ nm, $l = 133 ± 2$ nm) (32).

We have also used electrochemical template synthesis to prepare gold-silver iodide pair structures, with the metal and semiconductor layers either in contact or at variable interparticle distances. The synthesis of these paired particles is somewhat more complicated than that of the single Au nanorod or paired Au particle systems described above. Following step C of Fig. 4 again, a second silver layer is deposited onto the gold layer (4D'). The silver plating solution is then replaced with an aqueous potassium iodide solution, and the silver layer deposited above the gold layer is oxidized to form silver iodide (4E″) via the reaction

$$Ag_{(s)} + I^-_{(aq)} \rightarrow AgI_{(s)} + e^- \qquad (2)$$

If the entire silver layer is converted to AgI, the metal and semiconducting layers are in electrical contact. However, if only part of the silver layer is converted and the remainder is etched away with nitric acid, the result is two particles, one composed of Au and one of AgI, separated by a distance corresponding to the thickness of the uncoverted silver layer (4F″).

Figure 10 shows the UV/visible extinction spectra of three Au/AgI nanoparticle pair systems in porous AAO. The numbers next to the curves are the average spacing between the two particles, as estimated from TEM images. When the Au and AgI phases are far apart (top curve), the plasmon resonance band of the gold particles is relatively narrow, and the band edge of the AgI particles is well defined and shows the expected exciton peak (28). As the average interparticle spacing is decreased, the Au plasmon resonance band red-shifts and broadens, and the exciton peak at the AgI band edge becomes less distinct (middle curve). When the Au and AgI phases are in contact (bottom curve), the plasmon resonance band is further red-shifted and the exciton peak is not observed.

Based on Rayleigh limit spectral simulations for Au and AgI nanoparticles in contact or close proximity, the observed red shift of the Au plasmon resonance band does not appear to arise from simple electric dipole interactions between the two particles. The diminishing of the AgI exciton peak with decreasing distance was first thought to arise from an energy transfer between the two particles that results in lifetime broadening or dielectric screening arising from the nearby metal phase (30). However, we do not have direct evidence of these mechanisms.

Perhaps a simpler explanation, and one which implies a key limitation of the template synthesis method, is phase mixing or contamination. For example, the red shift and broadening of the Au plasmon resonance band may be explained by silver or iodide impurities which disrupt the Au lattice and decrease the electron mean-free lifetime (33,34). It is also known that exciton formation can be hindered by the presence of ionizable impurities in the semiconductor phase (35). The fact that we observe gradual trends in spectral features with changes in the average interparticle spacing may arise because the interparticle spacing can only be controlled in an *average* sense; when the average spacing is small, there may be a significant population of Au/AgI particle pairs that are actually in physical contact and thus prone to material contamination. As the average spacing increases, the fraction of particles in physical contact decreases, as does the incidence of contamination.

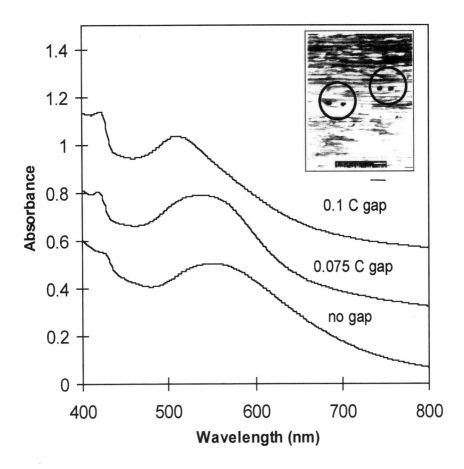

Fig. 10. UV/visible spectra of Au/AgI nanoparticle pairs in AAO (reprinted in part with permission from El-Kouedi, M.; Foss, C.A., Jr. The Journal of Physical Chemistry B 2000, 104, 4031–4037. Copyright 2000 American Chemical Society). (Top curve) Au and AgI particles at average edge-edge separation ca. 60 nm. (Middle curve) Separation ca. 30 nm. (Bottom curve) Au and AgI segments in contact. All spectra taken at normal incidence to AAO film surface, with incident electric field thus normal to Au/AgI pair axis. (Inset) TEM image of Au/AgI nanoparticle pair. Scale bar = 100 nm.

This interpretation seems consistent with preliminary studies in our laboratory that indicate that converting a small amount of the gold particle to Au_2S (via electrochemical oxidation of the Au phase in the presence of aqueous Na_2S) prior to Ag deposition provides a physical barrier to contamination. The inclusion of this layer partially restores the AgI exciton peak and prevents red-shifting of the Au plasmon resonance band even when the Au and AgI particles are in close proximity (36).

IV. CONCLUSIONS AND OUTLOOK

Electrochemical template synthesis provides a convenient, "low-tech" means for controlling metal particle structure at the nanoscopic level. Optical anisotropy, perhaps the most readily observed evidence for orientational ordering of nonspherical nanoparticles, is easily achieved. Template synthesized metal particles can also be sufficiently robust to endure removal from the template host and subsequent mechanical or field-induced alignment in other media.

If this chapter has emphasized anodic alumina films as template hosts, it is for good reason; porous AAO films are simple to prepare, their pores sizes are easy to dictate, and they are transparent over wide segments of the UV–near-IR range of the electromagnetic spectrum. Furthermore, because AAO films maintain their porous structure at elevated temperatures, annealing of the metal particles prepared within them is possible (27,37). High-temperature stability of AAO has also allowed for the preparation of carbon (38) and boron nitride (39) nanotubes via pyrolysis reactions. Li et al. (38) were also able to deposit cobalt within the carbon nanotubes, thus offering another level of metal nanoparticle complexity. Indeed, there are numerous other nonelectrochemical application of porous AAO films as template hosts, many involving extreme conditions that track-etched polymer membranes would not survive (40,41).

The primary drawback of electrochemical template synthesis remains its limited precision. While the typical distribution of rodlike particle radii in a given sample is 10% or less, parameters related to the thickness of the metal deposit (e.g., the rodlike particle length or interparticle spacing in paired structures) engender larger dispersion factors. The source of this problem is not clear, but for particles prepared in thick AAO films (Sec. II.B.2), roughness and irregularities on the side of the film contacting the plating solution and the resulting variations in metal ion transport efficiency may result in uneven metal deposition. This problem may be overcome with improvements in anodization techniques and a more thorough understanding of the pore formation process (42).

A problem related to particle length dispersion (in the thick AAO film open-pore method) is nonuniform particle placement within the film. Because the silver foundation thickness is not uniform (again, perhaps due to factors discussed in the previous paragraph), the gold particles deposited upon this foundation occur at widely varying distances from the AAO film surface. This "positional dispersion" is not reproducible from one sample to the next, and thus estimates of composite layer thickness and metal volume fraction can only be made via TEM examination of every sample (20). The problem of particle distribution within the AAO is not an issue if the metal nanoparticles are destined to be extracted and incorporated into other media. However, if optical theory-experiment comparisons in AAO/metal nanoparticle composites are the goal, uncertainties in composite nanostructure represent a serious challenge.

Another limitation of the electrochemical template synthesis method is the nature of the metals that can be incorporated into nanoparticle structures. First, the metals must be such that they can be deposited from solutions that do not damage the template host structure (this means, for example, avoiding pH extremes if the host is porous AAO or, for track-etched polycarbonate, organic solvents that dissolve the polymer). Second, open-pore methods of deposition involve an etching step that isolates or releases the final nanoparticle structures for subsequent study. The etchant attacks metal components such as Ag or Cu that provide an electrical contact and/or a spacing layer during electrodeposition. The problem is that it is difficult to actually incorporate these metals, or indeed any metal that is vulnerable to the etchant into the final nanoparticle structure. Thus, it is no accident that many electrochemical template synthesis studies to date have focused on nanoparticle structures of gold or platinum.

A challenge for future work is the development of new deposition/etching procedures that extend the range of metal elements that can be used to prepare nanoparticles. For example, one might envision a nonaqueous regime in which alkali metal or alkaline earth elements can form the electrodeposited foundations for subsequent deposition of non-noble metals. The foundation may then be etched by exposure to water, which would not damage nanoparticles composed of metals such as Ag, Cu, and Ni.

In spite of its current limitations, electrochemical template synthesis is likely to remain a viable approach to metal nanoparticle synthesis. Its low cost and simplicity make it a feasible strategy even in modestly equipped laboratories.

ACKNOWLEDGMENTS

The author acknowledges the support of the National Science Foundation (Early Faculty Development Award DMR 9625151) and the Petroleum Research Fund (PRF G Grant). The work originating in the author's laboratories was performed by past and present graduate students at Georgetown University, Dr. Nathir Al-Rawashdeh, Dr. Marie Sandrock, Mahnaz El-Kouedi, and Maryann Gluodenis. Georgetown undergraduates who contributed over the past seven years to anodic film preparation, template synthesis, and related computer modeling studies include Erich Wedam, Ann Kessinger, Carolyn Seugling, and John Michalik. The author is also grateful to Professors Janice Hicks (Georgetown) and Charles Pibel (American University) and to Dr. Franz Geiger (MIT) for useful discussion and assistance with the design and execution of the SHG experiments.

REFERENCES

1. M Faraday. Philos. Trans. 147:145, 1857.
2. G Mie. Ann. Phys. 25:377, 1908.
3. JC Maxwell-Garnett. Philos. Trans. R. Soc. A203:385, 1904.
4. See, for example, U Kreibig, M Vollmer, Optical Properties of Metal Clusters. Berlin: Springer-Verlag, 1995.
5. DR Gabe. Principles of Metal Surface Treatment and Protection Oxford: Pergamon Press, 1978, 2nd ed.
6. DGW Goad, M Moskovits. J. Appl. Phys. 49:2929, 1978.
7. A Andersson, O Hunderi, CG Granqvist. J. Appl. Phys. 51:754, 1980.
8. A Despic, V Parkhutik. In: Modern Aspects of Electrochemistry. J O'M Bockris, Ed. New York: Plenum, 1989, Vol 20.
9. RC Furneaux, WR Rigby, AP Davidson. Nature 337:147, 1989.
10. RC Furneaux, WR Rigby, AP Davidson. US Patent No 4,687,551, August 18, 1987.
11. RL Fleischer, PB Price, RM Walker. Nuclear Tracks in Solids. Berkeley: University of California Press, 1975.
12. CR Martin. Science 266:1961, 1994.
13. A Zagiel, P Natishan, E Gileadi. Electrochim. Acta 35:1019, 1990.
14. CK Preston, M Moskovits. J. Phys. Chem. 92:2957, 1988.
15. D Al-Mawlawi, CZ Liu, M Moskovits. J. Mater. Res. 9:1014, 1994.
16. CA Foss, Jr., MJ Tierney, CR Martin. J. Phys. Chem. 96:9001, 1992.
17. GL Hornyak. Doctoral dissertation, Colorado State University, 1997.
18. NAF Al-Rawashdeh, ML Sandrock, CJ Seugling, CA Foss, Jr. J. Phys. Chem. B. 102:361, 1998.
19. BMI van der Zande, MR Bohmer, LGJ Fokkink, C Schonenberger, Langmuir 16:451, 2000.
20. CA Foss, Jr., GL Hornyak, JA Stockert, CR Martin. J. Phys. Chem. 98:2963, 1994.
21. BMI van der Zande, MR Bohmer, LGJ Fokkink, C Schonenberger. J. Phys. Chem. B. 101:852, 1997.
22. BMI van der Zande, L Pages, RAM Hikmet, A van Blaaderen. J. Phys. Chem. B. 103:5761, 1999.
23. BMI van der Zande, GJM Koper, HNW Lekkerkerker. J. Phys. Chem. B. 103:5754, 1999.
24. CT O'Konski, K Yoshioka, WH Orrtung. J. Phys. Chem. 63:1558, 1959.
25. CT O'Konski. J. Phys. Chem. 64:605, 1960.
26. ML Sandrock, CF Pibel, FM Geiger, CA Foss, Jr. J. Phys. Chem. B. 103:2668, 1999.
27. ML Sandrock, CA Foss, Jr. J. Phys. Chem. B. 103:11398, 1999.
28. M El-Kouedi, CA Foss, Jr. J. Phys. Chem. B. 104:4031, 2000.
29. JD Klein, RD Herrick II, D Palmer, MJ Sailor, CJ Brumlik, CR Martin. Chem. Mater. 5:902, 1993.
30. M El-Kouedi, ML Sandrock, CJ Seugling, CA Foss, Jr. Chem. Mater. 10:3287, 1998.
31. BR Martin, DJ Dermody, BD Reiss, M Fang, LA Lyon, MJ Natan, TE Mallouk. Adv. Mater. 11:1021, 1999.

32. ML Sandrock. Doctoral dissertation, Georgetown University, 2000.
33. AH Ali, CA Foss, Jr. J. Electrochem. Soc. 146:628, 1999.
34. BNJ Persson. Surf. Sci. 281:153, 1993.
35. N Peyghambarian, S Koch, A Mysyrowicz. In Introduction to Semiconductor Optics. New Jersey: Prentice Hall, 1993.
36. M El-Kouedi, CA Foss, Jr. Unpublished results.
37. JC Hulteen, CJ Patrissi, DL Milner, ER Crosthwait, EB Oberhauser, CR Martin. J. Phys. Chem. B. 101:7727, 1997.
38. J Li, M Moskovits, TL Haslett. Chem. Mater. 10:1963, 1998.
39. KB Shelimov, M Moskovits. Chem. Mater. 12:250, 2000.
40. BB Lakshmi, PK Dorhout, CR Martin. Chem. Mater. 9:857, 1997.
41. CA Huber, TE Huber, M Sadoqi, JA Lubin, S Manalis, CB Prater. Science 263:800, 1994.
42. F Li, L Zhang, RM Metzger. Chem. Mater. 10:2470, 1998.

6
Nonlinear Optical Properties of Metal Nanoparticles

Robert C. Johnson and Joseph T. Hupp
Northwestern University, Evanston, Illinois

I. INTRODUCTION AND BACKGROUND

Nonlinear optics (NLO) is an interesting and active field of research and development where activity is now largely driven by a desire to progress toward real-world applications of novel technologies (1). A compelling emerging application of nonlinear optics is photonics, the photon-based analog of electronics, which utilizes nonlinear phenomena (e.g., sum- and difference-frequency generation and frequency doubling) for predictable modification, translation, and switching of optical signals. The fundamental physical principles of nonlinear photonics are well established. The challenge comes in discovering and/or designing materials that satisfy the technological requirements accompanying both proof-of-concept device demonstrations and practical applications. From this perspective, nonlinear photonics is an exceptionally multidisciplinary field, engendering the interest and efforts of engineers, chemists, physicists, and materials scientists. Eventual applications must successfully integrate fundamental (shedding light on the mechanisms of NLO response and enhancement) and applications-based (device design and materials synthesis) work. Accordingly, much of the chemical research in nonlinear optics and photonics has centered on development of new molecules, polymers, and other materials with favorable and programmable NLO properties. Almost all commercially available NLO devices, however, are still based on simple inorganic crystals such as potassium dihydrogen phosphate (KTP); such crystals are widely used, for example, as frequency doublers in laser applications. Nevertheless, organic molecules and polymers are potentially highly attractive for NLO purposes, offering superior mechanical properties, structural and physical tailorability, and a potentially wider range of optical properties. As the wealth of synthetic knowledge has grown, the search for novel, potent NLO materials has become correspondingly sophisticated and exotic.

Despite their remarkable optical properties, and though their existence has been known for nearly a century and a half, metal nanoparticles have not been included in this search until recently. While the linear optical properties of metal nanoparticles have been extensively investigated, the potential of these materials for nonlinear optical applications has been largely untapped. Nevertheless, because superior NLO responsivity is largely a matter of attaining superior electronic polarizability and hyperpolarizability (see below), one might expect metal nanoparticles, with their exceptionally high density of delocalized electrons, to be good-candidate NLO materials. This chapter summarizes recent studies of the nonlinear optical properties of metal nanoparticle solutions and suggests applications beyond the general goals of NLO research. While supported metal nanoparticles or nanoparticle films have been by comparison well investigated in this regard, these topics are not included here. Brief introductions to NLO and relevant experimental techniques are provided; the interested reader may consult the references for a more technical analysis.

II. NLO: THEORETICAL BACKGROUND

The effectiveness of NLO chromophores can be quantified and compared by tabulating NLO properties which describe the magnitude of the nonlinear response. Briefly, an intense electromagnetic field interacting with matter induces within the matter a dipole P_{ind} that can be expressed as

$$P_{ind} = \alpha_{ij}E_j + \beta_{ijk}E_jE_k + \gamma_{ijkl}E_jE_kE_l + \cdots \tag{1}$$

The subscripts represent the molecular or particle coordinates x, y, z; repeated subscripts indicate summation over the three components. E_j is the electric field component along the j axis. α, β, and γ represent the polarizability, first hyperpolarizability, and second hyperpolarizability tensors, respectively. α describes the linear response to the field, while β, γ, etc., describe the nonlinear response. β is the quantity in which the second-order response is manifested; from a chemical perspective, maximizing β is a primary objective. For dipolar molecules β is generally dominated by the component β_{zzz}, where z is the axis aligned with the molecular dipole. More detailed discussions about β are given elsewhere (2–4).

Perhaps the most common NLO phenomenon is second-harmonic generation (SHG), which is coherent light scattering at double the incident frequency. As with all even-order nonlinear optical processes, SHG is to a first approximation absent for species featuring a center of symmetry (5). SHG is also absent for *collections* of species when the species are configured such that signals (vector quantities) from individual moieties cancel each other. This includes, to a very good approximation, isotropic solutions of chromophores which individually are capable of SHG. [An important exception has been described by Eisenthal and co-workers, who showed that residual coherent SHG can be obtained from solutions which are macroscopically centrosymmetric but contain organized, ordered, and locally noncentrosymmetric chromophoric domains having dimensions comparable to, or in some cases even as small as ca. 10%, of the wavelength of incident light (6).]

Solution SHG measurements can be made via the EFISHG (electric-field-induced second-harmonic generation) technique, in which an external electric field is applied to orient dipolar solute molecules. Briefly, the orientation breaks the overall centrosymmetry of the solution, thereby permitting generation of a second-order signal. From a slightly different perspective, the sensitivity of SHG to symmetry reduction has been very effectively exploited by surface scientists, who have recognized that interface formation necessarily causes a loss of centrosymmetry. SHG can be used, therefore, as an interface-specific structural and analytical tool (7). Although this interesting field is beyond the scope of this chapter, a handful of studies of coherent SHG are described, including one involving a nanoparticle film which bears relevance to this chapter.

While coherent SHG is the preferred method for NLO analysis of films and interfaces, a more effective technique for solution species is hyper-Rayleigh scattering (HRS), which is incoherently scattered second-harmonic light (essentially "incoherent SHG"). While HRS was experimentally demonstrated in 1965 (8), not until decades later did it become popular as a means for obtaining first hyperpolarizabilities for solution species (9). At first glance, HRS would appear to be impossible: nonlinear scattering, for example, from dipolar molecules comprising a truly isotropic solution should lead to overall cancellation and no net signal. To a large extent this is true: HRS signals are typically several orders of magnitude less intense than coherent SHG signals from the same nonlinear chromophores because the doubling signals from individual, randomly oriented chromophores indeed do nearly completely cancel each other. Briefly, HRS works, despite the orientational randomization, because signals scale as the *variance* of the orientation of the chromophoric species with respect to the electromagnetic field (10). (Thus, although orientational randomization necessarily results in an average electromagnetic polarization of zero, the variance of the orientation is nonzero.) HRS also circumvents the need to orient the scattering entities within the sample and can thus be used to analyze, for example, solutions of ionic and nondipolar species, both of which are inaccessible to SHG experiments.

In general, the experimental hyper-Rayleigh scattered intensity ($I_{2\omega}$) is given by

$$I_{2\omega} = G \times B^2 \times (I_\omega)^2 \times \exp(-N_c \sigma_{c(2\omega)} l) \qquad (2)$$

where B is the (macroscopic) second-order susceptibility. For a two-component (solute plus solvent) solution, B^2 is

$$B^2 = \langle N_c \beta_c^2 + N_s \beta_s^2 \rangle \tag{3}$$

where the brackets indicate an average over all chromophore orientations. Subscripts c and s denote chromophore and solvent, respectively; G contains instrumental factors, collection efficiency, and local field factors; N represents concentration; I_ω is the intensity of the incident fundamental light; $\sigma_{c(2\omega)}$ is the absorption cross section at the second-harmonic wavelength; and l is the effective scattering pathlength. The exponential term accounts for self-absorption of the scattered photons.

Details of HRS setups and experimental procedures are provided elsewhere (10,11). The HRS signal is very weak, so a high-power laser source and appropriately sensitive detection are necessary. Depending on the wavelength of interest, typically the output of a nanosecond-pulsed Nd:YAG laser (1064 nm), a Nd:YAG-pumped optical parametric oscillator (broadly tunable within the visible and near-infrared spectral regions), or a femtosecond-pulsed titanium-sapphire laser (typically tunable from ca. 720–1100 nm) is used. If the solvent's β is known at the relevant wavelength, the HRS measurements can be internally referenced against the solvent. Alternatively, a different chromophore with a known β at the wavelength of interest can be used as an external reference. Since β_c usually exceeds β_s by a few to several orders of magnitude, the external reference method is typically less sensitive to small uncertainties in measurement and is thus inherently more accurate.

Several components of the β tensor can be extracted by performing second-order light-scattering experiments using incident fundamental frequencies ω_1 and ω_2 (for HRS experiments, $\omega_1 = \omega_2$) and intelligently choosing the polarizations of the incident and scattered light (12). One simple yet informative experimental variation of HRS is the measurement of D, the ratio of the intensities of vertically polarized and horizontally polarized HRS signals obtained by using vertically polarized incident light. D, which is approximately equal to β_{zzz}/β_{xzz}, can provide symmetry information when molecular scatterers are involved. For example, $D = 5.0$ for molecules with C_{2v} symmetry and $D = 1.5$ for purely octupolar species (13). Depolarization experiments on metal nanoparticles are described below.

III. GOLD, SILVER, AND COPPER NANOPARTICLES

Although theories of nonlinear scattering from metal nanoparticles are discussed later, a simplified treatment of the corresponding theory for *molecular* chromophores is summarized in Eq. (4), which is applicable to a hypothetical system characterized by a ground electronic state and a single charge-transfer excited state (denoted as 1 and 2, respectively), and an infinitely narrow charge transfer absorption band (14):

$$\beta = \frac{3\mu_{12}^2 \Delta \mu_{12} E_{op}^2}{2(E_{op}^2 - E_{inc}^2)(E_{op}^2 - (2E_{inc})^2)} \tag{4}$$

In the equation, μ_{12} is the transition dipole moment—a quantity closely related to the absorption intensity or oscillator strength, $\Delta\mu_{12}$ is the change in dipole moment upon optical excitation (proportional to the degree and distance of internal charge transfer accompanying excited-state formation), and E_{op} and E_{inc} are the energies of the optical transition and the incident light, respectively. This model predicts signal enhancements at wavelengths that are one- or two-photon resonant with an optical transition. It also predicts large signals for strongly allowed electronic transitions.

We and others reasoned that gold nanoparticles, with their spectacularly intense visible-region absorption bands (collective electron excitation or "plasmon" bands), might prove particularly effective as nonlinear optical chromophores. In HRS studies by Vance and co-workers with 13-nm, citrate-stabilized gold particles in aqueous solutions, an enormous hyperpolarizability—7×10^{-25} esu—was recorded with $E_{inc} = 12,200$ cm^{-1} = 1/(820 nm) (15). Similarly large hyperpolarizabilities have recently been reported by Galletto and co-workers (17). For comparison, β for the standard NLO benchmark compound, p-nitroaniline, is 34×10^{-30} esu at 1064 nm, while for water it is 0.56×10^{-30} esu. The comparison, of course, is somewhat misleading because a single gold nanoparticle contains thousands of atoms, while most molecular chromophores contain less than a hundred. Clearly, some sort of normalization is required in order to make meaningful comparisons. For photonic applications, perhaps the best figure of merit is β^2 per unit volume. (Recall that NLO responses scale as β^2 rather than β. If the candidate nonlinear chromophores are dipolar and are ultimately used in ways that require the application of an external field in order to attain alignment, and, therefore, net signal generation, then the most meaningful figures of merit could also include the dipole moment.) A slightly more convenient figure of merit—one which obviates the need to consider packing differences, bond-length differences, etc.—is β^2 per atom. As shown in Table 1, the normalized nonlinear scattering efficiency of the gold nanoparticle sample (partially resonant) exceeds by several orders of magnitude the off-resonance efficiencies of the best existing molecular NLO chromophores. In the molecular NLO chromophore literature, responses are typically reported in terms of β rather than β^2; therefore, Table 1 also includes $(\beta^2/\text{atom})^{1/2}$. For either figure of merit, it remains to be seen how great the nanoparticle/molecule disparity is under conditions where both sets of measurements are made off resonance; almost certainly, however, the disparity will be less.

Table 1

Scatterer	β (10^{-30} esu)	β^2/atom (10^{-60} esu)	$(\beta^2/\text{atom})^{1/2}$ (10^{-30} esu)
H$_2$O	0.56[a]	0.10	0.32
H$_2$N–C$_6$H$_4$–NO$_2$	34[b]	72	8.5
Me$_2$N-aryl-ethynyl-Zn(porphyrin)-ethynyl-aryl-NO$_2$	5000[a,c]	2.7 × 10^5	520
Ag (21 nm)	7 × 10^5 [b,d]	7 × 10^5	840
Au (13 nm)	7 × 10^5 [a,e]	8 × 10^6	2800

[a] 820-nm incident radiation.
[b] 1064-nm incident radiation.
[c] From Ref. 16.
[d] From Ref. 20.
[e] From Ref. 15.

What about nanoparticle size effects? At some point, as the size of the nanoparticle is increased, one would expect to see primarily residual coherent SHG, rather than HRS, thereby complicating the extraction of β^2 from the observed doubled-light intensities. Fortunately, there is a simple experimental diagnostic for distinguishing residual coherent SHG from HRS: signals for the former scale with the square of the chromophore concentration, while signals for the latter scale linearly. In any case, Galletto and co-workers have reported on size effects over the 5- to 22-nm-diameter particle range, using incident light at 1064 nm (17). Using β/atom as a figure of merit, they reported striking decreases in HRS response with increasing particle size. By using what we would argue is a more meaningful figure of merit, β^2/atom, the available data actually point to a significant *increase* in normalized efficiency with increasing particle size (Fig. 1). Notably over the same size range (a) the linear absorptivity, and therefore μ_{12} [see Eq. (4)], remain more or less constant on a per atom basis, and (b) the plasmon absorption maximum moves from 508 to 520 nm when the particle diameter is changed from 5 to 22 nm, perhaps slightly changing resonance effects.

What about wavelength effects? Qualitatively for 13-nm gold particles we find that nonlinear scattering occurs more efficiently with excitation at 1064 nm than at 820, 760, or 720 nm. This would be consistent with a two-photon enhancement effect due to resonance with a plasmon absorption band near 532 nm. More compelling evidence for plasmon resonance enhancement comes from closely related SHG measurements: excitation profiles for colloidal gold particles at the air-toluene interface, using a range of second-harmonic wavelengths tuned to include the plasmon absorption, show a narrow peak in the SHG profile centered near the absorption peak (18). Nanoparticles embedded in alumina and deposited

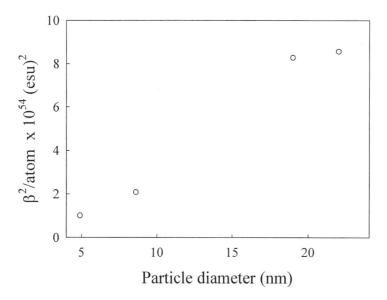

Fig. 1. Relative HRS response of gold nanoparticles of varying size; based on data from Ref. 17.

on silica behave similarly: the SHG excitation profile shows a peak centered near the plasmon absorption maximum (19). The combined experiments additionally show that nonlinear spectroscopic response also depends on the surrounding medium, at least in an energetic sense: the position of the plasmon peak for the supported particles is red-shifted compared to the solution particle (see Chapter I), and the SHG excitation profile is similarly red-shifted.

Can the phenomenon be observed with metal nanoparticles other than gold? The first reports of HRS from metal nanoparticles were for silver, not gold (20,21). Triest and co-workers reported seemingly modest HRS signals from colloidal silver nanoparticle suspensions: $\beta/\text{atom} \approx (0.7 \text{ to } 1.0) \times 10^{-30}$ esu, where the variability reflects differences in particle size. If the results are recast as $(\beta^2/\text{atom})^{1/2}$, the responses are rather more impressive: ca. 800×10^{-30} esu for the smallest particles examined (cf. Table 1). Notably, the particles examined by these investigators featured a narrow plasmon absorption centered at 400 nm and, therefore, exhibited absorbance at neither the fundamental (1064 nm) nor the second-harmonic (532 nm) wavelength of the experiment. Unpublished studies at Northwestern, based on wavelength-tunable excitation, reveal a strong two-photon enhancement effect due to plasmon resonance, with resonant $(\beta^2/\text{atom})^{1/2}$ values as large as $\sim 7000 \times 10^{-30}$ esu (22).

Further studies using 820-nm incident light show that copper nanoparticles also are capable of strong nonlinear responses, comparable to those of silver and gold (22). Presumably the strong response is due to partial plasmon resonance in both one- and two-photon senses. The copper particle spheres interrogated by HRS spectroscopy featured a broad, but intense, plasmon absorption band centered at $\sim$560 nm. In contrast, similarly sized tetrahedral platinum particles yielded no detectable hyper-Rayleigh scattering. The key difference between platinum and the coinage metals is presumably the absence of a visible-region plasmon band, needed for resonance enhancement.

IV. AGGREGATION EXPERIMENTS

The incoherent second-harmonic response from gold colloids is enhanced upon coagulation of the nominally spherical particles into larger random (and presumably nonspherical) aggregates. This has been shown by HRS experiments on gold with aggregation induced by addition of salt (15) or pyridine (23). Aggregation causes broadening of the plasmon peak and increases in extinction at longer wavelengths (600–800 nm). Under conditions of extreme aggregation the original plasmon absorption peak disappears essentially completely (Fig. 2). In view of the prime role played by the plasmon absorption in enhancing the particles' nonlinear optical response, it is tempting to conclude that the further enhancement accompanying aggregation is a consequence of plasmon band energy shifts and a

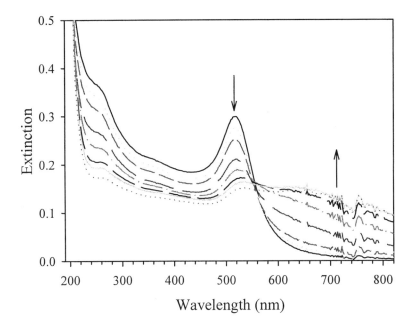

Fig. 2. UV-vis extinction spectra recorded after successive additions of 6.0 M NaCl to a solution of 13-nm gold nanoparticles (15). (Reproduced with permission from J. Phys. Chem. B., 1998, 102, 10091–10093. Copyright 1998 Am. Chem. Soc.)

resulting improvement in optical resonance. A more careful evaluation shows, however, that the large increase in HRS response occurs before the aggregation process has progressed far enough to cause noticeable changes in the linear extinction spectrum (15). Instead, HRS signal enhancement appears to be associated with small aggregate formation, with further aggregation having little additional effect. This is shown in a striking way in Fig. 3, from Vance and co-workers, where the linear scattering response (Rayleigh scattering, a rough measure of aggregate size) is plotted together with the second-order scattering response (hyper-Rayleigh response) as a function of the amount of aggregation agent added: HRS is clearly more sensitive to the formation of small aggregates than is Rayleigh scattering.

Why does the otherwise inconsequential formation of very small aggregates lead to such extensive enhancement of the metal nanoparticles' nonlinear scattering capabilities? Recall that the available *molecular* theory precludes dipole-based frequency doubling by centrosymmetric chromophores, such as spherical nanoparticles. If a particle aggregate behaves electromagnetically as a single collective chromophore rather than a collection of individual chromophores, then the overall symmetry and not just the symmetry of the component particles comes into play. Evidently, small aggregate formation causes a sufficient decrease in

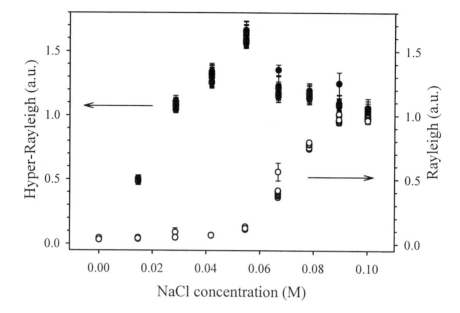

Fig. 3. Comparison of Rayleigh (open circles) and hyper-Rayleigh (filled circles) signals from 13-nm gold nanoparticles after aggregation induced by addition of 6.0 M NaCl solution (15).

collective symmetry (centrosymmetry) to permit the observation of an additional NLO response that would otherwise be cancelled. Furthermore, the reduction in chromophoric centrosymmetry upon small aggregate formation is sufficiently great that further aggregation, and further symmetry reduction, offer no additional advantage.

Although shifts in plasmon band energy and intensity appear not to play a dominant role in the studies of salt-induced aggregation summarized in Figs. 2 and 3, these effects might be important under other conditions (other excitation wavelengths, other aggregation agents, etc.) The formation of ellipsoidal aggregates rather than highly asymmetric aggregates has been observed for pyridine-induced aggregation (24). Note that a *perfect* ellipsoid, like a perfect sphere, possesses an inversion center and therefore should offer no advantage in a nonlinear scattering experiment. In any case, gold nanoparticle aggregates of approximately spheroidal shape display a substantially enhanced HRS response (23). They also display a broad extinction in the near IR not seen for the component spherical particles; this feature has been assigned as a longitudinal plasmon absorbance attributable to an electromagnetic response of each of the aggregates as a whole. (The precise shape and location of the longitudinal band depend on aggregate size and aspect ratio.) In any case, as Galletto and co-workers have noted, the enhancement in HRS upon formation of these aggregates might well be due to partial resonance of this new band with the 1064-nm incident radiation used in these experiments (23). Excitation profiles could prove helpful in establishing the extent to which longitudinal resonance effects are responsible for the HRS enhancement phenomenon. When coupled with available theory for deconvolving the linear extinction spectrum (25), the profiles should prove helpful in determining to what extent scattering versus absorption can contribute to HRS resonance effects. As one would expect, longitudinal plasmon resonance effects also are very much in evidence in the linear extinction spectra of rod-shaped gold particles. Templated synthesis of such particles with precise control over the particle dimensions has been reported (26,27), but no nonlinear optical experiments of suspensions of these interesting particles have yet been reported. Coherent SHG from aligned arrays of rod-shaped gold particles, however, has recently been reported (28).

The possible existence of yet another set of HRS enhancement schemes has been suggested by Triest and co-workers (20,21). They found, with silver nanoparticles, that adsorption of p-nitroaniline—itself a nonlinear chromophore—greatly increases the NLO response. They suggested that this could be the result of (a) a collective response from aligned dye molecules at the particle surface, (b) electromagnetic enhancement by the nanoparticle of the *molecular* chromophore's HRS response, perhaps in a fashion reminiscent of known mechanisms for surface-enhanced Raman scattering (29), or (c) fortuitous resonance with particle-dye charge-transfer transitions. (The authors also discussed the possibility of enhancement due simply to dye-adsorption-induced aggregation.) Mechanism (c) has an interesting precedent in studies by Liu and co-workers of residual coherent SHG from catechol-functionalized semiconductor particles (30). For mechanism (b) a combination of polarization and excitation-profile studies might be sufficient to establish the degree of applicability to the silver/p-nitroaniline system.

In the same study (20), these authors also reported that, in the absence of a chromophoric adsorbate, the normalized nonlinear scattering efficiency, expresses as β/atom, systematically increases with increasing particle diameter; over the admittedly limited size range examined (26–34 nm), the normalized HRS response appears to scale linearly with diameter. If the findings are recast in terms of β^2/atom (see above), the normalized dependence of the nonlinear scattering efficiency upon particle size of course becomes stronger.

V. CORE-SHELL PARTICLES

Core-shell nanoparticles are becoming increasingly popular due to their interesting electronic properties. The shell provides an extra size variable and allows for tunability of the composite particle properties. The syntheses of such particles have recently received increasing attention, and their linear optical properties have been addressed and correlated with theoretical prediction (25,31). However, the NLO properties of core-shell particles have not, to our knowledge, been seriously investigated. Recall that the visible-region optical properties of small-particle gold and silver colloids arise mainly from plasmon resonance effects, which in turn depend on polarization resonance effects whose energies are determined by dielectric functions for both the particles and their surroundings. One might expect the environmental dielectric modification associated with shell formation to have some impact, therefore, on the NLO properties of the colloids.

Preliminary HRS experiments on solutions of 13-nm gold core particles encased in a thin (4 nm) shell of silica show a collective signal roughly five times smaller than that of the unencapsulated particles (32). It should be noted that colloidal silica also exhibits a sizable HRS signal $[(\beta^2/SiO_2 \text{ unit})^{1/2} = 60-250 \times 10^{-30}$ esu for 10-nm colloidal particles (33)]; presumably the silica shell provides some of the HRS signal from the core-shell particles as well. Preliminary HRS depolarization ratio measurements (see NLO background section) yield a D value of 3.7 ± 0.4 for the silica-coated gold particles. While this does not approach the extremely large value reported for colloidal silica ($D_{silica} = 22 \pm 3$) (33), it is clearly larger than that of naked gold particles ($D_{gold} = 2.2 \pm 0.3$; see Fig. 4) (15), suggesting that the silica shell is indeed playing some role in the nonlinear scattering response of the composite species. The D values should not be interpreted as providing symmetry information about the particles; rather, they serve as a qualitative indicator that both core and shell components may be contributing to the observed HRS signal.

Why does shell formation have such a significant effect upon the nonlinear scattering efficiency? Despite the initial suggestion regarding electrostatic medium effects, the silica shells used in the preliminary HRS studies were too thin and perhaps too porous to alter the linear extinction spectrum. (Thicker shells, on the other hand, do induce the expected plasmon band red-shift and broadening.) A more prosaic explanation is that, under the conditions of the experiment, the silica shell prevents the gold particles from coalescing and hence eliminates intense HRS from residual low-symmetry aggregates otherwise present in the nominally nonaggregated sample. If the explanation is correct, then it indicates that the HRS signal enhancement accompanying aggregation and overall symmetry reduction is even greater than suggested above (i.e., ca. 40- to 50-fold rather than 8- to 10-fold).

While the silica shell provides protection against unintentional particle collision and aggregation, the core-shell particles can also be intentionally aggregated by addition of a difunctional linker ligand such as HBPA (hexane-1,6-bisphosphoric acid) (32). Preliminary work shows that when the acid groups bind the silica shells and cross-link the particles, they induce the familiar aggregation and color change. Linking also induces a substantial enhancement in the HRS response, showing that the aggregated nanoparticles can behave as collective low-

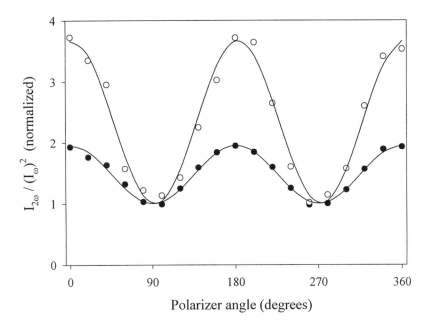

Fig. 4. Representative HRS depolarization experiments for naked 13-nm gold nanoparticles (filled circles, $D = 2.2 \pm 0.2$) and silica-coated 13-nm gold nanoparticles (open circles, $D = 3.7 \pm 0.4$). The curves are least-squares best fits of the data to $y = a\cos^2(x - c) + b$; the depolarization ratio D is equal to $(a + b) / b$. Both plots are normalized to signal = 1 at 90°.

symmetry chromophores despite physical separation (ca. 9 nm in this instance) due to the shell structures. The combined linear and nonlinear optical results also underscore a well-established but important characteristic of metal nanoparticles as chromophores. In comparison to molecular chromophores, metal particles display remarkably-long-range electromagnetic communication. Given the ready synthetic accessibility of core and shell structures of a variety of dimensions, further studies of the linear and nonlinear studies of these interesting composite particles and their aggregates could provide excellent tests of available theories regarding the distance dependence of particle-particle electromagnetic communication.

VI. DIMERS AND TRIMERS

While formation of random low-symmetry aggregates of colloidal metal particles clearly increases the incoherent second-harmonic response, a more detailed understanding can, in principle, be obtained by studying smaller aggregates of particles with known symmetry. Brousseau and co-workers have shown that appropriately designed, rigid organic linkers can be used as templates for assembling nanoparticles into well-defined clusters (34). Prototypical examples of templated dimers and trimers of gold nanoparticles are shown in Fig. 5. From the perspective of nonlinear optical behavior, the important differences between the various trimers, dimers, and free monomers are symmetry and interparticle separation. For assemblies of identically sized particles, trimers are the smallest aggregates that lack a center of symmetry. Furthermore, for the trigonally arranged trimers in Fig. 5, the component metal particles are characterized by a much smaller average separation distance than in the corresponding dimer structures (see figure caption). Among the questions addressable with these and related aggregates are the following: (1) Can the NLO response of metal nanoparticles be influenced by symmetry over dimensions approaching those of the particles themselves? (2) Over what interparticle separation distances in solution can metal nanoparticles communicate electromagnetically?

Available liner extinction spectra show only slight differences between monomeric, dimeric, and trimeric species. Preliminary nonlinear experiments (820- or 1064-nm excitation), on the other hand, show small HRS signal enhancements upon dimer formation and large enhancements upon trimer formation, with the largest enhancement accompanying the formation of the more compact trimer (35). The preliminary experiments provide further evidence for the roles of both symmetry reduction and chromophore proximity in magnifying nonlinear optical responses from metal nanoparticle aggregates. Polarized HRS excitation/detection measurements point to a further behavioral difference between trimers, on the one hand, and dimers and monomers, on the other. For both trimers, the depolarization ratio is 1.5 ± 0.1; for the dimers and monomers it is ca. 2.2. The nominally trigonal planar geometry of the trimers* is reminiscent of *molecular* NLO chromophores also possessing D_{3h} symmetry (e.g., crystal violet). The molecular species lack the ground-state dipole moment usually required for second-order nonlinear behavior, but overcome this deficit by utilizing a molecular octupolar moment (36). A key experimental signature for molecular octupolar scatterers is a D value of 1.5. It is tempting to interpret the values for the trimeric metal particles as also indicating a role for octupolar moments, although appropriate theory to support or refute the interpretation has yet to be reported. Clearly, however, the depolarization ratio is a diagnostic feature that merits further consideration by theorists (43).

---

*Note that all assumptions about symmetry in dealing with metal nanoparticles are necessarily approximations. The particles are crystalline, and the faceted surface structure reduces the precise symmetry. Such effects are presumed to be negligible on the bulk scale.

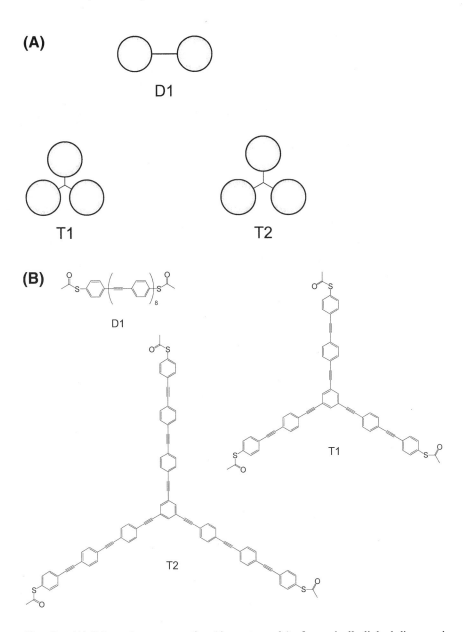

Fig. 5. (A) Schematic representation (drawn to scale) of organically linked dimer and trimers of 8-nm gold nanoparticles. Average interparticle spacings (surface to surface) for linked 8-nm-diameter spherical particles: (D1) 6.0 nm; (T1) 2.1 nm; (T2) 3.2 nm. (B) Chemical structures of the dimer and trimer linkers (34).

VII. LANGMUIR FILMS

An important non-solution-phase (or partial solution-phase) result that is relevant here has been reported by Collier et al. (37,38) The SHG response of Langmuir monolayers of small (≤4 nm) organic-ligand-capped silver nanoparticles to 1064-nm incident light was recorded continuously as the monolayers were compressed. The authors found that the change in susceptibility $\Delta\chi^{(2)}$ upon film compression displayed an exponential dependence upon average interparticle separation, indicative of distance modulation of particle-particle communication. The SHG signal was observed to increase sharply upon film compression, but with peaking followed by a dramatic falloff well before compression-induced collapse of the monolayer. The sharp decrease, seen for monolayers of 2.7-nm as well as 4-nm particles, was attributed to an insulator-to-metal transition within the monolayer. This interpretation was supported by linear reflectance data. As the authors note, these observations are an excellent example of the utility of SHG as a monitor of interparticle coupling.

VIII. THEORY

Second-harmonic generation from small metal particles was first investigated theoretically nearly two decades ago by Agarwal and Jha (39), before most of the experimental work on the subject had been performed. Enhancement of infrared absorption and Raman scattering from molecules adsorbed at metal surfaces had already been observed (40,41); the initial theoretical treatment

$$S = 192\pi^2 c |E_\omega|^4 \left(\frac{2\omega R}{c}\right)^4 \left\{ \left| \frac{e[1 - \varepsilon(2\omega)]/8\pi m\omega^2}{[\varepsilon(\omega) + 2][\varepsilon(2\omega) + 2]} \right|^2 + \frac{36}{5} \left| \frac{e[\varepsilon(\omega) - 1]/8\pi m\omega^2}{[\varepsilon(\omega) + 2]^2[2\varepsilon(2\omega) + 3]} \right|^2 \right\} \quad (5)$$

sought to explain such enhancement. Using a previously derived expression for the nonlinear polarization, and incorporating the familiar Mie solution for the local field at a spherical surface, the authors derived Eq. (5) for the total second-harmonic power, S, scattered by a small metal sphere (small compared with the wavelength of incident light). In the equation, m and e are the electron mass and charge, c is the speed of light, R is the sphere radius, E_ω is the incident intensity, and ε represents the dielectric constant of the material. The terms in brackets represent volume and surface contributions, respectively, to the scattered power. The authors further noted that (a) resonance enhancement is expected when Re($\varepsilon(\omega)$) + 2 ≈ 0 (as in standard Mie theory); (b) near resonance, the surface contribution ([ε(ω) + 2]$^{-4}$ dependence) is expected to dominate over the volume contribution ([ε(ω) + 2]$^{-2}$ dependence); (c) a two-photon resonance effect should also exist; and (d) all contributions should scale as the particle radius to the sixth power. Point (d) could also be expressed as a scaling of intensity with the square of the total number of atoms in the particle. For comparison, a log-log plot of the raw data (total β^2 versus particle radius) comprising the admittedly small experimental sample in Fig. 1 yields a slope of ~4.4, corresponding to an apparent increase in HRS signal intensity as total number of atoms per particle to the ca. 1.5 power, or roughly the square of the number of surface atoms per particle. Clearly, additional experimental work would be valuable.

Recalling the linear scaling of plasmon absorption intensity with number of atoms per particle (at least for smaller particles), the Agarwal-Jha theory also implies that HRS signals will scale as the square of the plasmon absorption coefficient or oscillator strength. The raw data comprising Fig. 1 suggest a scaling of HRS signal intensity as plasmon extinction coefficient to the power of ~1.5. (Interestingly, the corresponding theory for molecular chromophores predicts— based of course on different physics—a scaling with oscillator strength to the fourth power [see Eq. (4), where HRS signal intensity is proportional to β^2].)

Agarwal and Jha's work was later refined and extended by Hua and Gersten (42), who used a perturbation theory approach and a full Mie theory derivation to arrive at an analytical expression for an effective cross section for second-harmonic generation. Their results explicitly showed the importance of higher-mode (i.e., quadrupolar) contributions to the nonlinear scattering. [Note, however, that the second term in Eq. (5)—the "surface" term—is in fact a quadrupolar scattering term.] Scheme 1 illustrates in a qualitative way the idea of quadrupolar polarization of a small metallic sphere by incident electromagnetic radiation. Notably, quadrupolar second-harmonic generation is not subject to the noncentrosymmetry constraint associated with most dipolar SHG mechanisms. For very small particles, the Hua-Gersten theory—like the Agarwal-Jha theory—predicts a

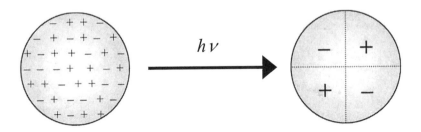

Scheme 1 Qualitative depiction of quadrupolar polarization of a metal sphere. Note the loss of centrosymmetry upon quadrupolar polarization.

radius-to-the-sixth-power (or number of atoms to the second power) dependence for the per-particle nonlinear scattering intensity. As the particle size increases, however, the sensitivity to size is predicted to weaken, ultimately dropping to a radius (r)-to-the-fourth-power dependence. Additionally, the shift toward r^4 dependence is predicted to occur more readily under near-resonance conditions. [Interestingly, the effect is paralleled by a shift from radius-cubed to radius-squared for the particle-size dependence of the linear absorption component of the plasmon extinction coefficient (25,42).] Preliminary modeling studies by Hua and Gersten for aluminum and silver nanoparticles (5–20 nm) showed multiple peaks in the calculated SHG cross section excitation profile corresponding to dipolar and quadrupolar plasmon resonance frequencies.

Recently, Dadap and co-workers treated the case of HRS from small spherical particles (43). Their theory emphasizes the role of the particle surface in creating the nonlinear response; presumably the theory, if further developed, will be useful in describing or predicting the nonlinear spectroscopic consequences (beyond simple dielectric medium effects) of specific particle surface modifications. Like Agarwal and Jha and Gersten and Hua, these workers note the importance of higher-order multipole effects and cite contributions from nonlocal dipole moment excitation and from local quadrupole moment excitation as the main contributors to the second-order signal. Interestingly, the Dadap treatment is not limited to metals, although sample calculations for aluminum particles were presented. From the sample calculations, they conclude that resonance enhancement should occur at excitation energies equaling E(plasmon) $\cdot$ $(3)^{-1/2}$, E(plasmon) $\cdot$ $(10)^{-1/2}$, E(plasmon) $\cdot$ $(12)^{-1/2}$, and E(plasmon) $\cdot$ $(2/5)^{+1/2}$. Like the earlier theories, the treatment of Dadap and co-workers predicts a limiting radius-to-the-sixth-power dependence for the per-particle HRS intensity. In addition, the theory treats second-harmonic radiation patterns as well as polarization effects, where the former apparently have yet to be explored experimentally. The authors state that the latter can be used to distinguish nonlocal dipole scattering contributions from quadrupolar contributions to the overall nonlinear response.

As new experimental findings emerge, corresponding needs and opportunities for further theory development will emerge. Some of the simpler existing experimental observations that a complete theory could help to explain are

1. The strong sensitivity of incoherent second-order nonlinear scattering responses to small-aggregate formation
2. An apparent scaling of near-resonant single-particle NLO signal intensity neither with particle area nor particle volume, but with a stronger function of particle size
3. The sensitivity of particle-cluster-based HRS signals to precise cluster symmetry and particle-particle separation distance
4. The existence of nonunit depolarization ratios

Nonlinear Optical Properties 159

We note that to some extent available theories already are capable of accounting for observations 2 and 4.

IX. APPLICATIONS

The notably different NLO responses for isolated versus aggregated metal nanoparticles, together with the comparative enormity of the responses, opens the door to many potential applications. The well-known red-to-blue color change of gold nanoparticles upon aggregation has been exploited in a sensitive colorimetric DNA sensing method, which uses oligonucleotide-coated nanoparticles designed to aggregate in the presence of a specific DNA base sequence (44). Accurate sensing of specific DNA sequences has clear technological potential for rapid disease screening. Vance and co-workers (45,46) have shown that similar sensing can be achieved with HRS as the reporter. The difference is that formation of very small aggregates (too small to induce a significant change in the linear extinction spectrum) can be detected. Under identical experimental conditions, the switch from linear to nonlinear solution-phase detection extends the DNA detection limit by a little less than an order of magnitude.

An important characteristic of the gold particle-based DNA detection system is the "melting" transition of the complementary DNA strands. In the presence of the correct complementary strand, the particles are linked. At sufficiently high temperatures enough of the hydrogen bonds connecting the strands are broken to unzip the oligonucleotides, "melting" the DNA and releasing the particles (44). Above the melting temperature, little HRS is observed. Below this temperature, substantial signals are seen. One application of the HRS detection scheme has been to evaluate the DNA concentration dependence of the "melting" temperature. Consistent with the underlying thermodynamic description, the "melting" temperature decreases as the concentration is lowered. In principle, the temperature dependence should provide qualitative or semiquantitative information about the corresponding enthalpy of complementary strand association/dissociation.

As noted earlier, modification of gold nanoparticles with alkanethiols and other surface passivating groups is straightforward and well documented. This allows for rational design of particles with specific surface properties. A potential practical application of such design is a sensitive technique for detecting toxic heavy metals in water. By decorating the surfaces of gold nanoparticles with 11-mercaptoundecanoic acid [HS-$(CH_2)_{10}$-CO_2H; MUA], the particles become efficient metal detectors in solution (32). The terminal acid functionality chelates certain divalent metal ions, including Pb^{2+}, Cd^{2+}, and Hg^{2+}—thereby inducing particle aggregation (shown schematically in Fig. 6.) Again, HRS is sensitive to this aggregation, and substantial HRS signal enhancement from 13-nm gold particles coated with MUA is seen in the presence of very small amounts of lead or

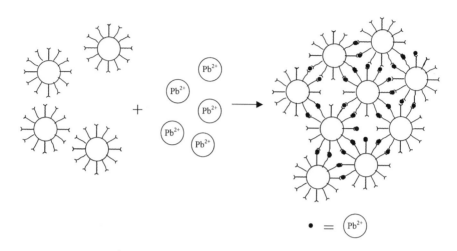

Fig. 6. Schematic representation (not drawn to scale) of metal-ion-induced aggregation of gold nanoparticles. The MUA(⇁)-decorated gold particles shown on the left are aggregated in the presence of Pb^{2+} in solution; the nonlinear (HRS) response of the resulting aggregates (shown on the right) is nearly an order of magnitude greater than that of the unlinked particles. Note that in this case Hg^{2+} and Cd^{2+} will also induce the aggregation and lead to signal enhancement.

cadmium. Addition of ethylenediaminetetraacetic acid—an extremely effective metal ion extraction agent—largely reverses the response. No enhancement is seen upon addition of zinc ions, consistent with the known poor affinity of Zn^{2+} for carboxylate ligands. Judicious tuning of the metal particle size and the receptor ligand's surface concentration could substantially increase analytical sensitivity, while manipulation of the ligand's chemical composition could enhance selectivity.

Finally, the enormous nonlinear scattering coefficients for metal nanoparticles featuring visible-region plasmon bands suggests that they could prove particularly useful in photonics applications. In contrast to metal macrostructures, the particles are characterized by good red and near-IR transparency, an important prerequisite in photonic devices because of the need to avoid absorption losses. Nevertheless, there are some obvious technical challenges in meaningfully utilizing metal nanoparticles as nonlinear optical components in photonic devices. One is to align or otherwise configure the particles so that efficient *coherent* SHG can be accomplished. [Recall that with HRS (incoherent SHG) individual particle responses largely cancel each other.] A related challenge will be to achieve the needed configuration without compromising red and near-IR transparency; recall that placement of particles in close proximity to each other typically leads to plasmon-enhanced red and near-IR region scattering and higher multipole absorption.

ACKNOWLEDGMENTS

We thank the MURI Program of the Army Research Office for financial support of this research through grant no. DAAG55-97-1-0133.

REFERENCES

1. PN Prasad, DJ Williams. Introduction to Nonlinear Optical Effects in Molecules and Polymers. New York: Wiley, 1991.
2. E Hendrickx, K Clays, A Persoons. Acc Chem Res 31:675–683, 1998.
3. A Willetts, JE Rice, DM Burland, DP Shelton. J Chem Phys 97:7590–7599, 1992.
4. P Kaatz, EA Donley, DP Shelton. J Chem Phys 108:849–856, 1998.
5. JA Giordmaine. Phys Rev 138:A1599–A1603, 1965.
6. H Wang, ECY Yan, E Borguet, KB Eisenthal. Chem Phys Lett 259:15–20, 1996.
7. RM Corn, DA Higgins. Chem Rev 94:107–125, 1994 and references therein.
8. RW Terhune, PD Maker, CM Savage. Phys Rev Lett 14:681–684, 1965.
9. K Clays, A Persoons. Phys Rev Lett 66:2980–2983, 1991.
10. K Clays, A Persoons. Rev Sci Instrum 63:3285–3289, 1992.
11. K Clays, A Persoons. Rev Sci Instrum 65:2190–2194, 1994.
12. M Kauranen, A Persoons. J Chem Phys, 104:3445–3456, 1996.
13. P Kaatz, DP Shelton. J Chem Phys 105:3918–3929, 1996.
14. JL Oudar, DS Chemla. J Chem Phys 66:2664–2668, 1977.
15. FW Vance, BI Lemon, JT Hupp. J Phys Chem B. 102:10091–10093, 1998.
16. L Karki, FW Vance, JT Hupp, SM LeCours, MJ Therien. J Am Chem Soc 120: 2606–2611, 1998.
17. P Galletto, PF Brevet, HH Girault, R. Antoine, M. Broyer. Chem Commun 1999: 581–582.
18. R Antoine, PF Brevet, HH Girault, D. Bethell, DJ Schiffrin. Chem Commun 1997: 1901–1902.
19. R Antoine, M Pellarin, B Palpant, M Broyer, B Prével, P Galletto, PF Brevet, HH Girault. J Appl Phys 84:4532–4536, 1998.
20. K Clays, E Hendrickx, M Triest, A. Persoons. J Mol Liq 67:133–155, 1995.
21. M. Triest. Ph.D. thesis, Katholieke Universiteit Leuven, Belgium, 1994
22. J Li, RC Johnson, JT Hupp. Unpublished work.
23. P Galletto, PF Brevet, HH Girault, R Antoine, M. Broyer. J Phys Chem B 103: 8706–8710, 1999.
24. CG Blanchard, JR Campbell, JA Creighton. Surf. Sci. 120:435–455, 1982.
25. KL Kelly, TR Jensen, AA Lazarides, GC Schatz, Chapter 4 this volume, and references therein.

26. BMI van der Zande, MR Böhmer, LGJ Fokkink, C Shönenberger. J Phys Chem B. 101:852–854, 1997.
27. CA Foss, MJ Tierney, CR Martun. J Phys Chem 96:9001–9007, 1992.
28. ML Sandrock, CD Pibel, FM Geiger, CA Foss, Jr. J Phys Chem B 103: 2668–2673,1999.
29. WH Yang, J Hulteen, GC Schatz, RP VanDuyne. J Chem Phys 104:4313–4323, 1996.
30. Y Liu, JI Dadap, D Zimdars, KB Eisenthal. J Phys Chem B 103:2480–2486, 1999.
31. For example, LM Liz-Marzán, M Giersig, P Mulvaney. Langmuir 12:4329–4335, 1996.
32. RJ Johnson, Y Kim, JT Hupp. Unpublished studies.
33. FW Vance, BI Lemon, JA Ekhoff, JT Hupp. J Phys Chem B 102:1845–1888, 1998.
34. LC Brousseau III, JP Novak, SM Marinakos, DL Feldheim. Adv Mater 11:447–449, 1999.
35. FW Vance, JP Novak, RC Johnson, JT Hupp, DL Feldheim. BI Lemon, LC Brousseau. J Am Chem Soc 122:12029–12030, 2000.
36. T Verbiest, K Clays, A Persoons, F Meyers, JL Bredas. Opt Lett 18:525–527, 1993.
37. CP Collier, RJ Saykally, JJ Shiang, SE Henrichs, JR Heath. Science 277:1978–1981, 1997.
38. Remacle F, Collier CP, Markovich G, Heath JR, Banin U, Levine RD. J Phys Chem B. 102:7727–7734, 1998.
39. GS Agarwal, SS Jha. Sol State Commun 41:499–501, 1982.
40. A Harstein, JR Kirtley, JC Tsang. Phys Rev Lett 45:201–204, 1980.
41. PN Sanda, JM Warlaumont, JE Demuth, JC Tsang, K. Christmann, JA Bradley. Phys Rev Lett 45:1519–1523, 1980.
42. XM Hua, JI Gersten. Phys Rev B 33:3756–3764, 1986.
43. JI Dadap, J Shan, KB Eisenthal, TF Heinz. Phys Rev Lett 83:4045–4048, 1999.
44. RA Reynolds, CA Mirkin, RL Letsinger. J Am Chem Soc 122:3795–3796, 2000.
45. FW Vance. Ph.D. thesis, Dept. of Chemistry, Northwestern University, 1999.
46. FW Vance, J. Storhoff, JT Hupp, CA Mirkin. Unpublished studies.

7
Electrochemical Synthesis and Optical Properties of Gold Nanorods

Chao-Wen Shih, Wei-Cheng Lai, Chuin-Chieh Hwang, Ser-Sing Chang, and C. R. Chris Wang
National Chung Cheng University, Min-Hsiung, Chia-Yi, Taiwan, R.O.C.

I. INTRODUCTION

The growth of the scientific literature pertaining to nanostructured materials has been remarkably fast in recent years. Nanotechnology as a fundamental science and its promise for great impact in applied technology have resulted in extraordinary interdisciplinary efforts. The unique properties of metal nanoparticle systems have attracted interest in a wide variety of fields, including but not limited to, catalysis (1), surface-enhanced Raman spectroscopy (2–4), and biosensing (5–6). In many of these applications, the observed electromagnetic effects depend strongly on the size and shape of the metal nanoparticles. As such, a natural focus of fundamental research and technological application is on synthetic methods which allow for the control of metal nanoparticle size and shape.

In fact, there are two major challenges in metal nanoparticle preparation. The first is indeed size and shape control; both the tuning of optical properties for practical applications and the evaluation of theories of these optical properties require uniformity in particle size and shape. The second challenge is surface modification. Adding a foreign functionality onto each single particle can directly lead to a multifunctioned nanoparticle and can offer the opportunity to study the mutual interactions between the new functionality and the core materials. In addition, control of the interaction between individual nanoparticles and their ultimate integration into superlattice structures or other devices will almost certainly require control over their surface chemistry.

Of all the synthetic approaches that have been developed to date for metal nanostructures, only a few are available that allow for shape control. On the basis of the means employed to achieve shape control, they can be divided roughly into two categories: The first is *template synthesis* (7), which involves deposition of metal nanostructure in a manner which prevents them from forming the thermodynamically favored spherical geometry. Widely used and less expensive template methods employ a porous host membrane such as anodic alumina (8–11) or polycarbonate (12). Metals deposited into the pores of these materials assume the pore geometry and orientation. More expensive and less widely available to most laboratories are the lithographic mask-based methods, which allow for the preparation of 2D structures on surfaces (13). The second method does not employ a preexisting template or mask but relies on the thermodynamically favored structures that result from the interaction of the metal particles' surface and some stabilizing reagent. For example, El-Sayed and co-workers have used polyacrylate stabilizers to produce tetrahedral and cubic platinum nanoparticles (14,15).

A simple unique electrochemical method for preparing large quantities of Au nanorods suspended in aqueous solution (16,17) has been recently developed. This method utilizes mixed cationic surfactants and offers the advantage of convenient control over the particles' dimensions. Several key ingredients and experimental parameters used in this method are discussed. Perhaps the only drawback at the present time is that the particle growth mechanism and its relation to the dynamics of the surfactant micelle structures are not fully understood. Nonetheless, as we intend to demonstrate, this method works well from the standpoint of synthesis.

The surface modification of Au nanorods to form silica coatings is attractive from at least two perspectives. First, silica-coated Au nanorods represent a model system for insulated nanowires. Second, coating the Au particle with silica opens up a wide range of surface modification chemistries that were not available for bare gold. In this chapter, we also discuss the experimental scheme for coating Au nanorods with a silica layer of desired thickness. Henceforth, we shall refer to silica-coated Au nanorods as Au nanorod@silica.

Before proceeding to the synthesis section, a word about the optical properties of metal nanorods is in order. For particles whose dimensions are small compared to the incident wavelength, the classical electrostatic model has been shown to be reasonably successful in predicting the absorption cross sections of metal nanostructures, where the surface plasmon (SP) resonances are the main spectral features (17). In the case of Au nanorods, the dominant SP band corresponds to the long-axis, or longitudinal, component (henceforth referred to as SP_l), while the transverse, or short-axis resonance, SP_t, is comparatively weak. We will outline electrostatic models suitable for the two experimental systems considered in this chapter: micelle-stabilized Au nanorods and the core-shell Au nanorod@ silica structures. A comparison of the calculated and experimental absorption spectra will then be made. We also demonstrate that the value of the dielectric constant of the silica shell can be extracted from the spectral measurements by observing the spectral shift of the SP_l band as a function of the shell thickness.

II. SYNTHESIS OF AU NANORODS IN AQUEOUS SOLUTION

A. Experimental Setup

We have described elsewhere a method for synthesizing suspensions of Au nanorods in aqueous solutions (16,17), wherein a mixed surfactant system was employed to define the size and shape of the nanorods. Since these first studies, we have developed an improved method (in terms of yield and control over size and shape). Briefly, the synthesis is performed within a simple two-electrode electrochemical cell. A gold metal plate (3 × 1 × 0.05 cm) and platinum plate (3 × 1 × 0.05 cm) serve as the anode and cathode, respectively. Both electrodes, fixed face-to-face at distance of 0.25 cm (determined by a Teflon spacer), are immersed into the electrolyte solution to a depth of ca. 1.5 cm. The electrolyte solution volume is typically 3 mL and contains 0.08 M of the cationic surfactant hexadecyltrimethylammonium bromide (C_{16}TABr, 99%, Sigma), and 12.6 mg of the more hydrophobic cosurfactant tetradodecylammonium bromide (TC_{12}ABr, >98%; Fluka). The C_{16}TABr-TC_{12}ABr system serves as the supporting electrolyte and as the stabilizer for the Au nanorods. The glass electrochemical cell containing the electrodes and mixed surfactant solution is then placed into an ultrasonic bath (Branson model 1210) whose water solution is held at 36°C. The cell temperature is allowed to equilibrate with the bath for 5 min.

Immediately prior to the electrolysis, 65 μL of acetone and 45 μL of cyclohexane are added to the electrolyte solution. Acetone is used to loosen the micelle framework, and cyclohexane is necessary for enhancing the formation of rodlike C_{16}TABr micelles (18). Meanwhile, one (or two) silver plate(s) of similar dimensions to the anode and cathode is (are) placed above the solution near the cathode and is (are) not connected to an electrode lead. The electrolysis is done under constant-current mode, with the typical setting being 5 mA. The electrolysis time is typically 20 min under constant sonication (again, the bath temperature is held at 36°C). Gold metal is then oxidized to form a soluble complex, which then diffuses to the Pt cathode where it is reduced. Also, the silver plates are gradually immersed into the electrolytic solution during the electrolysis to constantly provide fresh Ag surface (vide infra). The gold nanorods form in the solution with the aid of stabilizers directing their growth.

Following electrolysis, the solution is centrifuged (20 min at 6500 rpm at 25°C in a Hettich model D-78532) to remove a large fraction of the spherical Au particles that also form. The resulting supernate, which is rich in Au nanorods, is transferred to another centrifuge tube. A second centrifugation (20 min at 12,000 rpm at 25°C) yields flocculent precipitates of Au nanorods. The precipitates can be redispersed in deionized water to the desired concentration for further surface modification or transmission electron microscope (TEM) analysis.

Figure 1 shows a typical TEM image of Au nanorods prepared by this method and their corresponding absorption spectrum. The SP_l resonance, corresponding to the long-wavelength band in Fig. 1, is known to red-shift as the mean aspect ratio (nanorod length/diameter) increases. The smaller peak centered at 520 nm is due primarily to the spherical particles in the sample. The transverse plasmon resonance SP_t occurs near this region of the spectrum, but its intensity relative to the SP_l band is very small. A summary plot of our data concerning the measured SP_l λ_{max} values versus mean aspect ratio is shown in Fig. 2. A nearly linear trend is observed, which agrees nicely with the predictions of the classical electrostatic model (vide infra).

B. Temperature and Nanorod Yield

The yield of Au nanorods can be estimated by the relative heights of the two plasmon resonance bands, such as those seen in Fig. 1. The yield of nanorods relative to nanospheres is high, while the peak absorbance ratio **Abs**(SP_l)/**Abs**(520) is large. This ratio is a good indicator of the success of the synthetic routine. Figure 3 shows a plot of $\mathbf{A}_{max}(SP_l)/\mathbf{A}(520)$ versus electrolyte solution temperature. The optimum temperature appears to be ca. 36°C for nanorod yield. Meanwhile, from both SP_l band position and TEM images, we find that, at least within the range considered, the temperature has little effect on the mean aspect ratio of the nanorods. We suspect that the optimum temperature for nanorod yield is related to a balance between the kinetics of nanorod formation (favored by higher temperature) and the structural integrity of the cylindrical micelles (disfavored by higher temperatures). At high temperature, the micelle structures simply collapse and then drop their stabilizing capability. Therefore, the optimized temperature of 36°C is adopted in all of the Au nanorod synthesis.

C. Effect of Silver Ion Release on Mean Aspect Ratio

We find that the mean aspect ratios of the Au nanorods can be influenced by the appearance of silver ions during the electrolysis. As mentioned, we include one or two silver plate(s) in the electrolytic solution near the cathode. The immersed surface area of silver metal is gradually increased during the electrolysis. A spontaneous redox reaction between Ag metal and Au ion complexes occurs

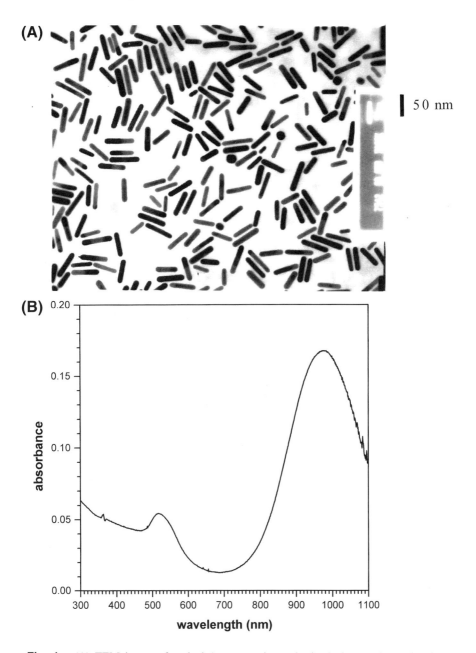

Fig. 1. (A) TEM image of typical Au nanorods synthesized via the electrochemical method. Scale bar represents 50 nm. (B) The absorption spectrum of the Au nanorod sample shown in (a).

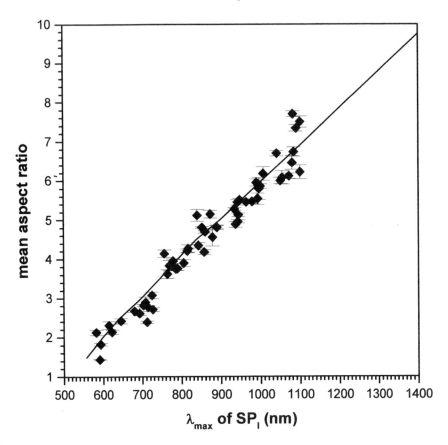

Fig. 2. Summary plot showing the relationship between the λ_{max}'s for the SP_l resonance bands and the mean aspect ratios of the Au nanorods. (Solid diamonds) Experimental λ_{max} values. (Solid line). Prediction of electrostatic limit theory.

simultaneously parallel with the electrolysis and nanorod formation. It leads to the Au deposition onto the Ag plate and, most importantly, the release of Ag^+ into the solution. Figure 4 shows a plot of $\lambda_{max}(SP_l)$ versus immersed Ag area. Recalling that $\lambda_{max}(SP_l)$ is related nearly linearly to the mean aspect ratio (Fig. 2), Fig. 4 clearly demonstrates the effect of silver plate immersion. While the mechanism of the Ag ion influence is not known at the present time, we suspect that the small amount of Ag^+ ions released consume Br^- of the micellar counterions, and then the subsequent instability of the micelles causes preferential 1-D growth of Au nanorods. It turns out that the total immersed surface area of Ag plate can efficiently tune the mean aspect ratios of the Au nanorods. The relevant parameters

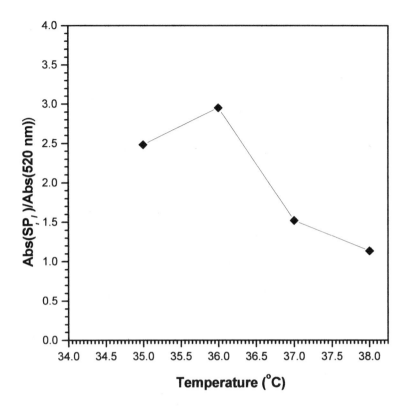

Fig. 3. Plot of the nanorod yield versus synthesis temperature. The yield is based on the spectral absorbance ratio $\mathbf{A}_{max}(SP_l)/\mathbf{A}(\lambda = 520 \text{ nm})$ (see text).

appear to be the amount of released Ag^+ ions and their release rate. The latter effect was evidenced by the result of directly placing the Ag metal with a certain total surface area inside the electrolytic solution. It showed no sufficient controllability over the particle aspect ratios. An alternative way of having Ag^+ effect is to directly inject Ag^+ into the solution. However, this did not result in easy control of the particle aspect ratios.

III. PREPARATION OF CORE-SHELL AU NANOROD@ SILICA PARTICLES

The core-shell Au nanorod@silica synthesis was first described elsewhere (19). An analogous coating procedure (17) has also been employed for Au nanorod embedded in a thin silica layer.

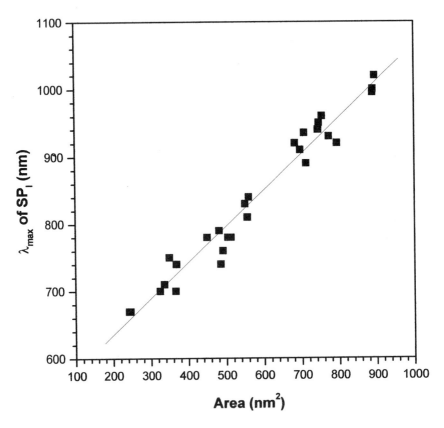

Fig. 4. Effect of the area of silver plate immersion on the $\lambda_{max}(SP_l)$ for Au nanorods. Note that the nanorod aspect ratio is linearly related to the $\lambda_{max}(SP_l)$ value (see Fig. 2).

A. Preparation Scheme

A gold nanorod solution (ca. 1 mL) is prepared by redispersing the centrifuged product as described above. The concentration is adjusted to the desired level by monitoring the absorbance at the SP_l band maximum. For example, an absorbance of ca. 2 (using a 1-cm-path-length cuvette) at $\lambda = 900$ nm was used for a sample of nanorods whose mean aspect ratio was 5. This solution was transferred to a polytetrafluoroethylene (PTFE) bottle for the silica-coating step.

The silane reagent was prepared by mixing (3-mercaptopropyl)trimethoxysilane (MPTMS, 97%; Fluka) with ethanol in a 1:1000 volume ratio (V_{MPTMS}: V_{EtOH}). Next, 40 µL of this reagent is added to 1 mL of the Au nanorod solution, and the mixture is stirred for 20 min with a magnetic stirrer. The stirring step ensures that the gold nanorods' surfaces are completely covered with the thiolate function of the MPTMS.

An aqueous silicate reagent solution was prepared by mixing 0.24 g of sodium silicate solution [$Na_2O(SiO_2)_{3-5}$; 27 wt%; Aldrich] with 50 mL of deionized water. Then 40 µL of this silicate solution (pH ≈ 9.5) is added to the MPTMS-modified Au nanorod solution with vigorous stirring. The resulting mixture (pH ≈ 7.5) is stirred for a certain reaction time period, at the end of which deionized water is added in a volume sufficient to dilute the original solution 2:1 and terminate the silica polymerization process. At this point the solution contains both Au nanorod@silica particles and spherical silica nanoparticles. Separation of these two particle types can be achieved by centrifugation. Centrifugation also allows for the removal of unreacted silicate without the need for dialysis (19).

Figure 5 shows a TEM image of Au nanorod@silica particles prepared as described above with a 24-hr silica polymerization step. Energy dispersive X-ray (EDS) analyses on individual particles confirm the presence of the elements Au, S, and Si. The thickness of the silica shell for this particular sample is about 11.7 ± 1.5 nm. As can be seen from the TEM image, the shell layer thickness is quite uniform.

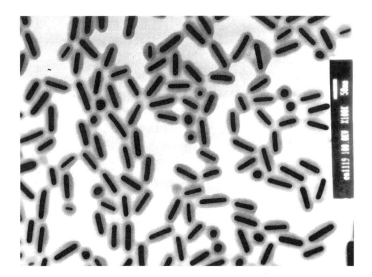

Fig. 5. TEM image of a typical purified sample of Au nanorod@silica particles. The scale bar represents 50 nm.

B. Control of Silica Layer Thickness

With the above preparation scheme, the thickness of the silica shell can be controlled by varying the reaction time of the coating process, in a manner entirely analogous to that used for spherical Au nanoparticles (19). Figure 6 (top) shows TEM images of two Au nanorod@silica samples, the left image of particles resulting from a 12-hr reaction time and the right image of particles which have undergone a 24-hr coating process. Figure 6 (bottom) shows that the dependence

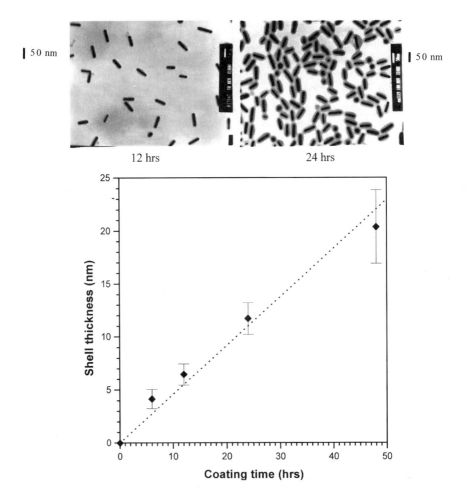

Fig. 6. (Top) TEM images of Au nanorod@silica nanoparticles after 12 hr (left) and 24 hr (right) of silica-coating reaction. Scale bar represents 50 nm. (Bottom) Dependence of mean shell thickness on silica-coating reaction time.

of shell thickness on reaction time is nearly linear. We note that it is possible to decrease the requisite reaction time for a given shell thickness by simply increasing the silicate reagent concentration.

IV. ABSORPTION SPECTRA OF AU NANORODS

A. Model Predictions for Single Rodlike Nanoparticles

The absorption spectra of metal nanoparticles show characteristic features which arise from SP resonances. The SP bands are known to be sensitive to particle shape and size. However, when the dimensions of the particles are small relative to the incident wavelength, the SP band position depends primarily on shape and only insignificantly on size. In this section we discuss the simple electrostatic limit of Mie scattering theory (20,21) and compare the predicted and experimental λ_{max} (SP_l) values. While somewhat more sophisticated (electrodynamic) scattering models are available, which address particle dimensions which are *not* negligible with respect to λ (22), we will show that for our systems the simple model suffices.

The interaction between the incident electromagnetic wave and a nonspherical metal nanoparticle can be described by a lossy induced dipole at the center of the particle. Again, if the particle's dimensions are small relative to the incident wavelength, the primary interaction can be assumed to be absorption; scattering terms can be assumed to be negligible. Furthermore, we can limit the discussion to an electric dipole and ignore higher-order terms, such as the magnetic dipole and electric quadrupole. The mean absorption cross section $\sigma(\lambda)$ for a small prolate metal nanoparticle, averaged over all orientations, is given by (23)

$$\sigma(\lambda) = \frac{2\pi\varepsilon_m^{1/2}}{\lambda} \mathrm{Im}\{\alpha_l + 2\alpha_t\} \qquad (1)$$

where α_l and α_t are the polarizabilities of the particle along its longitudinal and transverse axes, respectively. The dielectric function of the host medium is given by ε_m. The polarizabilities depend on the dielectric properties of the particle and host medium via

$$\alpha_{l,t} = \frac{V[\varepsilon_{l,t}(\omega) - \varepsilon_m]}{\varepsilon_m + [\varepsilon_{l,t}(\omega) - \varepsilon_m]P_{l,t}} \qquad (2)$$

where V and $\varepsilon_{l,t}(\omega)$ are the volume and frequency-dependent dielectric function of the particle, respectively. The parameter $P_{l,t}$ is the depolarization factor that depends on particle shape. The subscripts on the depolarization factor and dielectric function signify that they are formally dependent on the particle's orientation in the incident field. In these and all subsequent equations in this section, the comma between the l and t is taken as "or." The depolarization factors are given by (20)

$$P_l = \frac{1-e^2}{e^2}\left[\frac{1}{2e}\ln\left(\frac{1+e}{1-e}\right)\right] \quad (3)$$

$$P_t = \frac{1-P_l}{2} \quad (4)$$

where e^2 is related to the semimajor axis (a) and semiminor axis (b) via $e^2 = 1 - b^2/a^2$.

While we do not consider electrodynamic size corrections, which are needed when the particle dimensions become a significant fraction of λ, we do take into account mean-free path corrections, which are important when the particles are very small. When the particle dimensions approach the mean-free path of the metal's conduction electrons, the dielectric function diverges from the bulk case. Furthermore, since we are dealing with rodlike particles, the extent of deviation from bulk behavior depends on the direction of electron motion; electrons traveling parallel to the particle's transverse (semiminor) axis will experience a higher degree of confinement than those traveling parallel to the long axis. Thus, we have two sets of complex dielectric functions,

$$\varepsilon_{l,t}(\omega) = \varepsilon_{1,l,t}(\omega) + i\varepsilon_{2,l,t}(\omega) \quad (5)$$

where the subscripts 1 and 2 denote the real and imaginary components, respectively. The real and imaginary components are given by

$$\varepsilon_{1,l,t}(\omega) = \varepsilon^\infty - \left[\frac{\omega_p^2}{\omega^2 + \omega_{d,l,t}^2}\right] + B_1(\omega) \quad (6)$$

$$\varepsilon_{2,l,t}(\omega) = \left[\frac{\omega_p^2 \omega_{d,l,t}}{\omega(\omega^2 + \omega_{d,l,t}^2)}\right] + B_2(\omega) \quad (7)$$

where ε^∞ is the bulk dielectric at infinite frequency (~1 for most metals), ω_p is the plasma frequency [equal to 2.18×10^{15} Hz for Au (24)], and $B_{1,2}(\omega)$ represents the contribution of interband transitions arising from bound electrons. The interband terms can be determined from the bulk optical properties of the metal (25) and indeed are assumed to be unchanged from bulk values. The key term for the present discussion is $\omega_{d,l,t}$, which is the damping frequency, a term related to the electron mean-free path l.

As mentioned above, as the dimensions of the particle decrease, the confinement of the conduction electrons implies an effective mean-free path l_{eff}, which in turn is related to the damping frequency via

$$\omega_{d,l,t} = S\frac{V_F}{l_{\text{eff},l,t}} \quad (8)$$

where $l_{\text{eff},l,t}$ are the effective mean-free paths along the longitudinal and transverse axes, v_F is the velocity of the electrons at the Fermi energy [1.39×10^6 m/s for Au (24)], and S is a slope parameter between zero and unity which is assumed to be isotropic (26). The effective mean-free path can be related to the bulk mean-free path l_{bulk} [equal to 31.0 nm for Au (24)] and the particle dimensions by the expressions

$$\frac{1}{l_{\text{eff},l}} = \frac{1}{l_{\text{bulk}}} + \frac{1}{a} \tag{9a}$$

$$\frac{1}{l_{\text{eff},t}} = \frac{1}{l_{\text{bulk}}} + \frac{1}{b} \tag{9b}$$

Using the formalism just described, we calculated the plasmon resonance spectra for rodlike Au nanoparticles of various aspect ratios. The short-axis dimension was fixed at 10 nm, which is close to the experimental values. The calculated dependence of the $\lambda_{\text{max}}(\text{SP}_l)$ on particle aspect ratio is shown by the solid line in Fig. 2. It is clear that there is excellent agreement between theory and experiment.

While our experimental work herein focuses on Au nanorods, we have also calculated the absorption spectra of other metals to illustrate the dependence of the SP spectral features on the type of material. Figure 7 shows the spectra for various rodlike particles composed of Au, Ag, Cu, and Na in a host medium whose dielectric constant is that of water ($\varepsilon_m = 1.77$). A key point is that the SP_l resonance band positions are strongly dependent on aspect ratio but only weakly dependent on size (via electron damping effects). However, insignificant spectral shifts were indicated for the SP_t bands in four cases. For Au and Cu, the SP_t band intensity is negligible relative to the intensity of the SP_l band. However, for Ag and Na, the SP_t and SP_l bands are strong.

B. Experimental Spectra of Au Nanorod Suspensions

The success of the electrostatic model in predicting the shift in the SP_l band as a function of mean aspect ratio (Fig. 2) has implications for the interpretation of experimental spectra of nanorod systems, such as that in Fig. 1. If the quantitative validity of the $\lambda_{\text{max}}(\text{SP}_l)$ dependence on aspect ratio holds in the case of the relative intensities of the SP_t and SP_l bands, then we can expect (Fig. 7) that the SP_t band in the case of Au nanorods should be negligibly small relative to the SP_l band. It indeed has been further verified in designed separation experiments (27), in which pure Au nanorods and their absorption spectrum were obtained. We thus interpret that the SP band centered near $\lambda = 520$ nm in Fig. 1 is to arise from spherical Au nanoparticles in the sample and not from the SP_t band of the nanorods.

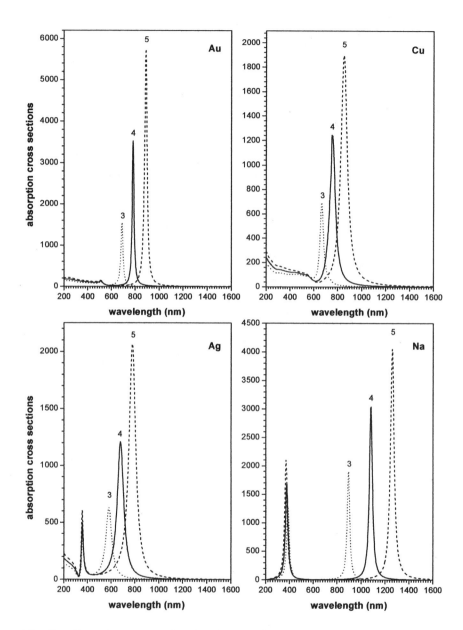

Fig. 7. Absorption spectra of prolate Au, Ag, Cu, and Na nanoparticles, electrostatic theory simulations. Aspect ratios = 3 (dotted lines), 4 (solid lines), and 5 (dashed lines). In all cases the host medium dielectric constant is assumed to be that of water ($\varepsilon_m = 1.77$). The parameter S is assumed to be unity in all cases (see text).

C. Model Predictions for Au Nanorod@Silica Core-Shell Particles

For silica-coated Au nanorods, we need to consider the perturbation introduced by the thin surface layer. The equation for the absorption cross section of a small coated prolate particle has the same form as for the uncoated particle [Eq. (1)]. The particle polarizability expressions, however, reflect the increased complexity:

$$\alpha_{l,t} = \frac{V(\varepsilon_s - \varepsilon_m)\{\varepsilon_s + [(\varepsilon_c - \varepsilon_s)(P^1_{l,t} - P^2_{l,t})]\} + f\varepsilon_s(\varepsilon_c - \varepsilon_s)}{\{[\varepsilon_s + (\varepsilon_c - \varepsilon_s)(P^1_{l,t} - fP^2_{l,t})][\varepsilon_m + (\varepsilon_s - \varepsilon_m)P^2_{l,t}]\} + fP^2_{l,t}\varepsilon_s(\varepsilon_c - \varepsilon_s)} \quad (10)$$

In Eq. (10), V is the total volume of the metal particle and its shell layer. The volume fraction occupied by the core (metal) material is given by f. The complex dielectric functions of the particle (core), shell, and host medium are given by ε_c, ε_s, and ε_m, respectively. For clarity, we have omitted the l,t subscripts for ε_c in Eq. (10); however, mean-free path effects are assumed to be operative as in the case of uncoated particles [see Eqs. (2), (5), and (6)]. The $P^1_{l,t}$ and $P^2_{l,t}$ terms are the depolarization factors for the core and shell structures, respectively. The core terms can be calculated from the metal particle's dimensions using Eqs. (3) and (4). For the shell layer, Eqs. (3) and (4) can still be used, but with the factor e^2 expressed as

$$e^2 = 1 - \frac{(b + t_s)^2}{(a + t_s)^2} \quad (11)$$

where t_s is the thickness of the shell layer.

A special case of Eq. (10) is the coated sphere, whose polarizability has been given elsewhere (23) as

$$\alpha = 4\pi (r + t_s)^3 \frac{(\varepsilon_s - \varepsilon_m)(\varepsilon_c + 2\varepsilon_s) + f(\varepsilon_c - \varepsilon_s)(\varepsilon_m + 2\varepsilon_s)}{(\varepsilon_s + 2\varepsilon_m)(\varepsilon_c + 2\varepsilon_s) + f(\varepsilon_c - \varepsilon_s)(2\varepsilon_s - 2\varepsilon)} \quad (12)$$

where r is the radius of the sphere. The corresponding absorption cross section for a coated sphere is

$$\sigma(\lambda) = \frac{2\pi \varepsilon_m^{1/2}}{\lambda} \text{Im}\{\alpha\} \quad (13)$$

Using the same size corrections for $\varepsilon_{l,t}(\omega)$ as before, simulations of the absorption cross sections of Au nanorods@silica were done, using Eq. (10) for various silica shell thicknesses. Figure 8 summarizes the effect of shell thickness, using two parameters we felt were meaningful indicators: $\Delta\lambda_{max}$ and $\sigma(\lambda_{max})$. The band-shift parameter is defined as $\Delta\lambda_{max} = \lambda_{max}(t_s) - \lambda_{max}(t_s = 0)$. Assuming

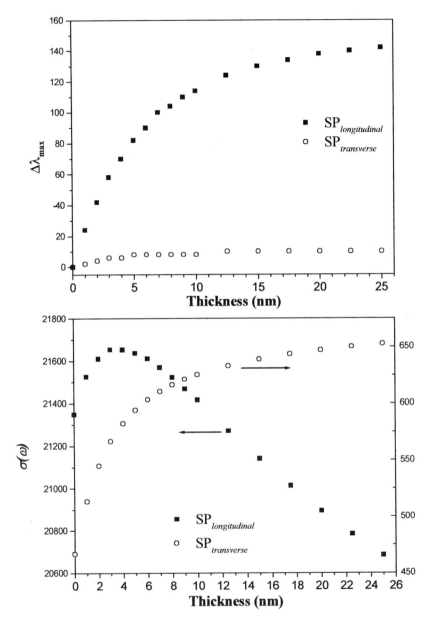

Fig. 8. Calculated dependence of the SP_l and SP_t spectral shift (top) and absorption cross section (bottom) for an Au nanorod on silica shell layer thickness. The aspect ratio of Au nanorod is set to 5.8, and the minor axis is 10 nm in calculations. The silica layer and host medium dielectric constants are assumed to be 2.46 and 1.77, respectively.

that the shell and host medium dielectric constants are 2.46 and 1.77, respectively, Fig. 8 demonstrates that we should expect both the SP_l and SP_t bands to red-shift with increasing silica shell layer thickness, with the effect on SP_l being much more pronounced.

The calculated effect of shell thickness on absorption cross section differs for the SP_l and SP_t bands. While $\sigma(\lambda_{max})$ for the SP_t bands rise with shell thickness, the cross section for the SP_l band initially rises with t_s and then drops dramatically after a maximum near $t_s = 3$ nm. The trends for the SP_l band arise from a trade-off between band oscillator strength and bandwidth, which both increase with increasing t_s. The initial rise of $\sigma(\lambda_{max})$ with t_s occurs because the oscillator strength increases more rapidly than does the bandwidth when the shell layer is thin.

D. Experimental Spectra of Au Nanorod@Silica Core Shell Particles

Figure 9 shows three absorption spectra of Au nanorods at different stages of silica shell layer formation. After modifying the Au nanorod surfaces with MPTMS, we observe a decrease in the $\sigma(\lambda_{max})$ for the SP_l band, but no significant shift in band position. After the silica layer is developed, the SP_l band shows a noticeable red shift. The change in band position and peak intensity (bare Au rods versus silica-coated rods) is in good agreement with the electrostatic limit prediction discussed in the last section.

While we initially used a value for the dielectric constant of glass [$\varepsilon_s = 2.46$, taken as the square of the refractive index (25)], we realized that this value may not apply to the silica layer formed in our experimental process. By comparing the $\Delta\lambda_{max}$ and t_s quantities obtained from the experimental data with the quantity calculated using Eq. (10), assuming different ε_s values, we are able to estimate the true ε_s value for the experimental systems. Figure 10 shows plots of $\Delta\lambda_{max}$ for the SP_l band versus t_s for various calculated spectra (open circles), where the only variation is the assumed ε_s value. The experimental $\Delta\lambda_{max}$ values, shown as the dark diamonds, correspond very closely to the curve calculated, assuming an ε_s value of 2.06. We note that this value is in close agreement with the value obtained for silica derived by the sol-gel process (28).

V. CONCLUSIONS

We have demonstrated that nonspherical metal nanoparticles can be prepared electrochemically in aqueous solution by using cationic surfactants. One-dimensional rodlike Au nanoparticles can be effectively synthesized with adjustable mean aspect ratios from 1 to 7. The mechanism is not clearly verified at the current

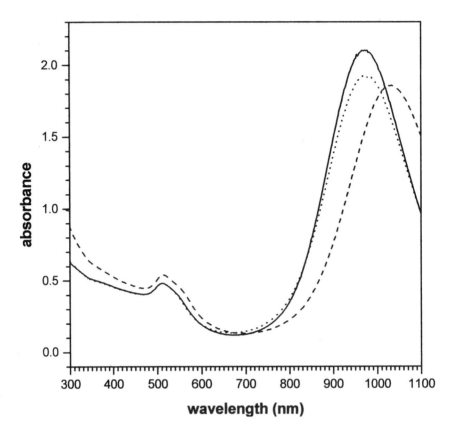

Fig. 9. Experimental absorption spectra of Au nanorods at different stages of silica coating. Micelle-stabilized Au nanorod (solid curve). Au nanorod coated with MPTMS (dotted curve). Au nanorod@silica structures after 24-hr coating reaction (dashed curve). For uncoated Au nanorods, $\lambda_{max}(SP_l) = 978$ nm.

time, but is assumed to be related to the cylindrical geometry of the micelle superstructures formed by these surfactants during the rod growth. The optical spectra of Au nanorods are in very close agreement with theoretical predictions. Their absorption spectral features pertaining to the surface plasmon resonances contain an intense SP_l band. Its position is subjected to a sensitive red shift as the particle's mean aspect ratio increases. This spectral shift can also be well described by the simple electrostatic model.

The coating of the Au nanorods with a silica shell, as an example of surface modification, can be achieved by using the rod-MPTMS-silica preparation

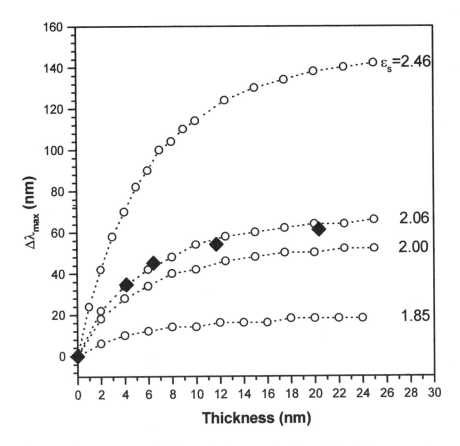

Fig. 10. $\Delta\lambda_{max}$ versus shell thickness for Au nanorods. Open circles depict curves calculated using electrostatic theory, assuming the shell dielectric constants shown at right. Solid triangles are experimental values. The aspect ratio of Au nanorod is set to 5.8, and the minor axis is 10 nm in calculations. The calculated spectrum for uncoated Au nanorod reflects correctly to the experimental SP_l resonance in Fig. 9.

sequence. The thickness of these shells is also controllable. Both theory and experiment show the SP resonance shifts to the red as the silica layer thickness increases. Meanwhile, the spectral shift of SP_l band for rodlike nanoparticles is much more sensitive compared to that of the SP band for spherical ones. The optical properties of these Au nanorod@silica structures are also in close accord with theory, with the agreement sufficiently precise as to allow for the back-calculation of the dielectric constant of the silica layer.

REFERENCES

1. JP Wilcoxon, A Martino, RL Baughmann, E Klavetter, AP Sylwester. Mater Res Soc Symp Proc 286:131–136, 1993.
2. (a) S Nie, SR Emory. Science 275:1102–1106, 1997. (b) SR Emory, S Nie, J Phys Chem 102:493–497, 1998.
3. RG Freeman, KC Grabar, KJ Allison, RM Bright, JA Davis, AP Guthrie, MB Hommer, MA Jackson, PC Smith, DG Walter, MJ Natan. Science 267:1629–1632, 1995.
4. B Vlcková, XJ Gu, M Moskovits. J Phys Chem B 101:1588–1593, 1997.
5. CA Mirkin, RL Letsinger, RC Mucic, JJ Storhoff. Nature 382:607–609, 1996.
6. R Elghanian, JJ Storhoff, RC Mucic, RL Letsinger, CA Mirkin. Science 277:1078–1081, 1997.
7. CR Martin. Science 266:1961–1966, 1994.
8. CA Foss, Jr., GL Hornyak, JA Stockert, CR Martin. J Phys Chem 98:2963–2971, 1994.
9. JC Hulteen, CJ Patrissi, DL Miner, ER Crosswait, EB Oberhauser, CR Martin. J Phys Chem B 101:7727–7731, 1997.
10. NAF Al-Rawashdeh, ML Sandrock, CJ Seugling, CA Foss, Jr. J Phys Chem B 102:361–371, 1998.
11. BMI van der Zande, MR Bohmer, LGJ Fokkink, C Schonenberger. J Phys Chem B 106:852–854, 1997.
12. C Schonenberger, BMI van der Zande, LGJ Fokkink, M Henny, C Schmid, M Kruger, A Bachtold, R Huber, H Birk, U Staufer. J Phys Chem B 101:5497–5505, 1997.
13. W Gotschy, K Vonmetz, A Leitner, FR Aussenegg. Opt Lett 21:1099–1101, 1996.
14. TS Ahmadi, ZL Wang, TC Green, A Henglein, MA El-Sayed. Science 272:1924–1926, 1996.
15. TS Ahmadi, ZL Wang, A Henglein, MA El-Sayed. Chem Mater 8:1161–1163, 1996.
16. YY Yu, SS Chang, CL Lee, CRC Wang. J Phys Chem B 101:6661–6664, 1997.
17. SS Chang, CW Shih, CD Chen, WC Lai, CRC Wang. Langmuir 15:701–709, 1999.
18. M Tornblom, U Henriksson. J Phys Chem B 101:6028–6035, 1997.
19. LM LizMarzan, M Giersig, P Mulvaney. Langmuir 12:4329–4335, 1996.
20. JA Creighton, DG Eadon. J Chem Soc Faraday Trans 87:3881–3891, 1991.
21. DS Wang, M Kerker. Phys Rev B24, 1777, 1981.
22. EJ Zeman, GC Schatz. J Phys Chem 91:634–643, 1987.
23. CF Bohren, DR Huffman. Absorption and Scattering of Light by Small Particles. New York: Wiley, 1983.
24. NW Ashcroft, ND Mermin. Solid State Physics. Philadelphia: Saunders, International Edition, 1988.
25. DR Lide, ed. CRC Handbook of Chemistry and Physics, 74th ed. Boca Raton: CRC Press, 1994, pp 12–113.
26. U Kreibig, L Genzel. Surf Sci 156:678–700, 1985.
27. GT Wei, FK Liu, SS Chang, CRC Wang. Anal Chem 71:1085–2091, 1999.
28. K Awazu, H Onuki. App Phys Lett 69:482–484, 1996.

8
Surface Plasmon Resonance Biosensing with Colloidal Au Amplification

Michael J. Natan
SurroMed, Inc., Palo Alto, California

L. Andrew Lyon
Georgia Institute of Technology, Atlanta, Georgia

I. INTRODUCTION

For more than a decade, biosensors based on surface plasmon resonance (SPR) have been commercially available (1–4), making real-time, tagless biomolecular interaction analysis accessible to those in biochemistry-related fields. In parallel with (and largely driven by) these commercial successes, there have been large increases in the number of academic pursuits involving biosensing and bioassays with SPR (5–9). The motivations for these studies are quite diverse. Most attempts have been aimed at lowering the accessible range of the sensor platform with respect to both molecular weight and concentration. However, methods by which SPR can be interfaced with other instruments to form hyphenated techniques have been developed (10–13). Reports of multichannel and imaging sensors for high-throughput applications are now available (5,14–17). Finally, despite the traditional tagless format practiced in SPR, a number of systems using particulate and molecular tags for sensitivity enhancement have been reported (7,16,18–21).

This chapter will detail the fundamentals and early application of one of these tag-amplified SPR methods: colloidal Au-enhanced biosensing (7,20,21). We first describe the basics of SPR sensing as it is traditionally practiced. A general discussion of how particulate materials modulate the SPR of smooth metal surface follows. Experimental results from nonbiological SPR studies of colloidal

Au-modified films are used to qualitatively illustrate the influence of colloidal Au on SPR curves. Finally, examples involving direct binding and sandwich immuno-assays demonstrate the utility of the method in bioassay development.

II. BACKGROUND

A. Surface Plasmon Resonance of Thin Metal Films

For a more mathematically rigorous discussion of surface plasmon resonance, the reader is referred to work concerning the optical properties of thin metal films (22). We will limit our discussion to the phenomena pertinent to biomolecular interaction analysis.

Originally described by Ritchie in the 1950s (23) surface plasmons are coherent fluctuations in electron density occurring at a "free-electron" metal/dielectric interface. Examples of free-electron metals (those with a lone electron in the valence shell) are Ag, Au, Al, and Cu, where Au is the most commonly used in biosensing. As is typical for surface-confined modes, the transverse component (perpendicular to the metal surface) of the electric field vector decays exponentially into the dielectric medium from a maximum value at the metal surface. With typical decay lengths being ~200 nm (at mid-visible excitation wavelengths), surface plasmons are exquisitely sensitive to changes in the properties of the interface. Specifically, changes in the thickness and/or refractive index of the medium in contact with the metal film results in a change in the propagation of the surface plasmon mode (9). This sensitivity can be described by the magnitude of the electric field propagating in the plane of the metal film:

$$E_x(x,z,t) = E_x^0(x) \exp(i\omega t - ik_z z) \tag{1}$$

where ω is the optical frequency, and the complex propagation constant along the z-axis is given by

$$k_z = k_z' + ik_z'' \tag{2}$$

It is this value (often referred to as K_{sp}, the wavevector of the surface plasmon) that is dependent on the refractive index at the interface.

$$K_{sp} = \frac{\omega}{c} \sqrt{\frac{\varepsilon_m \varepsilon_s}{\varepsilon_m + \varepsilon_s}} \tag{3}$$

In Eq. (3), ε_m and ε_s are the real parts of the metal and sample dielectric functions, respectively, and c is the speed of light in vacuum. Given this, it is apparent that radiation impinging upon the metal film with the same propagation wavevector can be used to excite this mode. The wavevector for a photon in the sample is

$$K_{light} = \frac{\omega}{c} \sqrt{\varepsilon_s} \tag{4}$$

Surface Plasmon Resonance Biosensing

However, a simple calculation shows that for a plasmon-supporting metal (large, negative ε_m), K_{sp} is always greater than K_{light}. Surface plasmons therefore cannot be excited by photons propagating in free space.

To overcome this problem, one must use illumination conditions that allow for modulation of the incident (momentum) wavevector magnitude. The most common method for accomplishing this was proposed originally by Kretschmann in 1971 (Scheme I) (24). In this technique, the thin metal film is optically coupled to a prism made of a high-refractive-index material (e.g., glass). Illumination of the film through this prism under total internal reflection conditions allows for excitation of the surface plasmon via an evanescent wave. Mathematically, the wavevector of the evanescent field is

$$K_{ev} = \frac{\omega}{c} \sqrt{\varepsilon_g} \sin \theta \tag{5}$$

where ε_g is the real part of the glass dielectric function and θ is the incident angle of excitation. From this we can see that light of a specific frequency can excite a plasmon of that frequency when θ satisfies the equality $K_{ev}=K_{sp}$. Experimentally, this is observed to result in a sharp minimum in a plot of reflectance versus incident angle. Because K_{sp} is dependent on the refractive index above the metal film, θ is likewise sensitive to this value. Furthermore, since the resonance condition

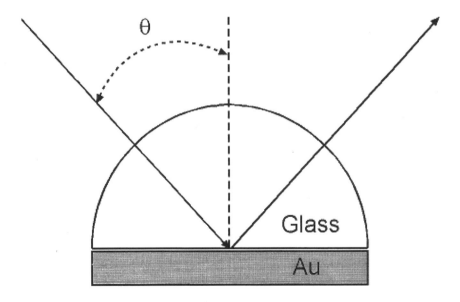

Scheme I. The Kretschmann SPR illumination geometry.

can now be experimentally observed, it can be correlated with the optical properties of the sample as indicated by Eq. (3); the position of the reflectance minimum is used to quantitatively determine the sample refractive index. A convenient method for the simulation and fitting of SPR curves based on Fresnel's equations was proposed by Hansen (25). Freeware based on that work has been developed by the group of Robert Corn at the University of Wisconsin and can be downloaded from the World Wide Web (http://www.corninfo.chem.wisc.edu/).

A number of practical experimental conditions, such as metal film thickness and incident wavelength, must be considered when designing an SPR experiment for optimal sensitivity (9,22,26). Figure 1 shows simulated curves demonstrating typical changes in the SPR resonance condition as a function of these variables. Panel (a) illustrates the change in SPR curves for a series of Au films of different thickness. Under illumination at 632.8 nm, the SPR curve progresses from a shallow, sharp minimum for an 80-nm-thick film, to a deep minimum at 47 nm, and finally to a shallow broad curve at 20 nm. Given these data, it is evident that the 47-nm-thick film is the optimal one for sensing applications (largest signal difference). These changes also give us physical insight into the mechanisms responsible for the SPR minimum. In the optimal (47 nm) film, a portion of the incident radiation reflects from the glass/Au interface, while a portion propagates through the film to the Au/sample interface and excites surface plasmons, which in turn radiate back into the metal film. The back-scattered radiation is then out of phase with the incident radiation, thereby resulting in a destructive interference process and, hence, a decrease in reflectance. For the thicker Au film, much of the wave propagating into the metal has decayed before reaching the Au/sample surface, resulting in a decrease in the magnitude of the back-reflected plasmon radiation. Conversely, the thin film displays a back-reflected field that is much greater in magnitude and therefore increases the reflectance of the ATR device near the plasmon angle.

Because the optical constants for thin metal films are varying functions of the incident wavelength, the photon energy can be used to change the shape and position of the SPR curve. Figure 1b illustrates this effect; curves for a 47-nm-thick Au film are shown at four different excitation wavelengths. We can see from these data that the SPR curve becomes sharper at longer wavelength, reflecting a decrease in the amount of damping experienced by the surface mode. Quantitatively, this damping is related to the mean-free path of electrons at the incident frequency and is therefore related to the decay length of plasmons at the Au/sample interface:

$$I(x) = \exp(-2k_x'' x) \tag{6}$$

in which the imaginary portion of the plasmon wavevector is

$$k_x'' = \frac{\omega}{c} \left(\frac{\varepsilon_m' \varepsilon_s}{\varepsilon_m' + \varepsilon_s} \right)^{3/2} \frac{\varepsilon_m''}{2(\varepsilon_m')^2} \tag{7}$$

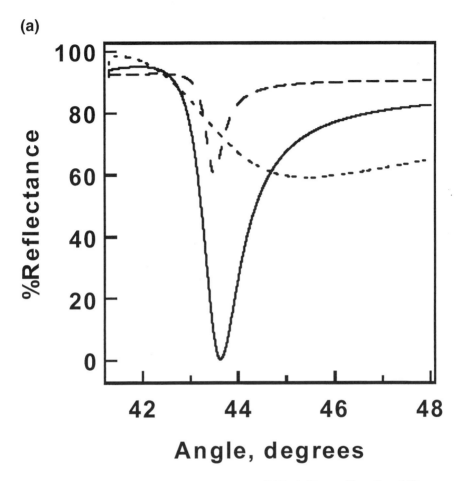

Fig. 1a. (a) Simulated SPR curves for 47-nm- (solid line), 20-nm- (dotted), and 80-nm- (dashed) thick Au films illuminated at 633 nm. (b) Simulated curves for a 47-nm-thick Au film at the indicated illumination wavelengths.

Equation (7) assumes that the real (′) and imaginary (″) portions of the metal dielectric function are frequency dependent. From these equations, it follows that decreases in the optical frequency will result in less plasmon damping and, hence, longer-range propagation of the plasmon mode [i.e., Eq. (6) decays more slowly]. This effect recently has been exploited by Corn's group, where IR-excited plasmons are used to obtain higher SPR sensitivity (15,27).

Taken alone, the above discussion of the surface plasmon effect suggests the numerous advantages of the technique with respect to biomolecular interaction

(b)

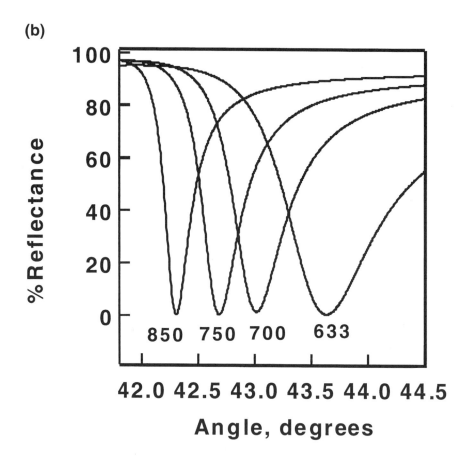

analysis. The technique is tagless; no enzyme, fluorophore, or radioisotope labeling is required for detection or amplification of a binding event. SPR can also be performed in real time, thereby allowing for kinetic as well as thermodynamic parameters to be determined. Finally, the technique is noninvasive since a "backside" illumination geometry is employed; the source beam does not have to travel through the sample solution during operation. This decreases any interference that could arise from turbid or strongly absorbing solutions. Despite these advantages, it is still desirable in some applications to utilize "tagging" to either amplify or differentiate observables in the SPR experiment. In our case, we have used metallic (Au) nanoparticles as protein tags in order to amplify the resultant SPR signal (7,20,21). The next section briefly describes the physics behind SPR perturbations via the introduction of nanoscale roughness.

B. Surface Plasmon Resonance of Roughened and Nanoparticle-Modified Metal Films

When a metallic roughness feature is introduced to the thin metal film, it presents a site for conversion of the nonradiative surface plasmon into a radiative electromagnetic mode (22). For very small roughness factors (<1 nm rms), this is observable with a simple photodiode directed toward the metal/air interface in the laser/prism/(reflectance) detector plane. However, under such conditions, the roughness does not result in a large change in the reflectance curve. As the roughness features become larger, they present large perturbations to the SPR dispersion relations and, hence, the shape of the reflectance curve. The main culprits in the dramatic changes that are observed are multiple scattering phenomena. Specifically, plasmon emission is not only observed to radiate from the roughness sites into the dielectric medium, but can also couple back into the metal film, where it is detected along with the specularly reflected light (7,20–22). Under these conditions, the angles where the back-scattered radiation destructively interferes with the incident radiation are very different than in the smooth thin-film approximation. Indeed, at very high roughness factors, complete extinction of the reflected beam is never observed due to the wider angular distribution of scattered light emitted from the roughness site.

Roughness can be introduced via a number of methods. Optically transparent undercoatings have been used prior to metal film evaporation in order to create well-defined roughness features (28). A simpler, but less easily controlled, route involves the variation of the metal evaporation rate; slower rates result in smoother films. In our research, we reasoned that monodisperse metallic nanoparticles would enable the introduction of well-defined "roughness" sites to thin metal films. Furthermore, the roughness is tunable in a manner unlike that obtained by more traditional methods, as both the size and the spacing of the features can be controlled (29–31). The following sections represent a review of our early investigations into Au nanoparticle-modulated SPR of thin Au films and the application of this method to immunoassay amplification.

III. COLLOIDAL AU MODULATION OF SURFACE PLASMON RESONANCE

A very simple architecture was employed to investigate the influence of particle size and spacing on the resultant SPR signal. Scheme II illustrates this sample arrangement. A 47-nm-thick Au film deposited via thermal evaporation onto BK7 glass coverslips serves as the SPR substrate. Monolayers of 2-mercaptoethylamine (MEA) were then deposited onto the Au surface from ethanolic solutions of the mercaptan. The amine-coated surfaces were exposed to Au hydrosols for a fixed period of time, resulting in the formation of a colloidal Au submonolayer,

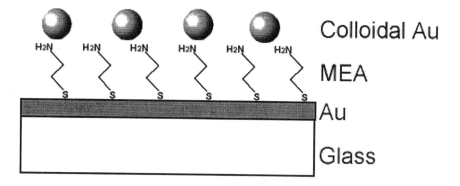

Scheme II. "Designed" roughness with colloidal Au.

the density of which was controlled by varying the immersion time and/or the hydrosol concentration. Tapping-mode AFM (TM-AFM) images of representative colloid-modified surfaces are shown in Fig. 2. The random arrangement of particles on the surface is typical of particles adsorbed via a random process under conditions where the particle-particle interaction potential is repulsive (29–31). A loosely packed submonolayer of particles is thus obtained, with no observable

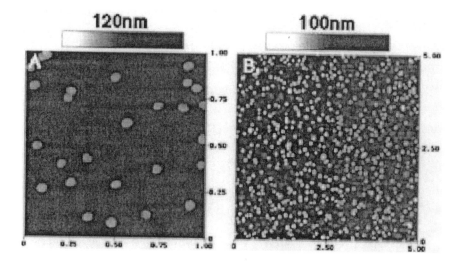

Fig. 2. Representative 1 μm × 1 μm (A) and 5 μm × 5 μm (B) TM-AFM images of 30-nm-diameter colloidal Au immobilized on evaporated Au. Colloid exposure time was 30 min. Z-dimension scale bars are as indicated.

multilayer formation at these coverages. This provides for a consistent film morphology between different particle sizes; multilayer formation would hopelessly complicate the analysis of the SPR data. These films were interrogated using an instrument arranged as in Scheme III. Excitation of the surface plasmon was performed in the Kretschmann geometry (24) with a HeNe laser (632.8 nm, 5 mW maximum power, Melles Griot). After being focused onto the prism/sample assembly, the intensity of the reflected beam was measured with a silicon photodiode detector (Thor Labs). Rotation of the sample and detector with respect to the incident laser was accomplished with a rotation stage assembly consisting of two high-resolution (0.001°) servodrive stages (Newport) that are mounted together such that their axes of rotation are collinear. A more detailed discussion of this instrumentation is given elsewhere (7).

Representative SPR curves of a 47-nm-thick, MEA-coated Au film are shown in Fig. 3 prior to and following 60-sec and 60-min exposures to a ~17 nM sol of 11-nm-diameter colloidal Au. The SPR curve of the bare Au film displays a sharp reflectance minimum at an angle of 43.7°. However, the Au nanoparticle-modified films have significantly shifted plasmon angles, broadened curves, and an increased minimum reflectance. As described above, these changes reflect a modulation of the surface plasmon dispersion relationship. In contrast to the angle shifts induced by smooth dielectric (e.g., polymer) films (9), however, these curves are due to a more complex mechanism of plasmon damping that involves the multiple scattering effects described above. In this manner, the overall changes in curve position and shape are enhanced by the presence of a discontinuous metal layer that mimics the effects of large-scale film roughness. As one might expect, the magnitude of the SPR shifts are observed to increase with increasing particle size. Figure 4 illustrates the particle size effect for submonolayer coverages of particles ranging in diameter from 30 to 59 nm. Even at these low

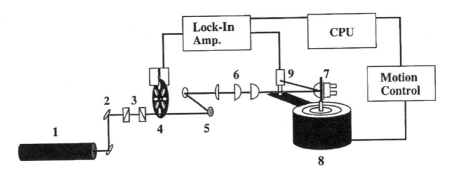

Scheme III. Surface plasmon resonance apparatus. Reprinted with permission from Anal. Chem. 1998, 70, 5177–5183. Copyright 1998 American Chemical Society.

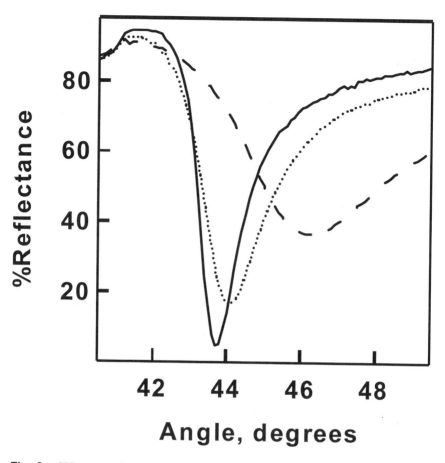

Fig. 3. SPR curves for a 2-mercaptoethylamine-coated 50-nm evaporated gold film modified with 11-nm-diameter colloidal Au for varying exposure times: 0 sec (solid line), 60 sec (dotted), and 60 min (dashed).

coverages (1.2% to 4.5% of a monolayer), enormous SPR perturbations are observed. Based on the results obtained with 11-nm-diameter particles, it was expected that similar coverage effects would be observed for larger particles. These effects are illustrated in Fig. 5 for 35- and 45-nm-diameter particles; as the coverage (induced roughness) increases, the curve broadens, shifts, and becomes shallower. Figure 6 shows the compiled plasmon angle data for all four particle sizes over a range of coverages. It is interesting to note that these curves deviate from linearity, suggesting that the traditional linear dependence of angle on the number of immobilized species (e.g., proteins) does not necessarily hold for colloidal Au

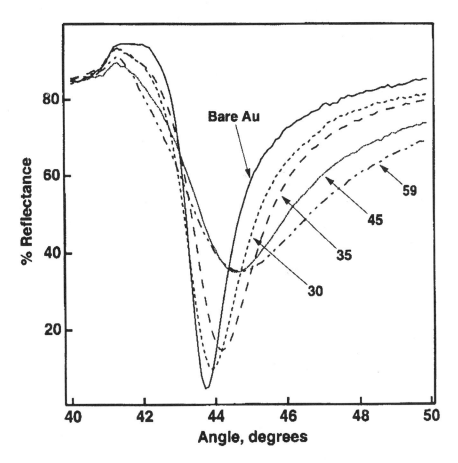

Fig. 4. SPR curves of a 47-nm-thick Au film coated with 2-mercaptoethylamine (solid line) and a submonolayer of colloidal Au (1.3 ± 0.15 × 10^9 particles/cm^2). Particle diameters are as indicated on the graph.

particles. Similarly, the maximum change in reflectance at any one angle is not a purely linear function, but tends to reach a plateau at high coverage (Fig. 7). Again, the nonlinearities of these curves are due to the change in plasmon damping mechanism. Linear behavior can be expected when the damping is due to modulation of the plasmon velocity by a change in refractive index (as in protein adsorption). In the case of nanoparticle adsorption, the plasmons are converted from nonradiative to radiative modes by the sharp discontinuities in film morphology that are produced by the nanoparticles, thereby producing a much larger damping effect.

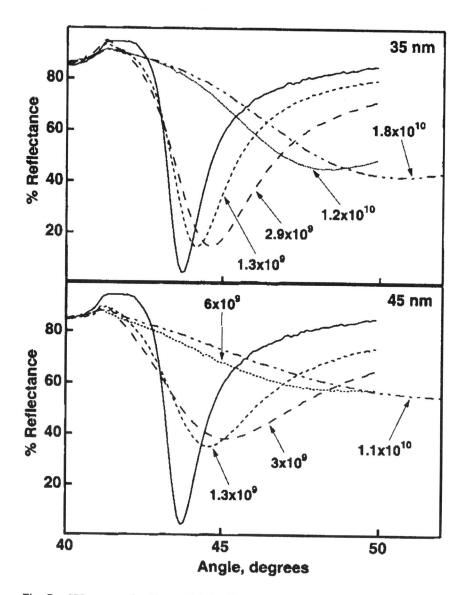

Fig. 5. SPR curves of a 47-nm-thick Au film coated with 2-mercaptoethylamine (solid line) and a layer of 35-nm-diameter colloidal Au (top) and 45-nm-diameter colloidal Au (bottom). Surface concentrations (in particles/cm^2) are as labeled.

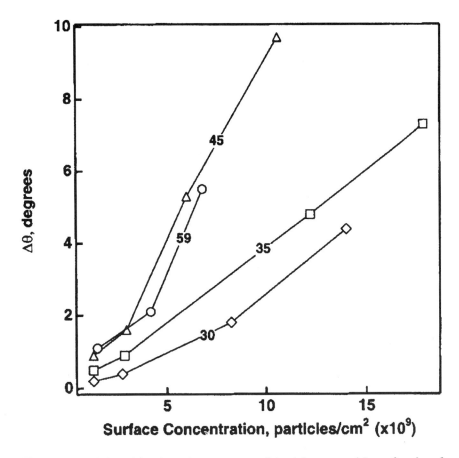

Fig. 6. Modulation of the plasmon resonance condition (plasmon angle) as a function of particle number density at four different colloidal Au diameters (as indicated on the plot).

IV. AMPLIFIED IMMUNOSENSING WITH COLLOIDAL AU TAGS

Because of the large changes in the plasmon curve that are brought about by metal nanoparticle adsorption, it was reasoned that such particles may be useful as amplification tags in a bioassay application. To evaluate the utility of such a method, two colloidal Au:protein conjugate architectures were investigated (Scheme IV). The first structure involves direct binding of an antigen:Au bioconjugate to an antibody-modified surface. The effect on the SPR curve due to Au

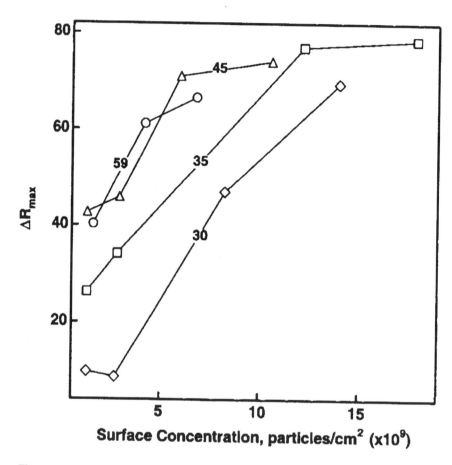

Fig. 7. Maximum SPR reflectance change (determined from difference curves) plotted as a function of surface coverage for colloidal Au diameters of 30, 35, 45, and 59 nm.

adsorption via an intervening protein layer (as opposed to a short-chain alkanethiol) can be evaluated with this architecture. A more bioanalytically relevant structure is shown in the second example; an antibody-derivatized surface is exposed to free antigen and then a secondary antibody:Au conjugate. This traditional sandwich immunoassay format allows for evaluation of colloidal Au tags in a standard antigen detection mode.

Representative curves for SPR detection of antibody-antigen binding are shown in Fig. 8. Exposure of an antibody-coated Au film [γ-chain-specific monoclonal goat antihuman immunoglobulin G (α-h-IgG(γ)] to the appropriate antigen

Surface Plasmon Resonance Biosensing

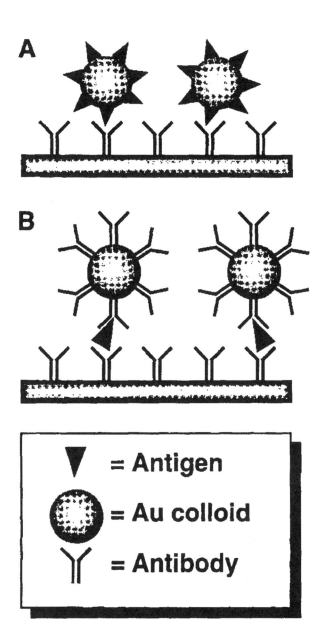

Scheme IV. Particle-enhanced immunosensing architectures. Reprinted with permission from Anal. Chem. 1998, 5177–5183. Copyright 1998 American Chemical Society.

[human immunoglobulin G (h-IgG)] results in a small shift in plasmon angle. However, a 1.5° shift and an observable increase in minimum reflectance and curve broadening are observed upon exposure of an antibody-coated surface to a 10-nm-diameter colloidal Au:antigen conjugate (h-IgG:Au). This large enhancement of the plasmon angle shift represents a >15-fold sensitivity increase over the nonenhanced assay. Figure 9 illustrates a similarly enhanced signal for the sandwich immunoassay format. Incubation of an antibody-modified film with a solution of h-IgG yields a small (0.04°) but detectable shift in plasmon angle. Exposure of the antibody:antigen coated film to a secondary antibody for the antigen [F_c-specific monoclonal goat antihuman immunoglobulin G (α-h-IgG(F_c)] results in a similarly small plasmon angle shift. Amplification of this signal is realized through the replacement of the free antibody with a colloidal Au:α-h-IgG(F_c) conjugate [α-h-IgG(F_c):Au]. Again, a large increase in plasmon angle, minimum reflectance, and curve breadth are observed. In this case, a 28-fold increase in plasmon angle is afforded through the use of protein:Au colloid conjugates in the sandwich immunoassay. Interrogation of colloidal Au-amplified protein binding can also be performed in a real-time monitoring mode. Figure 10 shows reflectance time courses for the immobilization of both the free and Au-bound secondary antibodies to antibody:antigen coated Au surfaces. Again, the magnitude of the signal is significantly greater in the case of Au enhancement. Finally, ultrasensitive detection of protein binding has been realized by this method. Under conditions where the antigen concentration is too low for unamplified detection of binding, the colloidal Au amplification step provides an easily detectable signal (Fig. 11). Figure 12 illustrates the effect of decreasing the antigen concentration further. As expected, the observed colloidal Au-induced shifts decrease as the solution antigen concentration is decreased from 3.0 μM to 6.7 pM. The lowest concentration of h-IgG investigated, 6.7 pM, yields an easily detectable shift of 0.33°. Together, these results suggest that Au enhancement may have utility in the detection of either low protein concentrations *or* low-molecular-weight (<500 g/mol) species that are very difficult to detect by traditional SPR methods.

In the case of direct (MEA-mediated) adsorption of Au particles, a nonlinear dependence of plasmon shift on particle coverage was observed. In order to evaluate whether this nonlinearity was present in the case of protein-mediated adsorption, AFM was used to determine the particle coverage over a wide range of *antigen* concentrations. The thus-obtained relationship between colloid coverage and plasmon angle shift is shown in Fig 13. In this case, a quasi-linear relationship is obtained, with smaller plasmon angle shifts accompanying lower colloid coverages. While it is clear that the relationship is not well-behaved enough to enable quantitative analysis at low antigen concentrations, the rough linearity suggests that such quantitation may be possible with better control over the surface attachment chemistries.

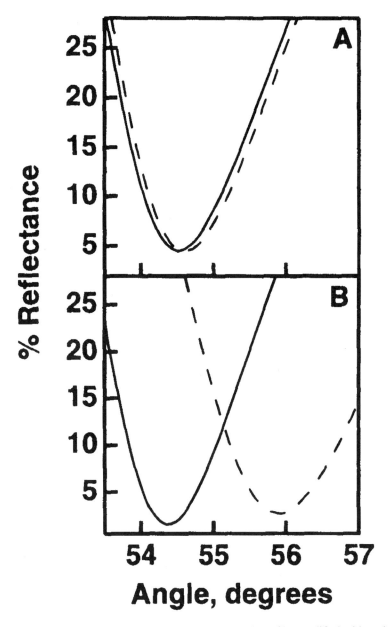

Fig. 8. In situ SPR curves of (A) an evaporated Au film modified with anti-IgG (solid line) and then exposed to a 1.0 mg/mL solution of antigen (human-IgG) (dashed line). (B) A film modified with anti-IgG (solid line) and then exposed to a solution of an electrostatic conjugate between human-IgG and 10-nm-diameter colloidal Au.

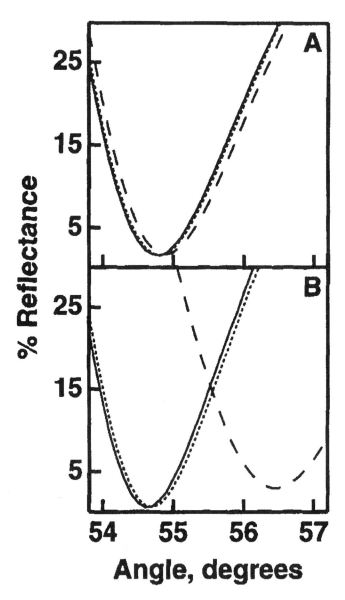

Fig. 9. In situ SPR curves of (A) an evaporated Au film modified with antibody (solid line) followed by sequential exposure to a 0.045 mg/mL solution of antigen (dotted line) and a 8.5 mg/mL solution of the secondary antibody, antihuman IgG (Fc-specific) (dashed line). (B) A film modified with antibody (solid line) followed by sequential exposure to a 0.045 mg/mL solution of antigen (dotted line) and a solution of secondary antibody: 10-nm-diameter Au conjugate (dashed line).

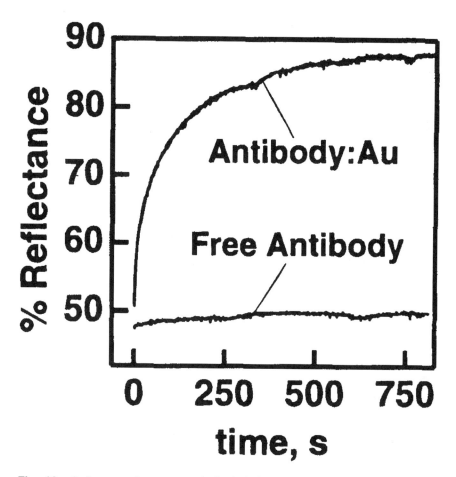

Fig. 10. Reflectance timecourses obtained during exposure of an antibody:antigen coated surface to a 17 nM solution of secondary antibody:colloidal Au conjugate (top curve), and a 8.5 mg/mL solution of the unmodified secondary antibody (bottom curve).

V. SUMMARY

Despite the attractiveness of tagless SPR assays from the standpoint of ease and speed of analysis, routine detection of low-molecular-weight components remains a challenge in the field. We have presented one possible solution to this problem: colloidal Au enhancement. This technique benefits from the long history of colloidal Au:protein bioconjugates and the more recent developments in oligonucleotide conjugates. Nearly any type of biologically relevant macromolecule can

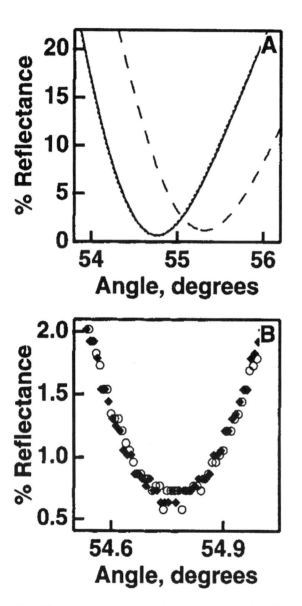

Fig. 11. In situ SPR curves of (A) an evaporated Au film modified with antibody (solid line) followed by sequential exposure to a 0.0045 mg/mL solution of antigen (dotted line) and a solution of secondary antibody:colloidal Au conjugate (dashed line). (B) Inspection of the plasmon minima for the antibody-modified Au surface before (circles) and after (diamonds) exposure to the 0.0045 mg/mL solution of human IgG. Within experimental error, there is no difference between the two curves.

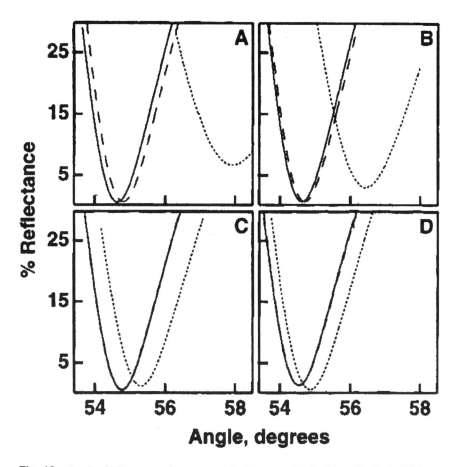

Fig. 12. In situ SPR curves of evaporated Au films modified with antibody (solid line) followed by sequential exposure to solutions of antigen (dashed lines) and secondary antibody:colloidal Au conjugate (dotted lines). The panels correspond to solution human IgG concentrations of (A) 3.0 µM, (B) 0.3 µM, (C) 3.0 nM, and (D) 6.7 pM.

be immobilized to the nanoparticle surface in an active form. Furthermore, standard sandwich and displacement immunoassay formats can be used with this method, thereby eliminating the need for new instrumentation or assay procedures. Indeed, similar methods have since been applied to ultrasensitive detection of DNA hybridization by both SPR and quartz crystal microgravimetry. The technique has also enhanced the contrast of SPR imaging of multianalyte arrays. Further investigations of Au-amplified bioassays in terms of practical application *and* fundamental phenomena will allow for continued expansion of the technique's utility.

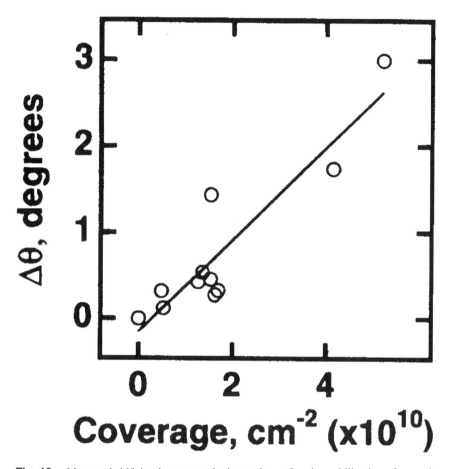

Fig. 13. Measured shift in plasmon angle due to the surface immobilization of secondary antibody:colloidal Au conjugate as a function of particle coverage (as determined by TM-AFM).

ACKNOWLEDGMENTS

This work was made possible through the support of NSF, NIH, and the Alfred P. Sloan Foundation.

REFERENCES

1. S Lofas. Pure Appl. Chem. 67:829, 1995.
2. RL Rich, DG Myszka. Curr. Opin. Biotechnol. 11:54, 2000.

3. J Homola, SS Yee, G Gauglitz. Sens. Actuator B-Chem. 54:3, 1999.
4. P Schuck. Curr. Opin. Biotechnol. 8:498, 1997.
5. AJ Thiel, AG Frutos, CE Jordan, RM Corn, LM Smith. Anal. Chem. 69:4948, 1997.
6. W Hickel, D Kamp, W Knoll. Nature 339:186, 1989.
7. LA Lyon, MD Musick, MJ Natan. Anal. Chem. 70:5177, 1998.
8. CE Jordan, RM Corn. Anal. Chem. 69:1449, 1997.
9. AG Frutos, RM Corn. Anal. Chem. 70:449A, 1998.
10. JR Krone, RW Nelson, D Dogruel, P Williams, R Granzow. Anal. Biochem. 244:124, 1997.
11. RW Nelson, JR Krone J. Mol. Recognit. 12:77, 1999.
12. RW Nelson, JR Krone, O Jansson. Anal. Chem. 69:4363, 1997.
13. RW Nelson, JR Krone, O Jansson. Anal. Chem. 69:4369, 1997.
14. CEH Berger, TAM Beumer, RPH Kooyman, J Greve. Anal. Chem. 70:703, 1998.
15. BP Nelson, AG Frutos, JM Brockman, RM Corn. Anal. Chem. 71:3928, 1999.
16. LA Lyon, WD Holliway, MJ Natan. Rev. Sci. Instrum. 70:2076, 1999.
17. CE Jordan, AG Frutos, AJ Thiel, RM Corn. Anal. Chem. 69:4939, 1997.
18. PT Leung, D Pollardknight, GP Malan, MF Finlan. Sens. Actuator B-Chem. 22:175, 1994.
19. T Wink, SJ van Zuilen, A Bult, WP van Bennekom. Anal. Chem. 70:827, 1998.
20. LA Lyon, DJ Pena, MJ Natan. J. Phys. Chem. B 103:5826, 1999.
21. LA Lyon, MD Musick, PC Smith, BD Reiss, DJ Pena, MJ Natan. Sens. Actuator B-Chem. 54:118, 1999.
22. H Raether. In: Springer Tracts in Modern Physics, Vol. 111 (Hohler, G., ed.). Springer-Verlag, New York, 1988.
23. RH Ritchie. Phys. Rev. 106:874, 1957.
24. E Kretschmann. Z. Phys. 241:313, 1971.
25. WN Hansen. J. Opt. Soc. Am. 58:380, 1968.
26. HE Debruijn, RPH Kooyman, J Greve. Appl. Optics 31:440, 1992.
27. AG Frutos, SC Weibel, RM Corn. Anal. Chem. 71:3935, 1999.
28. D Hornauer, H Kapitza, H Raether. J. Phys. D1:L100, 1974.
29. KC Grabar, RG Freeman, MB Hommer, MJ Natan. Anal. Chem. 67:735, 1995.
30. RG Freeman, KC Grabar, KJ Allison, RM Bright, JA Davis, AP Guthrie, MB Hommer, MA Jackson, PC Smith, DG Walter, MJ Natan. Science 267:1629, 1995.
31. KC Grabar, PC Smith, MD Musick, JA Davis, DG Walter, MA Jackson, AP Guthrie, MJ Natan. J. Am. Chem. Soc. 118:1148, 1996.

9
Self-Assemblies of Nanocrystals: Fabrication and Collective Properties

Marie-Paule Pileni
Université Paris et Marie Curie (Paris IV), Paris, France

I. INTRODUCTION

Self-assembled nanocrystals have attracted an increasing interest over the last five years (1–36). The level of research activity is growing seemingly exponentially, fueled in part by the observation of physical properties that are unique to the nanoscale domain. The first two- and three-dimensional superlattices were observed with Ag_2S and CdSe nanocrystals (1–4). Since then, a large number of groups have succeeded in preparing various self-organized lattices of silver (5–16), gold (15–27), cobalt (28,29), and cobalt oxide (30,31). With the exception of CdSe (2) and cobalt (28,29) nanocrystals, most superlattice structures have been formed from nanocrystals whose surfaces are derivatized with alkanethiols.

By varying the experimental conditions of nanocrystal deposition, a variety of organized structures have been achieved. For example, when silver (14,15), gold (17), and CdS (32) nanoparticles are deposited from oil suspensions containing low particle concentrations, circular domains of monolayer nanoparticle coverage are observed, surrounded by regions of bare substrate. Under other deposition conditions, large "wires" composed of silver nanoparticles have been observed (9), in which the degree of self-organization varies with the length of the alkyl chains coating the particles (33).

It has been demonstrated that three-dimensional (3D) superlattices of nanoparticles are often organized in a face-centered-cubic (FCC) structure (1–6, 10). In other cases, the particles pack in a hexagonal lattice (8,34). Interestingly, it has been recently demonstrated that the physical properties of silver (6,35) and cobalt (28,29) nanocrystals organized in 2D and 3D superlattices differ from those of isolated nanoparticles. Indeed, certain collective properties are observed, and in the case of silver nanoparticles these properties depend on whether the superlattice is a square or hexagonal network (36).

In this Chapter we discuss recent work from our laboratory in which self-organized 2D and 3D superlattices are formed from colloidal nanocrystals (which we will also refer to as "nanoparticles," or simply "particles"). In the case of 3D structures, we will use the more general term "aggregate" synonymously. We also discuss the optical and electrical properties of these self-organized systems.

II. SYNTHESIS OF NANOCRYSTALS

In recent years we have exploited the properties of micellar aggregates to synthesize nanoscopic metal and semiconducting particles. While a fundamental discussion of micelle formation and properties is beyond the scope of this chapter, background in micelle theory as it applies to our work can be found elsewhere (37–46). In our laboratories nanocrystals are synthesized using either reverse or normal micelles (47,48). The reverse micelles are based on diethyl sulfosuccinate, which is usually referred to as Aerosol OT, or, in the case of the sodium salt, Na(AOT). The solvent is either hexane or isooctane. The normal micelles we have used are based on sodium dodecylsulfate, commonly referred to as NaDS or SDS.

Key to our approach is the use of metal functionalized surfactants, $M^{n+}(AOT)_n$ in the case of reverse micelles and $M^{n+}(DS)_n$ in normal micelles, where M is the metal ion precursor to the nanocrystal material. While the micelle aggregate can influence the size and shape of the nanocrystal, addition of a surface derivatizing agent such as citrate or trioctyl phosphine is often necessary for the separation and purification of the particles. We have also made extensive use of alkanethiols, which not only allow us to derivatize the surface of the nanocrystals, but manipulate the interparticle interactions that are important factors in the 2D or 3D aggregate structure.

A. Synthesis of Silver Sulfide Nanocrystals (1–3)

Ag_2S nanocrystallites are synthesized by mixing two 0.1 M Na(AOT) micellar solutions (hexane solvent), each having the same water content, but with one containing 8×10^{-4} M sodium sulfide (Na_2S) and the other 8×10^{-4} M Ag(AOT). Brownian motion leads to collisions between the two types of micelles, and in some events there is an efficient exchange of water pool contents. After a few minutes, the reaction of Ag^+ and S^{2-} leads to nanosized Ag_2S particles. Significantly, the average size of the particles increases linearly with water content. The standard deviation in the distribution of particle sizes is about 30%, relative to the average size.

Pure alkanethiols, $C_nH_{2n+1}SH$ (where n is an even number varying from 6 to 18) are added (1 µL per mL of micelle solution) to the reverse micelle solutions containing Ag_2S nanocrystals. After evaporation of the solvents at 60°C, the resulting solid is washed with ethanol and filtered. The silver sulfide nanocrystals coated with the alkanethiols are then dispersed in heptane, forming an optically clear solution. This procedure does not allow for the extraction of the largest particles (ca. 10 nm) from the reverse micelles. Thus, the extraction engenders a size selection resulting in a smaller average diameter, as well as a decrease in the relative distribution of sizes from 30% to 14%.

B. Synthesis of Silver Nanocrystals (5,6)

As described above in the case of Ag_2S nanocrystals, the synthesis of Ag nanoparticles involves the mixing of two reverse micelle solutions, each with the water content parameter $w(=[H_2O]/[AOT])$ fixed at 40. The total AOT concentration in the first solution is distributed as 30% Ag(AOT) and 70% Na(AOT). The second solution contains Na(AOT) and 0.07 M hydrazine (N_2H_2). After the solutions are mixed and the Ag particles formed from the reduction of Ag^+ by hydrazine, addition of dodecanethiol (1 µL per mL of micelle solution) results in the coating of the particles with a dodecanethiolate layer, and subsequent flocculation. The solution is then filtered to isolate the particles, which can be easily redispersed in hexane.

After the dodecanethiol extraction step, the size distribution is reduced from ±40% to ±30%. Since the polydispersity is still large at this point, size-selective precipitation (SSP) is employed to reduce the size distribution. SSP is a well-known technique for separating mixtures of copolymers and homopolymers that occur during the synthesis of sequenced copolymers, and has also been used for the extraction of nanosized crystals (49). SSP utilizes a mixture of two miscible solvents, each differing in their ability to dissolve the surfactant alkyl chains. For example, alkanethiolate-coated silver particles are highly soluble in hexane but poorly soluble in pyridine. Thus, pyridine is added in steps to a hexane solution containing coated silver nanoparticles. When the hexane/pyridine volume ratio is about 50%, the solution becomes cloudy, indicating agglomeration of the largest silver particles. The larger particles agglomerate first because of their stronger van der Waals interactions. This solution is centrifuged, and the fraction of solution rich in agglomerate is separated. The resulting supernatant is small-particle rich. The silver particles in the agglomerate-rich fraction can be redispersed via addition of hexane to form clear solution. By repeating this procedure several times on the supernatant, very small particles can be obtained, with low polydispersity (15%).

C. Synthesis of Cobalt Nanocrystals (28,29)

A reverse micellar solution containing 0.25 M Na(AOT) and 0.02 M Co(AOT)$_2$ is mixed with a 0.25 M Na(AOT) micellar solution also containing 0.02 M sodium borohydride (NaBH$_4$). Both micellar solutions possess the same water content parameter ($w = 10$), which implies a water radius of $R_w = 1.5$ nm (42). The synthesis is carried out in air. After mixing, the solution remains optically clear and its color turns from pink to black, indicating the formation of colloidal particles. Under anaerobic conditions, the cobalt particles are extracted from the reverse micelles by covalent attachment of either trioctyl phosphine (29) or lauric acid (50), and then redispersed in pyridine or hexane, respectively. The covalent-attachment-based extraction step results in a size selection similar to that seen in the Ag$_2$S system discussed earlier. The surface modification of the cobalt particles also significantly improves their stability when exposed to air. Cobalt nanoparticles prepared in this manner can be stored without aggregation or apparent oxidation for at least one week.

D. Synthesis of Cobalt Ferrite Nanocrystals

Normal micellar solutions containing 5.25×10^{-4} M Co(DS)$_2$ and 7×10^{-3} M Fe (DS)$_2$ are mixed and then maintained at 28.5°C. Next, 0.44 M dimethylammonium hydroxide [(CH$_3$)$_2$NH$_2$OH] is added to the micellar solution while it is vigorously stirred. After 2 hr, a magnetic precipitate is observed and subsequently separated from the solution by centrifugation. The precipitate is washed with 0.01 M nitric acid and then redispered in hexane. A solution of 0.015 M sodium citrate (Na$_3$C$_6$O$_7$H$_5$) is added, and the solution is stirred at 90°C for 30 mins. Acetone is added to precipitate the particles again, and the solid is washed with copious amounts of acetone. After air drying, the resulting powder of citrate-coated cobalt ferrite nanocrystals can be redispersed in aqueous solution. The resulting solution is a neutral magnetic fluid.

The nanocrystals prepared according to the above procedure show the X-ray diffraction lines of a spinel phase with lattice constant $a = 8.41$ Å. This value of a is consistent with that of bulk cobalt ferrite (51). The diffraction peaks themselves are too broad to allow for the distinction between inverted and normal spinel phases. The elementary composition, as determined by energy dispersive X-ray spectroscopy (EDS), is 95% iron and 5% cobalt in mass. No Fe^{2+} is detected by Mossbauer spectroscopy. From these data, the formula of the nanocrystals is Co$_{0.13}$Fe$_{2.58}$_$_{0.29}$O$_4$, where _ represents a cationic vacancy. The particles can be used in either powder form or, once dispersed in aqueous solution, at the concentration desired by the user. A drop of such solutions can be deposited on a carbon grid in the presence or absence of an external magnetic field. In our studies, a field of 1.8 Tesla (T) is employed (vide infra).

III. FABRICATION OF 2D SUPERLATTICES

A. Monolayers of Silver Sulfide Nanocrystals (1,3,4,33,52)

1. Influence of the Substrate

In this section we discuss the self-assembly of Ag_2S nanocrystals on three different substrates, namely amorphous carbon, highly ordered pyrolytic graphite (HOPG), and molybdenum sulfide (MoS_2). As we will demonstrate, the substrate can have a strong influence on the structure of the superlattice of nanocrystals (52).

By using a dilute solution of Ag_2S nanocrystallites (particle volume fraction $\phi \leq 0.01\%$), monolayers of nanoparticles can be formed. Monolayer formation on amorphous carbon-coated TEM sample grids involves adding low-volume-fraction solutions drop by drop, allowing for solvent evaporation between drops. Large ribbon-like formations of monolayers are observed (Fig. 1A), with the rest of the grid covered with a lower density of nanocrystals. Within the ribbon domains, the nanocrystals are arranged in a compact hexagonal network, with an average (surface-surface) interparticle separation distance of 1.8 nm (1B). The ribbon-like formations are obtained for various nanocrystal sizes (from 2 to 6 nm). The length and width of the ribbons depend on the nanocrystal size (4). In most cases the length of the ribbons exceeds 100 mm.

Scanning electron microscope (SEM) images of 5.8-nm Ag_2S nanocrystals on HOPG are shown in Figs. 2A–C. The brightest regions correspond to 3D aggregates. The 3D aggregates have a somewhat oblate shape, with widths from 4 to 8 μm and heights less than 1 μm. The gray regions in the SEM images correspond to nanocrystal monolayers, while the darkest regions are holes in the monolayer structure. The 3D aggregates are always surrounded by breaks in the monolayer structure. Holes also occur within the large monolayer regions (2A).

Figure 3 shows atomic force microscope (AFM) images of the monolayer regions. The inset in the low-resolution image shown in Fig. 3A depicts the vertical profile of the line scan shown in the upper left of the image. The line scan was deliberately taken over a path that contained a hole (black), a flat plain (gray), and a small 3D aggregate (white). The peak in the line scan shown in the inset of Fig. 3A indicates the formation of very small aggregates whose range from 100 to 200 nm and whose height is 10 to 15 nm (corresponding to bi- and trilayers of Ag_2S nanocrystals). Since the depths of the holes are about 6 nm, and since particles are not observed within the holes when high-resolution scans are performed, the line scan is highly suggestive of a dense monolayer of Ag_2S nanoparticles on HOPG.

Figure 3B shows a high-resolution image of the flat plain region encircled in Fig. 3A. The Ag_2S nanocrystals are organized in local hexagonal network domains. The average center-to-center distance of the particles is ca. 8 nm. Given the diameter of the particles (determined from TEM images), the average surface-to-surface separation is about 2 nm. The images shown in Fig. 3A and 3B could,

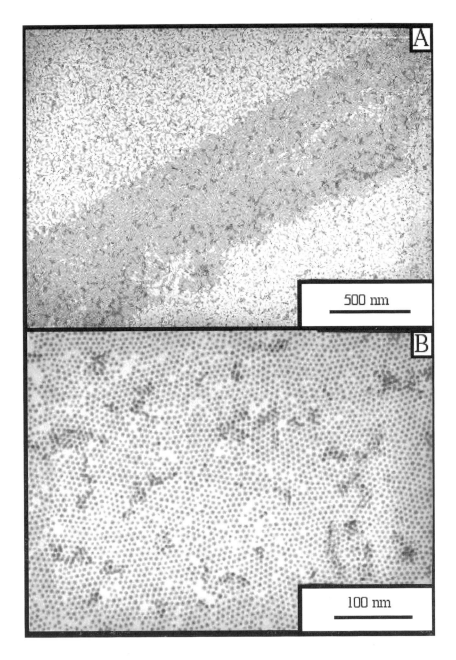

Fig. 1. TEM images of a monolayer of 5.8-nm-diameter silver sulfide nanocrystals deposited on amorphous carbon, observed at different magnifications.

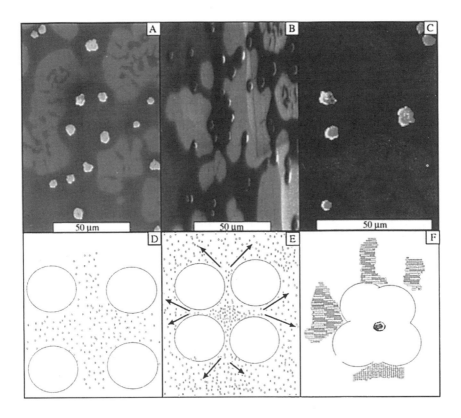

Fig. 2. SEM images (A, B, C) of 5.8-nm-diameter silver sulfide nanocrystals deposited on HOPG (from a 10^{-6} M hexane solution at $T = 20°C$). Schematics of mono- and multilayer formation of nanocrystals as a result of solvent bubble growth and coalescence (D, E, F).

in principle, result from dense multilayer structures. However, taking into account the overall similarity between the TEM and AFM images (52), as well as the flat vertical trace observed over a large range of surface line scans, our tentative conclusion is that a dense monolayer of Ag_2S nanocrystals is formed on HOPG.

The superlattice structures seen in the TEM, SEM, and AFM images above can be rationalized on the basis of interparticle and particle-substrate interactions and solvent-substrate interaction (wetting). Recent work in our laboratories has shown that the attractive force between two Ag_2S nanocrystals is 1.8 $\times$ 10^{-6} dynes (52). The force between a Ag_2S nanocrystal and the HOPG substrate is repulsive (-3.7×10^{-6} dynes) (52). Significantly, the heptane solvent on HOPG forms a nonzero contact angle (5–10°), which means that as evaporation occurs the solvent film becomes unstable and droplets begin to form (52).

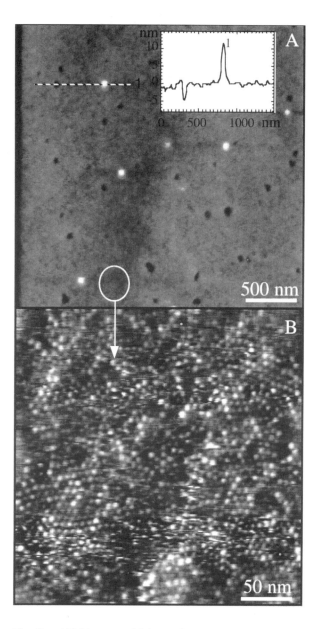

Fig. 3. AFM images of 5.8-nm-diameter Ag_2S nanocrystals deposited on HOPG substrate (A) Low magnification showing flat plains with isolated holes (black) and 3D aggregate (white). Line scan profile showing depth of holes and height of aggregates (inset in A). (B) High magnification of plains region (encircled in 3A) showing close-packed structure.

This process is an example of the well-known Marangoni effect (53). The dynamics of these droplets play a key role in the evolution of Ag_2S nanocrystal assembly on HOPG.

Consider the cartoon schematics in Fig. 2D–F accompanying the SEM images in Fig. 2A–C. Immediately after the solution containing the nanocrystals is deposited on the substrate, the solvent begins to evaporate and droplets form. The Ag_2S nanocrystals themselves are fully solvated (or "dressed") by the heptane, which prevents their assembly into dense structures (Fig. 2D). As the droplets grow and begin to merge, some of the Ag_2S particles (which are still mobile because of the thin solvent layer present on the HOPG surface) are expelled away from the merge center (see arrows in Fig. 2E). These dressed particles form compact monolayer islands, whose density increases after all of the solvent evaporates and interdigitation of the alkyl chains on the Ag_2S particles occurs.

Other particles are caught in the center of the droplet merge point. The pressure exerted on these particles by the droplet menisci is large, and, while a monolayer initially forms continued droplet coalescence engenders the formation of a 3D structure (Fig. 2F). This process may be viewed as analogous to the collapse of Langmuir films when the lateral pressure on the monolayer is too high. The 3D structure of dressed particles dries out as the solvent evaporates, and thus interdigitation of the particles' alkyl chain coating occurs in three dimensions instead of two. The sizes of the 3D aggregates of Ag_2S are similar, which implies that the merge regions between growing solvent droplets are also similar in size.

A key point implied by the scheme in Fig. 2D–F is that the formation of monolayer and 3D aggregates occurs essentially simultaneously. It should also be pointed out that very small 3D aggregates also form in the monolayer regions (Fig. 3A), but via simple interparticle attraction rather than confinement and compression due to droplet growth.

When the HOPG substrate is replaced with MoS_2, the assembly of Ag_2S nanocrystals results in a very different pattern. Figure 4 shows SEM images of 5.8-nm Ag_2S nanocrystals deposited on MoS_2 using a procedure similar to that used in the case of HOPG. In contrast to what is seen on HOPG, the MoS_2 substrate shows large interconnected domains of monolayers surrounding hole regions (light and dark regions, respectively). Three-dimensional aggregates can be seen on the surface of the monolayers. These aggregates are smaller than those on HOPG, with widths in the 1.6- to 2.3-μm range. SEM analysis does not allow for the determination of aggregate height. However, AFM studie indicates that the average height is ca. 150 nm.

The contrasting Ag_2S assembly structures on HOPG and MoS_2 can be attributed to differences in particle-substrate and solvent-substrate interaction (52). First, the contact angle formed by heptane on MoS_2 is too small to be measured (in other words, heptane wets MoS_2 better than it wets HOPG). Thus, droplets due to thin solvent layer instability are less likely to form. Second, and

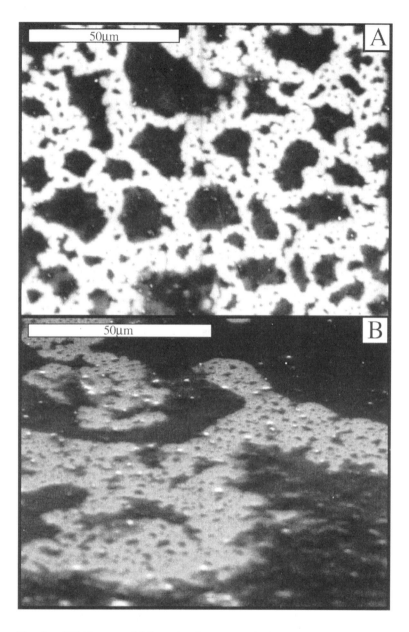

Fig. 4. SEM images of 5.8-nm Ag_2S nanocrystals deposited on MoS_2 substrate from a 10^{-6} M hexane solution at $T = 20°C$), observed at normal incidence (A) and at 60° tilt (B). Note the open web structure in 4A, compared to the closed monolayer structure in Figs. 2A–C.

perhaps most significant, is that the force between an Ag_2S nanoparticle and the MoS_2 substrate is, in contrast to HOPG, attractive (ca. 2.2×10^{-5} dynes). When the heptane solution of Ag_2S nanoparticles comes into contact with the MoS_2 substrate, the particles collide with the surface and remain fixed. The placement of these particles on the surface is random because of Brownian motion prior to adhesion.

As the solvent evaporates, the movement of Ag_2S particles in response to droplet growth and coalescence does not occur to a great extent because of the attractive particle-substrate interaction. Thus, the localized but very dense monolayers seen on HOPG are absent on MoS_2. Nonetheless, the particles do diffuse in two dimensions; attractive particle-particle interactions and capillary forces due to the thin solvent layer combine to promote the formation of small monolayer islands. The smaller size of the few 3D aggregates seen on MoS_2, is also consistent with attractive particle-substrate interactions.

2. Influence of the Alkanethiol Chain Length

In our laboratory most of the nanocrystals formed in micelles cannot be extracted via surface derivatization with hexanethiol, octanethiol, hexadecanethiol, or octadecanethiol (henceforth abbreviated C_6, C_8, C_{16}, and C_{18} thiols, respectively). Thus, they can only be removed from the oil environment along with the micelle-forming surfactant.

The instability of the particles coated with shorter-chain thiols (C_6 and C_8 thiols) may be attributed to the disordered nature of the alkyl chains; short-chain thiols do not form a crystalline environment and possess a large number of gauche defects. Thus, the surface layer on the nanoparticles may resemble more closely free alkanes in a liquid state than it does a self-assembled monolayer. Such behavior has been well documented for gold particles (54,55). While we cannot assume that the behavior of short-chain thiols on gold is identical to that on Ag_2S particles, the similar behavior of self-assembled monolayers on Au(111) and Ag(111), in spite of the large differences in chemisorption energy (56–58), leads us to tentatively conclude that the source of particle instability in the Au and Ag_2S systems is the same.

Extraction of the particles from their micelle environment is also not successful when very-long-chain alkanethiols (C_{16} and C_{18} thiols) are used. The hydrophobicity of these compounds is certainly high, and thus they tend to remain solubilized in the nonpolar medium (59). This could, in principle, prevent the long-chain thiols from reacting with the Ag_2S surfaces. However, the dynamic character of the reverse micelles is such that reactions can occur between hydrophobic and hydrophilic reactants (60). Apparently, the solvation energy of the long-chain thiols in the nonpolar solvent is sufficient to overcome the energy of the thiol-Ag_2S surface reaction.

The intermediate-chain-length thiols (C_8, C_{10}, and C_{14} thiols) appear to be the best suited for particle derivatization, extraction from the reverse micelles, and redispersion into hexane. TEM images of Ag_2S particles coated with these different thiols show that self-organized hexagonal networks of the particles form, whose interparticle spacings change systematically with alkanethiol chain length (33). However, the edge-edge separation distances do not correspond exactly to the distances one might expect from the alkyl chain lengths. For example, Table 1 summarizes the interparticle spacings (edge-edge distances, henceforth referred to as d_{pp}) measured from TEM images. For comparison, the distance ($2L$) calculated from the empirical equation of Bain et al. (61), which assumes an *all-trans* zigzag conformation, is also shown. It is clear that the experimental d_{pp} values are considerably shorter than twice the chain length of the surface thioalkane groups.

In the case of C_8 and C_{10} thiol surface groups, d_{pp} is longer than L but shorter than $2L$. In the case of the C_{12} thiol, d_{pp} is about equal to L. For C_{14}-thiol-derivatized particles, d_{pp} is actually shorter than L. These trends can be explained by the change in chain conformation and the various defects that depend on chain length (54). For example, the number of gauche defects near the particle surface decreases as the alkyl chain length increases. However, the incidence of gauche defects near the terminal (methyl) end of the chain increases with chain length. Hence, the shortest chains (C_8) behave almost as free alkanes (62) and are thus not prone to 2D organization or interdigitation. As the chain length increases, interdigitation between alkyl chains on adjacent particles occurs, leading to a d_{pp} value that approaches the length of a single chain, L (C_{10} and C_{12} thiol systems). In the case of C_{14} thiol surface groups, the increase in gauche defects near the chain terminal end allows pseudorotational motion of the chain about the R-S bond axis. This explains why the d_{pp} value is smaller than L for the C_{14} thiol system.

There are two key observations in the study of alkanethiol chain length effects. First, the interparticle spacing d_{pp} is dependent on chain length, but the

Table 1. Variation of the Average Distance (d_{pp}) between Alkanethiolate-Coated Nanocrystals Organized in a Hexagonal Network

	Alkanethiolate system			
	C_8	C_{10}	C_{12}	C_{14}
Experimental d_{pp} (nm)	1.4	1.6	1.75	1.8
L_{calc}[a] (nm)	1.27	1.52	1.77	2.03
$2L_{calc}$[a] (nm)	2.54	3.04	3.54	4.06

[a] From L (nm) = 0.25 + 0.127n, where n is the number of CH_2 groups in the chain. See Ref. 61.

dependence is not linear. Second, using very-short-chain alkanethiol groups (regardless of the difficulties inherent in the extraction step) to reduce interparticle spacing is not likely to yield successful results. Short-chain compounds simply are not prone to 2D assembly and behave more as free alkanes.

B. Monolayers of Silver and Cobalt Nanocrystals

1. Silver Nanocrystals

The 2D network formed after a drop of a low-concentration solution of silver particles (2.5×10^{-5} M) is placed on an HOPG substrate. Since the interparticle spacing in the hexagonal network is ca. 1.8 nm, arguments similar to those used in the interpretation of the Ag_2S nanoparticle studies can be made that essentially total interdigitation of alkyl chains on adjacent particles is occurring.

2. Self-Organization of Cobalt Nanocrystals

Cobalt nanocrystals coated with trioctylphosphine also form a hexagonal 2D network on amorphous carbon. However, in contrast to results seen for Ag_2S particles on amorphous carbon, the hexagonal network contains many bare regions. The size distribution in the cobalt nanocrystals is ca. $\pm$ 14%, similar to that in the Ag_2S particle studies. Since size distribution is a major factor in controlling the monolayer structure, one might expect that the cobalt nanoparticle network should resemble the Ag_2S networks in Fig. 1–4. The fact that the network in the cobalt nanocrystal case is more open may be due to the magnetic properties of the particles, or simply because the surface groups are different (trioctylphosphine instead of dodecanethiolate). Studies on these systems are in progress.

C. Organization of Nanocrystals in Rings and Hexagon

When nanocrystal-containing solutions are deposited on cleaved amorphous graphite held fixed by anticapillary tweezers and the solvent is allowed to completely evaporate, interesting patterns of nanocrystals are observed. Figure 5 shows TEM images of the circle structures formed from a variety of dodecanethiol-coated silver (5A), cobalt (5B), cobalt ferrite (5C), and cadmium sulfide (5D). Since only the cobalt and cobalt ferrite particles are magnetic, it can be concluded that the circle patterns have nothing to do with magnetic properties. Individual particle shape is apparently also not a critical factor, since the CdS nanocrystals are triangle shaped. We observe similar structures on amorphous carbon-coated TEM sample grids.

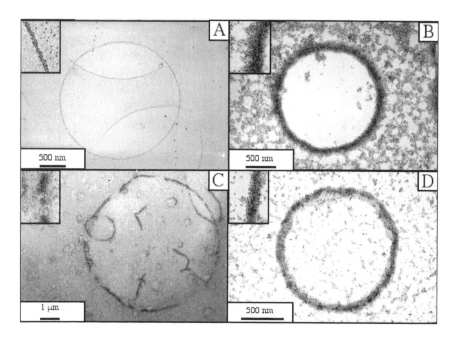

Fig. 5. TEM images of various nanocrystals deposited on cleaved amorphous graphite using an anticapillary tweezer. Silver (A), cobalt (B), cobalt ferrite (C), and triangular cadmium sulfide (D) nanocrystals.

IV. SUPRACRYSTALS: FCC LATTICES OF NANOCRYSTALS

Unlike some of the 2D structures discussed above, whose form depends in part on forces exerted by growing solvent droplets on flat substrate, the formation of a 3D lattice does not require external forces. The 3D lattice structure is due to the interplay of van der Waals attraction forces and hard-sphere repulsive interactions. Because these forces are isotropic, the best particle arrangement is achieved by maximizing the density of what we shall term a "supracrystal." The initial (2D) monolayer thus is hexagonally close packed. Three-dimensional growth could, in principle, result in a hexagonal (HCP) or face-centered-cubic (FCC) structure. However, we observe FCC structures with fourfold symmetry. In this section, we give an overview of the 3D supracrystal systems we have prepared in our laboratories. At the outset it is important to note that these supracrystal assemblies are obtained only after the nanocrystal size distribution has been reduced significantly via the methods discussed in Secs. II.A and II.B.

A. Silver Sulfide Supracrystal Assemblies

1. Ag$_2$S Nanocrystals Coated with Dodecanethiol

1. As discussed earlier, 3D structures of dodecanethiol-coated Ag$_2$S nanocrystals form during solvent evaporation on amorphous carbon and HOPG (Fig. 2). Fourfold symmetry, which can be attributed to the [001] plane of an FCC lattice, can easily be seen. The center-to-center distance between two nanocrystals along the [010] plane is ca. 11 nm. The average particle diameter is 5.8 nm, and the shortest center-to-center distance is 7.8 nm, leaving a 2-nm edge-to-edge separation that is consistent with alkyl chain interdigitation (see Sec. IV.A.2). The supracrystal structure is stable under ambient conditions for two months.

2. It has already been demonstrated that it is possible to prepare and isolate particles of different sizes (47,48). Using the same deposition procedure as that used for the 5.8-nm-diameter particles, 3D supracrystals of Ag$_2$S nanocrystals whose average diameters are 3 nm and 4 nm have been prepared. Upon increasing the nanocrystal concentration in hexane to ca. 6×10^{18} particles per mL solution, larger polycrystalline aggregates containing a large number of defects as dislocations are formed.

2. Effect of Alkyl Chain Length on the Formation of Supracrystals of Ag$_2$S Nanoparticles

For Ag$_2$S particles of virtually any size deposited with high concentration solutions (ca. 6×10^{18} particles per ml solution) on amorphous carbon, large aggregates will form. Nanocrystals coated with C$_8$ and C$_{10}$ thiols form well-faceted aggregates with fourfold symmetry, as observed for the C$_{12}$ thiol-coated particles discussed in the last section. Conversely, C$_{14}$-thiol-coated Ag$_2$S particles do not form well-defined supracrystals. Nanocrystals coated with C$_6$ thiol cannot be extracted from micelles.

The difference in self-assembly behavior when C$_{14}$ alkanethiols are used is related to the d_{pp} edge-to-edge spacing parameter discussed earlier. For alkyl chain lengths shorter than C$_{14}$, the d_{pp} value is either similar (for example, $n = 12$) or slightly longer ($8 \leq n \leq 12$) than the calculated chain length, assuming complete *trans* (zigzag) conformation. The interdigitation of alkyl chains in these systems leads to a dense packing of the nanocrystals, which in turn results in well-defined 3D lattices. In the case of the C$_{14}$ alkyl chains, the small d_{pp} value implies strong interparticle van der Waals attraction, in spite of the presence of a large number of gauche defects and poor interdigitation. We therefore conclude that well-defined 3D supracrystals require both strong interparticle interactions and alkyl chain interdigitation.

B. Supracrystals of Silver Nanoparticles (5,6)

Figure 6A shows a TEM image of aggregates formed from 5-nm silver particles coated with dodecanethiol on an amorphous carbon substrate. High-resolution TEM (Fig. 6B) shows that the silver nanoparticles appear to be arranged in either cubic or hexagonal structures. The transition from one phase to the other is abrupt and analogous to polycrystalline atomic lattices, wherein each nanocrystal domain has a different orientation. In a perfect hexagonal crystal, patterns of fourfold symmetry should not be observed. However, if one considers an FCC lattice, fourfold symmetry may be observed at certain tilt angles (we confirmed this by actually changing the tilt angle in the TEM studies). Thus, the "pseudohexagonal" structure seen in Fig. 12B can be attributed to the stacking of the {110} plane of an FCC lattice, and the features with fourfold symmetry arise from stacking of the {011} planes.

The cell parameter of the aggregate can be determined from either the fourfold symmetry or pseudohexagonal patterns in the TEM. In the former case, the cell parameter is 9 ± 1 nm. From the pseudohexagonal pattern, we find that the lattice constants (a,b,c) are not equivalent as they should be in a perfect hexagonal lattice. We find that $a = b = 9$ nm and $c = 6.6$ nm. This discrepancy may be attributed to a tetragonal distortion of the FCC structure, which was recently predicted from molecular dynamics simulations of alkylthiolate-coated gold nanoparticles (34).

STM images of 3D aggregates of silver nanoparticles on an Au(111) substrate largely confirm the interpretation of the TEM images. For example, in Fig. 6C, the fourfold symmetry arising from the stacking of {011} planes is clearly visible. The line trace (see inset) from the dark (low) to bright (high) regions of the image indicates clearly the stacking of monolayers of silver nanocrytals. Several such line scans were obtained, and the distance between two layers is consistently found to be in the 2- to 2.8-nm range. This distance is smaller than the total diameter of one coated nanocrystal (6.1 nm = 4.3 nm + 1.8 nm) and thus indicates that the particles of one layer sit in the center of the triangle formed by particles in an adjacent layer. The change in height of the line scan as it crosses nearly three layers of nanocrystals is consistent with interdigitation and zigzag conformation of the alkyl chains. Employing a similar deposition procedure, but with slower solvent evaporation, well-defined FCC supracrystals of silver nanoparticles can be obtained (64).

C. Supraaggregates of Ferrite Nanocrystals (65)

By depositing a 5-mL solution containing cobalt ferrite nanocrystals (0.0093 mass percent) on an amorphous carbon-coated TEM grid and allowing the solution to evaporate, an open mesh of randomly aggregated particles is obtained

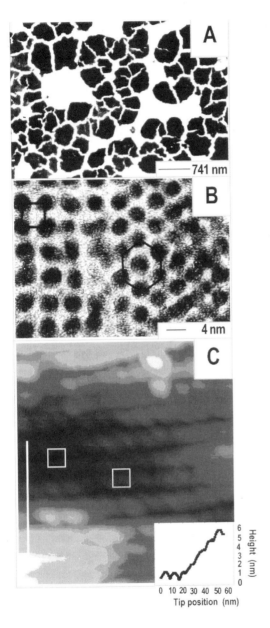

Fig. 6. Supracrystals of 5-nm silver nanocrystals. (A) Low-magnification TEM image. (B) High-resolution TEM image showing both hexagonal and fourfold symmetry. (C) AFM image of Ag supracrystal showing layered structure and source of fourfold patterns in TEM image.

(Fig. 7A). The cobalt ferrite sample used contained a narrow particle size distribution, with the average diameter based on measurements of 500 particles, being 10.7 ± −0.3 nm (66). Using solutions with a higher particle concentration of 0.17 mass percent, a higher particle density random aggregate structure results (see SEM image in Fig. 7B). It should be noted that the individual particles do not coalesce.

When the higher-concentration solution is deposited on amorphous carbon in the presence of a 1.8-T magnetic field is applied parallel to the substrate, long needle-like structures form (see SEM image in Fig. 7C). The higher magnification inset of Fig. 14C shows that the structure is quite regular. By tilting the sample in the electron beam, it is observed that a large number of nanocrystal layers are aligned with the magnetic field.

Other groups have observed self-organization of nanoparticles that is influenced by external magnetic fields. Depending on experimental conditions, needle-like structures (67), thin films with ordered structure (68), or microdrops (69) have been obtained.

V. COLLECTIVE PROPERTIES OF NANOCRYSTALS ORGANIZED IN 2D AND 3D SUPERLATTICES

A. Optical Properties of Monolayers

The basic theory of the optical properties of metal nanoparticles is discussed in Chapter 1 of this volume. In the UV-visible spectral range, the broad absorption bands of metal nanocrystals are due to plasmon resonance excitations or interband transitions. The spectra of dilute solutions of metal nanocrystals can be modeled quite well with Mie theory (70–76). Both the experimental (6,77) and simulated (75,78) absorption spectra show a decrease in the plasmon resonance band intensity and increase in bandwidth with decreasing particle size. The experimental peaks shapes are nearly Lorentzian, and discrepancies at higher energies are due to interband transitions in the experimental system ($4d$–$5sp$) (79). When silver nanocrystals are organized into a 2D lattice, the plasmon resonance peak is shifted to energies lower than what is obtained for dilute solutions of isolated particles. The absorption spectrum of dodecanethiol-coated 5-nm-diameter silver nanoparticles in the original hexane solution (before the particles are deposited on the surface) is similar to that of particles removed from the surface and redispersed in hexane. When the same nanocrystals are deposited in 2D and 3D lattices, spectral shifts are observed. The shift is more pronounced in three dimensions than in two. Since the original hexane solution spectrum and that of the redispersed particles are virtually the same, the spectral shifts associated with the 2D and 3D assemblies must come from interparticle electromagnetic interactions and not from some physical or chemical transformation associated

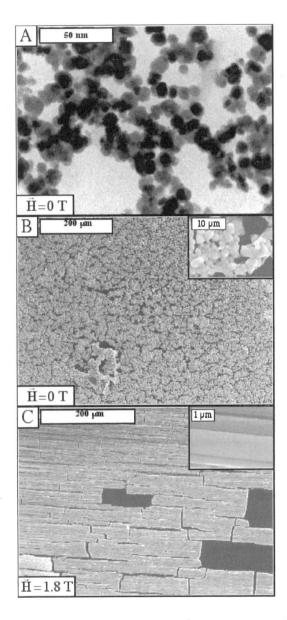

Fig. 7. Cobalt ferrite nanocrystals deposited on amorphous carbon. (A) TEM image of particles deposited from low-concentration solution (0.0093 mass percent in water) in the absence of an external magnetic field. (B) SEM image of random structure resulting from deposition with higher-concentration solution (0.17 mass percent in water). (C) Cobalt ferrite nanocrystals deposited in the presence of a 1.8-T field applied parallel to the substrate.

with their deposition. The bandwidth of the 2D plasmon resonance peak is ca. 1.3 eV, considerably larger than that of the particles in dilute solution (ca. 0.9 eV). Similar behavior is observed for silver nanocrystals of various sizes (from 3 to 8 nm).

UV/vis polarization spectroscopy can reveal information about interparticle electromagnetic interactions. In s-polarization, the electric field vector is oriented parallel to the plane of the substrate at all incidence angles θ. Plasmon resonance modes with components polarized perpendicular to the plane of the substrate are not seen when the incident light is s-polarized (80). On the other hand, p-polarized light, whose electric field is parallel to the plane of incidence can probe plasmon resonance excitations whose components are either parallel or perpendicular to the substrate. For 2D assemblies of metal nanocrystals on planar substrates, p-polarized light is thus ideal for probing interparticle interactions (81).

Figure 8 shows the UV-vis polarization absorption spectra of a hexagonal 2D assembly of 5-nm silver nanocrystals on a HOPG substrate. In s-polarization, the absorption spectra are virtually independent of incidence angle θ (Fig. 8A) and show a plasmon resonance band centered at 2.9 eV, which is similar to that seen in isolated silver nanocrystals. The degree of band asymmetry is also similar to that seen in isolated particles. However, in p-polarization a second band appears at higher energy as the incidence angle is increased (Fig. 8B). At θ = 60° the two peaks are well defined: the first is close in energy (2.8 eV) to the absorption maximum for isolated particles (2.9 eV), but the second is centered at ca. 3.8 eV. These data are reproducible and not sensitive to coverage effects.

To examine if polarization spectra such as those in Fig. 8B are due to a self-organized lattice, similar spectral studies were done on 5-nm silver nanocrystals randomly distributed on a HOPG substrate. In s-polarization at θ = 60°, the UV/v is spectrum shows a single plasmon resonance band centered at 2.7 eV. With p-polarized light at the same incidence angle, the plasmon resonance band is split into two peaks, one slightly higher and one slightly lower in energy than the peak in the s-polarization spectrum.

The TEM image shows that while the silver nanoparticle sizes are quite uniform, a few particles are touching and perhaps even to the point of coalescence. This interaction may account for the peak splitting. However, the energy difference between the two peaks is very small compared to the case of the p-polarization spectrum of a hexagonal 2D lattice (Fig. 8B).

It can be concluded that the high-energy band at 3.8 eV in Fig. 8B is due to the self-organization of the silver nanocrystals into a hexagonal network. The position of the peak can be explained in terms of local field effects; each nanocrystal experiences the electric field of the incident light plus the dipolar fields of the particles in its vicinity. Calculated spectra (35) for finite-sized clusters under s- and p-polarization are in qualitative accord with the results presented here.

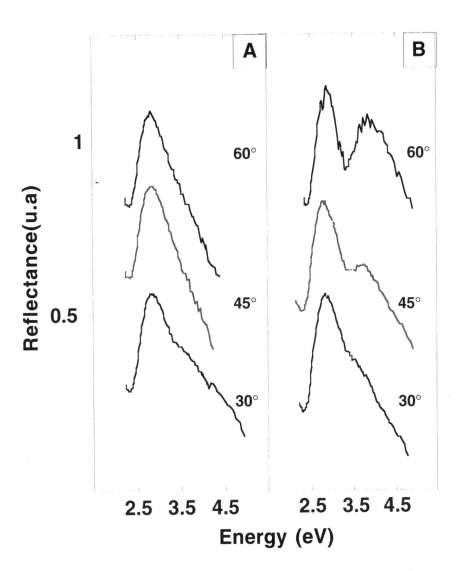

Fig. 8. UV/vis polarization absorption spectra of 2D hexagonal lattice of 5-nm-diameter silver nanocrystals on HOPG substrate: (A) *s*-Polarization, (B) *p*-Polarization. Numbers next to curve indicate incidence angle.

B. Collective Optical Properties of Particles Self-Assembled in Multilayers

The UV/vis spectrum of 3D aggregates shows a ca. 0.25 eV shift toward lower energy relative to the peak associated with silver particles in dilute solution. The bandwidth of the plasmon resonance peak in the 3D lattice spectrum is actually smaller (0.8 eV) than that of the isolated particles. As in the case of the 2D lattice systems, we also redispersed into hexane solution the particles composing the 3D lattice and obtained spectra virtually identical to that of the original hexane solution used to deposit the particles. Thus, the spectral changes (the peak shift and decrease in bandwith) must arise from the interaction of the silver nanoparticles in the 3D lattice and not some physical or chemical transformation.

According to theory (35,70), an increase in the dielectric constant of the medium surrounding a metal nanoparticle will cause the plasmon resonance maximum to shift to lower energies. For particles organized in multilayers, each silver particle is surrounded by 12 other particles, whereas in a monolayer each particle has only 6 near neighbors. Since the presence of other silver particles effectively increases the dielectric constant of the medium surrounding the particle in question, it is reasonable to expect that the plasmon resonance band for a 2D monolayer of silver nanoparticles should be red-shifted from that of an isolated silver particle, and that of a 3D structure red-shifted further still. The experimental data are entirely consistent with this model.

The fact that the plasmon resonance bandwidth for the 3D superlattice is smaller than that of the 2D monolayer and even the isolated silver nanoparticles is very interesting, and may be due to an increase in the electron mean-free path. The increase in mean-free path may in turn result from tunneling of electrons between particles. If true, this result would be rather surprising, since the dodecanethiol coatings result in a ca. 2-nm separation between the particles' surfaces, a distance that we would expect to preclude tunneling. In the next section we discuss the impact of electron tunneling on current-voltage $[I(V)]$ curves for these self-assembled systems.

C. Electron Transport Properties of Isolated Nanocrystals, and Nanocrystals Organized in 2D and 3D Superlattices (82)

When a single silver nanoparticle is deposited on a gold 111 substrate, the scanning tunneling spectroscopy measurement indicates a double-tunnel junction (Fig. 9A). Upon increasing the applied bias voltage V, the capacitor elements (defined by the tip-particle interface and particle-substrate interface) are charged up, and the detected current I is inititally close to zero. Above a certain threshold voltage, electrons can tunnel through the interfaces and the current increases with the

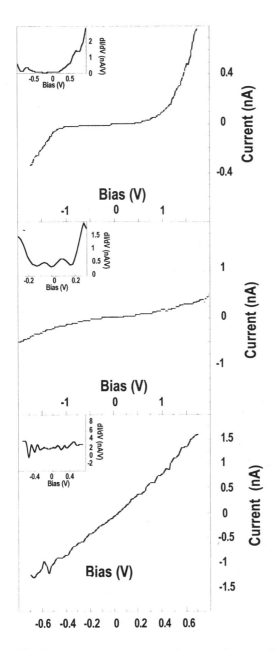

Fig. 9. Current-voltage curves for 5-nm-diameter silver nanoparticles on Au(111) substrate obtained with STM method: (A) Isolated silver nanocrystal. (B) 2D lattice. (C) FCC supracrystal. Insets in each case are corresponding derivative curves (dI/dV versus V).

applied voltage. A plot of dI/dV versus V (see inset of Fig. 9A) clearly shows that the derivative reaches zero at 0 V. The nonlinear profile of the $I(V)$ curve and zero dI/dV at zero bias voltage are characteristic of the well-known Coulomb blockade effect discussed in the Chapter 1.

The voltage range over which there is zero current is ca. 2 V and indicates that the dodecanthiolate ligands on the silver nanoparticles are sufficiently good electrical insulators that they act as tunneling barriers between the particles and the underlying substrate. It is also significant that no Coulomb staircase is observed in Fig. 9A; the present STS experimental setup employed a constant-current mode, which imposes a tip-particle distance to maintain a 1-nA current at a bias voltage of −1 V. The tip-particle distance in this case does not correspond to that needed for the observation of a Coulomb staircase. Since the particle-substrate distance is fixed by the dodecanethiolate coating, both tunnel junctions are characterized by fixed parameters. Similar Coulomb blockade behavior has been observed elsewhere (83,84).

Figure 9B shows the $I(V)$ curve for silver nanoparticles self-organized in a 2D superlattice on a Au(111) substrate. For large voltage bias, both positive and negative, the current is an order of magnitude lower than that observed for isolated particles. The Coulomb gap is small (ca. 0.45 V, compared to the 2 V seen in the isolated particles), and the overall $I(V)$ curve is more linear than that seen in Fig. 9A. This indicates an increase in the ohmic contribution to the current. In other words, the tunneling contribution to the total current decreases, and more conductive pathways between particles are established. The derivative curve shown in the inset of Fig. 9B indicates a metallic conduction behavior with $dI/dV \neq 0$ at zero bias voltage. From the $I(V)$ and dI/dV curves, it can be concluded that when the particles are arranged in a 2D lattice, the tunneling current exhibits both metallic and Coulomb contributions. This indicates that lateral tunneling between adjacent particles is very important and contributes to the total electron transport process.

The 2D lattice results presented here are in good agreement with data published elsewhere. For example, Rimbert et al. (85) measured at low temperature (40 mK) the $I(V)$ curves for aluminum island structures on aluminum substrates (linked by an Al_2O_3 spacer) when they were isolated and when they were organized in 2D lattices. The $I(V)$ curves they obtained were different for isolated islands and 2D lattices, with the latter system indicating interisland connections. Ohgi et al. studied the electron transport properties of 5-nm gold nanocrystals coated with dihexanethiolates (corresponding to a 1.4-nm ligand barrier) (86). They also found that the Coulomb gap decreases with increasing coverage of the particles on the substrate. Markovich et al. found similar effects in a Langmuir trough experiment on propanethiolate-coated silver nanoparticles; as the monolayer compression reduced the interparticle spacing to ca. 6 Å, the monolayer showed a metallic response to an applied voltage (87). The difference between these previous works and the present data is that the average interparticle spacing was either not controlled or rather small. In our studies the average distance between particles is large (1.8 to 2 nm) and the nanocrystals are self-organized in a close-packed hexagonal network. With the ordered packing, lateral electron transport takes place through insulating ligand shells even at these larger distances.

When silver nanocrystals are assembled in a 3D FCC structure, the $I(V)$ curve shows a linear ohmic behavior (Fig. 9C). The dI/dV curve is essentially flat, indicating metallic behavior without Coulomb staircases (inset 9C). Note that the voltage scale in Fig. 9C is smaller than that in Fig. 9A and B; the absolute current in the 3D structure is much higher than in the isolated silver particle and 2D monolayer structure. The ohmic behavior cannot be attributed to the coalescence of the particles. We have shown through TEM studies that the silver particles remain spherical (5,6,35). Furthermore, as discussed, the redispersal of the silver particles in the FCC structure into hexane solution leads to absorption spectra that are virtually the same as the original hexane solutions of silver nanocrystals used to deposit the assemblies. Similar behavior has been observed when HOPG is used as a substrate in place of gold (6,35).

Thus, we conclude that the FCC structure of the superlattice induces an increase in the tunneling rate via a decrease in resistance between the particles. The electron tunneling between adjacent particles becomes a major contribution to conduction, and the Coulomb blockade effect in the $I(V)$ curves is inhibited. The mechanism may involve an enhanced dipole-dipole interaction along the vertical (z) axis. When subjected to a voltage bias, the Fermi level of the individual nanocrystals is also perturbed. The details remain to be uncovered, but it is clear that a supercrystal of coated metal nanoparticles can behave as a metal.

D. Collective Magnetic Properties of Cobalt Nanocrystals

A comparison of the magnetic properties of isolated magnetic nanocrystals and 2D hexagonal assemblies reveals cooperative effects in the latter system. Figure 10A shows the magnetization curves for a 0.01% volume fraction solution of cobalt nanocrystals dispersed in hexane at a temperature of 3 K. While saturation magnetization for bulk cobalt is 162 emu/g, the value for the dilute cobalt nanoparticle solution is estimated to be ca. 120 emu/g (the estimation is based on an extrapolation of curves of H/M versus H). Saturation occurs at applied fields above 2 T. For the nanoparticle solution, the ratio of remanence to saturation magnetization (M_r/M_s) is equal to 0.45, and the hysteresis field is 0.11 T.

Figure 10B shows the magnetization curves for cobalt nanocrystals deposited on an HOPG substrate, for magnetic fields applied parallel (solid curve) and perpendicular (dashed curve). In the case of the 2D assembly, saturation magnetization is equal to 120 emu/g and occurs at ca. 0.75 T. Compared to the dilute solution, the coercive field decreases to 0.06 T. The observed changes cannot be

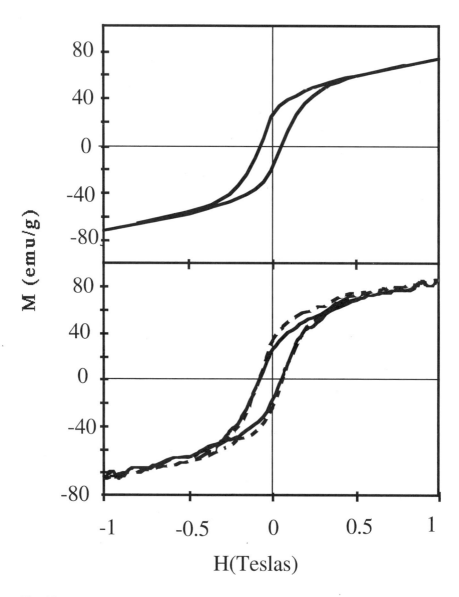

Fig. 10. Hysteresis magnetization loops obtained at 3 K for cobalt nanocrystals. (A) Dilute solution of particles in hexane (0.01% volume fraction in hexane). (B) Cobalt nanocrystals deposited on HOPG and dried under argon to prevent oxidation. (Solid curve) Magnetic field parallel to the substrate. (Dashed curve) Magnetic field perpendicular to substrate.

Self-Assemblies of Nanocrystals

attributed to coalescence of the nanocrystals, since TEM images taken over large areas of the sample show no evidence of this. Among the possible explanations for the change in magnetic properties (isolated particles versus 2D assemblies) is exchange coupling between adjacent particles. We exclude this mechanism because the edge-to-edge separation (2 nm) between cobalt particles is too large. However, magnetic dipolar coupling between particles may occur and lead to enhanced magnetization. There are two scenarios for dipole coupling enhancements. The first is an enhancement due to the long-range order of the 2D lattice and collective "flips" of the magnetic dipoles. The second model assumes that the distribution of cobalt particles on the substrate is random and that there exist local regions of high cobalt particle volume fraction. The two sets of hysteresis loops shown in Fig. 10B can help us distinguish between these two possibilities.

When the magnetic field is parallel to the substrate, the M_r/M_s ratio is 0.60 and the hysteresis loop is squarer than that obtained for particles in solution. When the field is perpendicular to the substrate, the loop is less square, and the M_r/M_s ratio decreases to 0.40. These results show that for a given saturation magnetization, the remanence magnetization markedly varies with the orientation of the magnetic field. We have recently calculated the dependence of the magnetization curves on field orientation, the simulations assuming 2D lattices of spheroidal particles of uniaxial symmetry with their easy axes randomly oriented (50,88). The easy axis is the direction favored by the magnetocrystalline anisotropy. The simulated curves resemble the experimental curves in Fig. 10B, showing variations when the magnetic field is applied parallel and perpendicular to the substrate. From these theory-experiment comparisons, it seems reasonable to conclude that the collective magnetic properties observed when the cobalt particles are arranged in a 2D lattice are due to an increase in magnetic dipole-dipole interactions.

VI. CONCLUSIONS

We have shown that metal and metal oxide or sulfide nanocrystals can self-assemble to form 2D and 3D superlattices, provided that care is taken to maximize monodispersity. In the case of 2D networks of nanocrystals, the final structure is determined by particle substrate interactions as well as interparticle interactions. The latter factors are, in turn, determined largely by the structure of the structure of the surface modifying group (e.g., alkanethiolate chain length and dynamics). The optical absorption spectra of metal nanoparticle systems show systematic changes as measurements are made on dilute solutions of particles, 2D structures, and 3D structures. The electronic and magnetic properties are also dependent on the degree of self-assembly; we have observed properties that appear to arise from the collective response from nanocrystals arranged in ordered lattices, as opposed to random aggregates.

ACKNOWLEDGMENTS

I would like to thank Drs. A. Courty, L. Motte, T. Ngo, C. Petit, V. Russier, and A. Taleb for their fruitful contributions to this work.

REFERENCES

1. L Motte, F Billoudet, MP Pileni. J Phys Chem 99:16425, 1995.
2. CB Murray, CR Kagan, MG Bawendi. Science 270:1335, 1995.
3. L Motte, F Billoudet, E Lacaze, MP Pileni. Adv Mater 8:1018, 1996.
4. L Motte, F Billoudet, E Lacaze, J Douin, MP Pileni. J Phys Chem B 101:138, 1997.
5. A Taleb, C Petit, MP Pileni. Chem Mater 9:950, 1997.
6. A Taleb, C Petit, MP Pileni. J Phys Chem B 102:2214, 1998.
7. SA Harfenist, ZL Wang, MM Alvarez, I Vezmar, RL Whetten. J Phys Chem 100:13904, 1996.
8. SA Harfenist, ZL Wang, RL Whetten, I Vezmar, MM Alvarez. Adv Mater 9:817, 1997.
9. SW Chung, G Markovich, JR Heath. J Phys Chem B 102:6685, 1998.
10. BA Korgel, S Fullam, S Connely, D Fitzmaurice. J Phys Chem B 102:8379, 1998.
11. BA Korgel, D Fitzmaurice. Adv Mater 10:661, 1998.
12. ZL Wang, SA Harfenist, RL Whetten, J Bentley, ND Evans. J Phys Chem B 102:3068, 1998.
13. ZL Wang, SA Harfenist, I Vezmar, RL Whetten, J Bentley, ND Evans, KB Alexander. Adv Mater 10:808, 1998.
14. PC Ohara, JR Heath, WM Gelbart. Angew Chem Int Engl 36:1078, 1997.
15. T Vossmeyer, S Chung, WM Gelbart, JR Heath. Adv Mater 10:351, 1998.
16. K Vijaya Sarathy, G Raina, RT Yadav, GU Kullkarni, CNR Rao. J Phys Chem B 101:9876, 1997.
17. PC Ohara, DV Leff, JR Heath, WM Gelbart. Phys Rev Lett 75:3466, 1995.
18. S Murthy, ZL Wang, RL Whetten. Philos Mag Lett 75:321, 1997.
19. MJ Hostetler, JJ Stokes, RW Murray. Langmuir 12:3604, 1996.
20. A Badia, L Cuccia, L Demers, F Morin, RB Lennox. J Am Chem Soc 119:2682, 1997.
21. RL Whetten, JT Khoury, MM Alvarez, S Murthy, I Vezmar, ZL Wang, CC Cleveland, WD Luedtke, U Landman. Adv Mater 8:428, 1996.
22. M Brust, D Bethell, DJ Schiffrin, CJ Kiely. Adv Mater 9:797, 1995.
23. J Fink, CJ Kiely, D Bethell, DJ Schiffrin. Chem Mater 10:922, 1998.
24. CJ Kiely, J Fink, M Brust, D Bethell, DJ Schiffrin. Nature 396:444, 1998.
25. LO Brown, JE Hutchison. J Am Chem Soc 121:882, 1999.
26. XM Lin, CM Sorensen, KJ Klabunde. Chem Mater 11:198, 1999.
27. TG Schaff, MN Hafigullin, JT Khoury, I Vezmar, RL Whetten, WG Cullen, PN First, C Gutierrez-Wing, J Ascensio, MJ Jose-Ycaman. J Phys Chem B 101:7885, 1997.
28. C Petit, A Taleb, MP Pileni. J Phys Chem B 103:1805, 1999.
29. C Petit, A Taleb, MP Pileni. Adv Mater 10:259, 1998.

30. JS Yin, ZL Wang. J Phys Chem B 101:8979, 1997.
31. JS Yin, ZL Wang. Phys Rev Lett 79:2570, 1997.
32. S Maenosono, CD Dushkin, S Saita, Y Yamaguchi. Langmuir 15:957, 1999.
33. L Motte, MP Pileni. J Phys Chem B 102:4104, 1998.
34. WD Luedtke, U Landman. J Phys Chem 100:13323, 1996.
35. A Taleb, V Russier, A Courty, MP Pileni. Phys Rev B 59:13350, 1999.
36. V Russier, MP Pileni. Surf Sci 425:313, 1999.
37. DJ Mitchell, BW Ninham. J Chem Soc Faraday Trans 2 77:601, 1981.
38. DF Evans, DJ Mitchell, BW Ninham. J Phys Chem 88:6344, 1984.
39. DF Evans, DJ Mitchell, BW Ninham. J Phys Chem 90:2817, 1986.
40. SA Safran, LA Turkevich, PA Pincus. J Phys Letters 45:L69-L74, 1984.
41. M Kotlarchyk, JS Huang, SH Chen. J Phys Chem 89:4382, 1985.
42. MP Pileni, ed. Reactivity in Reverse Micelles. Amsterdam:Elsevier, 1989.
43. M Zulauf, HF Eicke. J Phys Chem 83:480, 1979.
44. I Lisiecki, P Andre, A Filankembo, C Petit, J Tanori, T Gulik-Krzywicki, BW Ninham, MP Pileni. J Phys Chem 103:9168, 1999.
45. I Lisiecki, P Andre, A Filankembo, C Petit, J Tanori, T Gulik-Krzywicki, BW Ninham, MP Pileni. J Phys Chem 103:9168, 1999.
46. C Petit, TK Jain, F Billoudet, MP Pileni. Langmuir 10:4446, 1994.
47. MP Pileni. J Phys Chem 97:6961, 1993.
48. MP Pileni. Langmuir 13:3266, 1997.
49. WL Wilson, PF Szajowski, LE Brus. Science 262:1242, 1993.
50. V Russier, C Petit, J Legrand, MP Pileni. Phys Rev B 62:3910, (2000).
51. J Smit, HPJ Win. Adv Electron Phys 6:53, 1954.
52. L Motte, E Lacaze, M Maillard, MP Pileni. Langmuir 16, 3803 (2000).
53. A Oron, SH David, SG Bankoff. Rev Mod Phys 69: 931, 1997.
54. MJ Hostetler, JJ Stokes, RW Murray. Langmuir 12:3604, 1996.
55. A Badia, W Gao, S Singh, L Demers, L Cuccia, L Reven. Langmuir 12:1262, 1996.
56. A Ulman. Chem Rev 96:1533, 1996.
57. GM Whitesides, JP Mathias, CT Seto. Science 254:1312, 1991.
58. GK Jennings, PE Laibinis. Langmuir 12:6173, 1996.
59. PE Laibinis, MA Fox, JP Folkers, GM Whitesides. Langmuir 7:3167, 1991.
60. F Michel, MP Pileni. Langmuir 10:390, 1994.
61. CD Bain, EB Troughton, YT Tao, J Evall, GM Whitesides, RG Nuzzo. J Am Chem Soc 111:321, 1989.
62. JR Heath, CM Knobler, DV Leff. J Phys Chem B 101:189, 1997.
63. M Maillard, L Motte, T Ngo, MP Pileni (submitted).
64. A Courty, C Fermon MP Pileni. (submitted for publication).
65. T Ngo, MP Pileni. Adv Mater. 12:276 (2000).
66. The size distribution was determined using a log-normal method. See, for example, R Kaiser, G Miskolczy. J Appl Phys 41:1064, 1970.
67. B Jeyadevan, I Nakatani. J Mag Mag Mater 201:62, 1999.
68. HC Yang, IJ Jang, HE Hong, JM Wu, YC Chiou, CY Hong. J Mag Mag Mater 201:313–316, 1999.
69. HC Yang, IJ Jang, HE Hong, JM Wu, YC Chiou, CY Hong. J Mag Mag Mater 201:317, 1999.

70. CF Bohren, DR Huffman. Absorption and Scattering of Light by Small Particles. New York: Wiley, 1983.
71. KP Charle, F Frank, W Schulze. Ber Bunsenges Phys Chem 88:354, 1984.
72. G Mie. Ann Phys 25:377, 1908.
73. JA Creighton, DG Eaton. J Chem Soc Faraday Trans 2 87:3881, 1991.
74. H Hovel, S Fritz, A Hilger, U Kreibig, M Vollmer. Phys Rev B 48:18178, 1993.
75. BNJ Persson. Surface Science 281:153, 1993.
76. R Ruppin. Surface Science 127:108, 1983.
77. C Petit, MP Pileni. J Phys Chem 97:12974, 1993.
78. MA Alvarez, JT Khoury, TG Schaaf, MN Shafigullin, I Vezmar, RL Whetten. J Phys Chem B 101:3706, 1997.
79. T Yamaguchi, M Ogawa, H Takahashi, N Saito, E Anno. Surface Science 129:232, 1983.
80. SW Kennerly, JW Little, RJ Warmack, TL Ferrell. Phys Rev B 29:2926, 1984.
81. PA Bobbert, JV Leiger. Physica A 147:115, 1987.
82. A Taleb, F Silly, O Gusev, F Charra, MP Pileni. Adv Mat. 12:119, (2000).
83. C Petit, T Cren, D Roditchev, W Sacks, J Klein, MP Pileni. Adv Mater 11:1358, 1999.
84. U Simon. Adv Mater 10:1487, 1998.
85. AJ Rimbert, TR Ho, J Clarke. Phys Rev Lett 74:4714, 1995.
86. T Ohgi, HY Sheng, H Nejoh. Appl Surf Sci 130:919, 1998.
87. G Markovich, CP Collier, JR Heath. Phys Rev Lett 80:3807, 1998.
88. V Russier, C Petit, J Legrand, MP Pileni. Appl Surf Sci (in press).

10
Electrodeposition of Metal Nanoparticles on Graphite and Silicon

Sasha Gorer, Hongtao Liu, Rebecca M. Stiger, Michael P. Zach, James V. Zoval, and Reginald M. Penner
University of California, Irvine, Irvine, California

I. INTRODUCTION

Metal nano- and microparticles are technologically important electrocatalysts. For example, nanoparticles having a composition of $Pt_{0.5}Ru_{0.5}$ are the most efficient catalysts for the oxidation of methanol in the direct methanol fuel cell (1), silver nanoparticles are excellent catalysts for the reduction of oxygen in basic aqueous solutions, and nanoparticles of pure platinum are efficient catalysts for oxidizing a variety of molecules, including oxygen (in acidic solutions) (2) and many organic molecules [among these organic acids (3,4), and alkenes (5)]. On semiconducting TiO_2 surfaces, platinum nanoparticles have also been employed as photocatalysts to effect the light-driven splitting of water into H_2 and O_2 on semiconducting TiO_2 surfaces.

Ironically, although metal nanoparticles are important electrocatalysts, electrodeposition has rarely been employed by electrochemists to prepare metal particles. Instead, nanometer-scale metal particles have usually been synthesized by impregnation of a catalyst support (e.g., porous carbon) with an aqueous metal salt solution followed by drying and gas-phase reduction of the dispersed salt at high temperature in H_2. A second route to metal nanoparticles involves the evaporative deposition of metal atoms onto a surface from a hot filament source in vacuum ["physical vapor deposition (PVD)"]. Impregnation is popular because it provides a straightforward method for producing catalytically active metal nanoparticles in high-surface-area support media. PVD, which is employed primarily for fundamental investigations of catalysis, is capable of producing metal particles of high purity that are also narrowly distributed in diameter. One reason electrodeposition has been virtually excluded from consideration is because it has been impossible to produce dispersions of metal particles exhibiting a reasonable degree of size monodispersity.*

Why is it difficult to produce dimensionally uniform (6) structures electrochemically? There are two main reasons: First, the nucleation of metal particles is often "progressive." This means that new particles are formed continuously during the application of a voltage pulse. If nucleation is progressive, the number density of particles on the surface increases as a function of time, and late nucleating particles are small compared with early nucleating particles and broad particle size distributions are obtained. In selected particle growth experiments, nucleation can be temporally discrete† or "instantaneous" instead of progressive. In these experiments, more subtle mechanisms are responsible for the development of particle size dispersion. The most important of these we have termed (6) *interparticle coupling*. The growth of two particles located in close proximity on the surface becomes coupled after sufficient growth time has elapsed. Once coupled, these two particles share a flux of metal ions from solution, and for this reason these two particles grow at a slower rate than metal particles located in isolation on the same surface. In order for interparticle coupling to contribute to the development of particle size dispersion, a distribution of nearest-neighbor distances must exist on a surface. This is usually the case: on most surfaces nucleation is spatially random. We believe interparticle coupling is the most important mechanism for the development of particle size dispersion in electrodeposition experiments in which nucleation is instantaneous.

In this chapter we focus on the electrodeposition of mesoscale‡ metal particles that are uniform in size (RSD_{dia}§ $< 15\%$). Since dimensional uniformity is of central importance, we shall limit our attention to electrodeposition experiments in which nucleation is instantaneous. We begin by describing the "zero-order" experiment, which involves the electrodeposition of metal nanoparticles on graphite surfaces using large overpotentials ($\eta \approx -500$ mV) and dilute metal plating solutions ($[M^{n+}] = 1.0$ mM). These experiments demonstrated that metal

---

* Throughout this chapter the terms "dimensional uniformity" and "size monodispersity" applied to the particle diameter refer to the proximity of the relative standard deviation of the particle diameter (RSD_{dia}, the ratio of the mean diameter to the standard deviation of the diameter) to zero.
† "Temporally discrete" nucleation refers to a separation in time between the nucleation and growth phases of particle formation.
‡ "Mesoscale" will refer to a critical dimension from 10 Å and 1.0 μm.
§ The relative standard deviation of the particle diameter. RSD_{dia} is calculated by dividing the mean particle diameter by the standard deviation of the diameter.

nanoparticles could be electrodeposited with some size selectivity. In addition, they proved to us that size dispersion develops even when nucleation is instantaneous. A deeper understanding of the other mechanisms at work required the analysis of computer simulations of particle growth using the Brownian dynamics method. This exercise alerted us to the existence of interparticle coupling and suggested electrodeposition strategies that might prevent interparticle coupling from degrading particle uniformity. Finally, we describe three recently developed strategies for electrodepositing metal nanoparticles that are narrowly dispersed in diameter.

II. VOLMER-WEBER METAL NANOPARTICLE DEPOSITION USING LARGE OVERPOTENTIALS

Metal electrodeposition on gold, platinum, or silver electrodes typically proceeds via a Stransky-Krastanov (SK) mechanism if the overpotential is -20 mV or more (7). In SK growth processes, the initial deposition of metal occurs via an atomic layer-by-layer process beginning with an underpotentially deposited (UPD) monolayer. At some critical thickness of the electrodeposited layer, islands that are more than one atomic layer in height begin to form atop this continuous metal multilayer.

A much different mechanism is seen if the electrode surface exhibits a low surface free energy. Two such surfaces are graphite and hydrogen-terminated silicon. The wetting of these coordinatively saturated surfaces by electrodeposited metal is energetically unfavorable, and the layer-by-layer deposition of metal normally seen on gold and platinum is never observed. Instead, multilayer metal islands, or particles, are promptly formed. Although this Volmer-Weber (7) mechanism of deposition had been recognized by high-vacuum-surface scientists for years (8), the first clear example of VW growth in an electrodeposition experiment was reported by Zoval et al. (9) in 1995. In this paper, involving silver electrodeposition on graphite, it was reported that three-dimensional metal particles were observed using noncontact atomic force microscopy (NC-AFM) even when the total quantity of electrodeposited metal amounted to 1/100 of a single atomic layer.* This constituted clear evidence for Volmer-Weber growth. Qualitatively similar observations were reported for the deposition of other metals, including platinum (10), copper (11), cadmium (12,13), and zinc (14) on graphite. For all of these metals, the electrodeposition of metal from a 1.0 mM solution of M^{n+} (i.e., Ag^+, Cd^{2+}, etc.) using $\eta = -400-500$ mV for 10–100 ms produced metal

---

* The electrodeposition of submonolayer quantities of metal required the use of extremely short plating pulses having a duration of 10–100 ms and dilute metal plating solutions ([Mn^+] ≈ 1.0 mM).

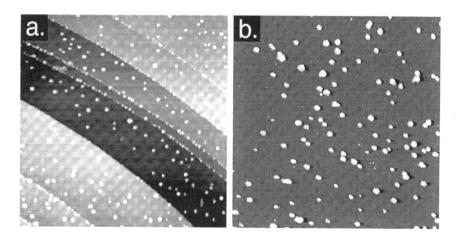

Fig. 1. Noncontact atomic force microscope (NC-AFM) images of graphite and silicon surfaces following the electrodeposition of metal nanoparticles. (a) A 3 × 3 μm image of platinum nanocrystals having a mean diameter of 52 Å on graphite (From Ref. 10). (b) A 3 × 3 μm image of silver nanocrystals having a mean diameter of 75 Å on Si(100) (From Ref. 15). (Reproduced with permission from Langment, 1999, 15, 790. Copyright 1999, Am. Chem. Soc.)

nanoparticles at a coverage of 5×10^8–5×10^9 cm^{-2}. NC-AFM images of these surfaces revealed that metal nanoparticles nucleated in a pseudorandom fashion on the graphite surface: as shown in Fig. 1a, step edges were densely nucleated with metal nanoparticles whereas the nucleation density on atomically smooth terraces was approximately 5×10^8 cm^{-2}. The nucleation density for platinum on graphite is independent of the deposition time over the range of deposition times from 10 − 100 ms. This observation suggests that nucleation on graphite is instantaneous.* Corroborating this hypothesis are measurements of the mean particle diameter as a function of deposition charge which are consistent with the growth of a fixed number of particles (10).

Similar behavior is seen for the electrodeposition of silver onto hydrogen-terminated Si(100) [henceforth, H-Si(100)] from acetonitrile-based electrolytes (15). An NC-AFM image of an H-Si(100) surface on which 0.037 monolayer of silver has been electrodeposited is shown in Fig. 1b. In contrast to graphite, preferential nucleation at step edges was not observed for silver deposition on Si(100), and a buildup of the nucleation density—indicative of progressive nucleation—was observed during the first 20 ms of growth (15).

---

* "Instantaneous" nucleation refers to a scenario in which the time interval during which nucleation occurs is much shorter than the subsequent particle growth phase.

On graphite and silicon surfaces, the standard deviation of the particle diameter, σ_{dia}, increases as a function of the deposition time. This trend is apparent in the histograms in Fig. 2. For experiments conducted at silicon surfaces, there is nothing surprising about this trend since nucleation is progressive. On graphite surfaces, on the other hand, nucleation is instantaneous and a narrowing of the particle size distribution as a function of time is expected. The origin of distribution narrowing was explained by Reiss (16) and LaMer (17) in the late 1950s and can be understood as follows: the growth law for an individual hemispherical metal particle is $r = kt^{1/2}$, where r is the particle radius, k is a collection of constants which depend on the identity of the deposited material, and t is the growth

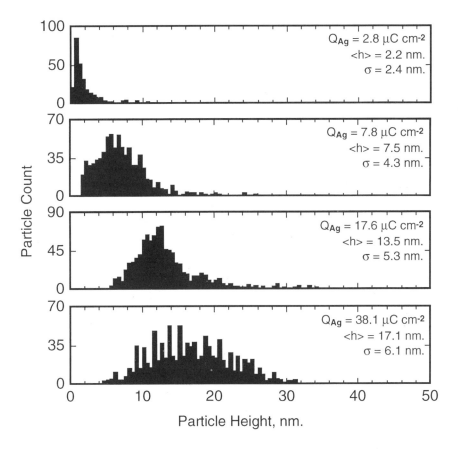

Fig. 2. Histograms of silver particle heights derived from NC-AFM images. These silver nanoparticles were electrodeposited onto hydrogen-terminated Si(100) surfaces from an acetonitrile-based plating solution (From Ref. 21). (Reproduced with permission from R. Stiger, B. Craft, and R. M. Penner, Langmuir, 1999, 15, 790. Copyright 1999, Am. Chem. Soc.)

duration. The growth rate is therefore $dr/dt = 0.5kt^{-1/2}$, a quantity which is larger for small particles than for large ones. Thus, a distribution of particles becomes narrower as a function of time because particles at the small end of this distribution "catch up to" the particles on the high end of the distribution as growth is propagated in time. Colloid chemists have exploited this fact to prepare extremely size monodisperse suspensions of particles (metal, semiconductor, polymer, etc.) for the last 50 years (excellent reviews of colloid growth are provided in Refs. 18 and 19). Although this literature pertains to the growth of colloidal suspensions, in principle these arguments are equally valid for particles which are confined during growth to a flat surface (20). With regard to metal particle growth on graphite surfaces, then, the obvious question is, "why do particle size distributions broaden as a function of time?"

III. UNDERSTANDING PARTICLE SIZE DISPERSION IN ELECTROCHEMICAL VOLMER-WEBER PARTICLE GROWTH

The origin of distribution broadening for the electrodeposition of metal particles on graphite could originate in the chemistry of these particles or in the physics of the deposition experiment. If particle chemistry is the culprit, then one might not expect to see distribution broadening for all metals; however, this is invariably the case, and distribution broadening is seen for metals ranging from reactive (e.g., zinc) to unreactive (e.g., platinum).

Thus, closer scrutiny is brought to bare on the physics of particle growth. The first question to ask in this regard is whether interparticle interactions are likely to influence the growth rate of particles. Another way to pose this question is to ask, "does the depletion layer for neighboring particles on the surface over lap"? The answer to this question depends on the radius of the depletion layer as a function of time, $r_d(t)$, which can be calculated as follows. Under steady-state conditions the concentration at the surface of a hemispherical electrode, $C(r)$, obeys the relationship (21)

$$C(r) = C^* \left[1 + \frac{r_o}{r} \right] \tag{1}$$

where C^* is the bulk concentration of a species undergoing electrolysis at the electrode surface, and r_0 is the electrode radius. Equation (1) assumes that this electrolysis reaction is diffusion controlled [i.e., $C(r_0) = 0$]. Under these conditions the time dependence of the radius is given by (22)

$$r_0(t) = \frac{(2DC^* Mt)^{1/2}}{\rho^{1/2}} \tag{2}$$

Electrodeposition of Metal Nanoparticles

where D is the diffusion coefficient, M is the atomic weight of the metal, and ρ is the density of the metal. Equation (2) also assumes that growth is diffusion controlled. Combining Eq. (1) and Eq. (2) provides an expression for the radius of the depletion layer at the surface of a growing metal particle, r_d, as a function of growth time. If r_d is defined by $C(r_d)/C^* = 0.95$, then $r_d = r_0/0.05$. The deposition time at which depletion layers for neighboring metal particles can be expected to overlap, t_c, can be calculated by equating the distance between nearest neighbors, a, and the quantity $2r_0 + 2r_d$. For a hexagonal array of particles, for example,

$$t_c = \frac{1}{\sqrt{3}(42r_0)^2} = \frac{1}{\sqrt{3}(42)^2}\left[\frac{\rho}{DC^*MN}\right]$$
$$= 3.2730 \times 10^{-4}\left[\frac{\rho}{DC^*MN}\right] \quad (3)$$

where the particle areal density, N, is given by $2/\sqrt{3}a^2$. Domains of coupled and uncoupled growth, delineated by t_c, are shown graphically for silver as a function of N in Fig. 3. The coordinates of growth time and N corresponding to the experiments discussed earlier (indicated by a rectangle labeled 1) are located within the region in which coupled growth is expected. It is therefore certain that interparticle interactions are important in these experiments. What role do these interparticle interactions play in the genesis of particle size dispersion?

Computer simulations can provide an answer to this question. We have employed Brownian dynamics simulations to simulate particle growth in the case where particles nucleate instantaneously, grow at a diffusion-controlled rate, and are confined during growth to a flat surface. We explicitly consider growth durations that are long compared with the interaction time, t_i, between nearest neighbors on the surface, and we have modeled a range of experimentally relevant nucleation densities. In these simulations, ensembles of up to 200 hemispherical metal particles were grown atop a 10^{-9}-cm^{-2} planar surface. At the beginning of the simulation, these particles are single atoms and the 0.5-ms duration of the simulation permits growth to a mean diameter of 3 nm from a 10^{-3} M "solution" of metal ions. Since the number of nuclei in each simulation is fixed at the beginning of the simulation, nucleation is rigorously instantaneous. Each metal particle in these ensembles was explicitly modeled so that the development of size dispersion for the ensemble could be monitored as a function of the deposition time.

Typical plots of the deposition current and the standard deviation of the particle diameter are shown for three different nucleation densities in Fig. 4. Data for three particle densities, corresponding to 5, 20, and 100 nanoparticles on a surface having a total area of 10^{-9}-cm^{-2}, are presented. Several features of these simulated data are worth noting. The onset of interparticle coupling is signaled by a peak in the current-versus-time transient. The location of this peak in

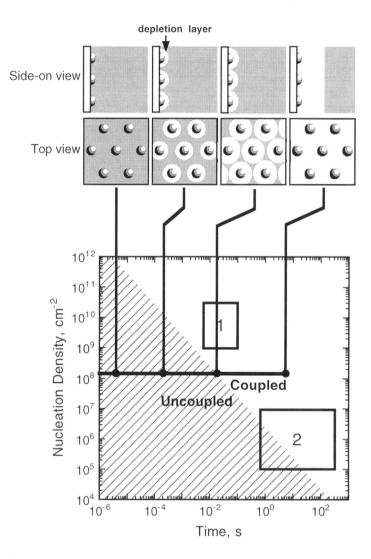

Fig. 3. Plot of the critical time (corresponding to the onset of interparticle coupling) as a function of nucleation density of particles and the deposition time. The critical time indicated in this plot has been calculated for electrodeposition experiments involving the diffusion-controlled growth of particles arranged in a hexagonal array. The rectangle labeled 1 indicates the region of parameter space relevant to the experiments described in Sec. II. The rectangle labeled 2 indicates the parameters encountered in the slow growth experiments discussed in Sec. IV.B. Shown at top is a schematic diagram depicting the spatial relationship between the depletion layers developed at neighboring particles in the array.

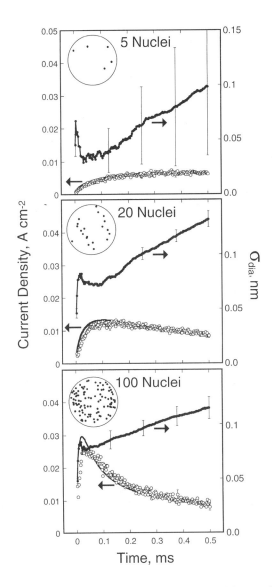

Fig. 4. Current density (open circles) and the standard deviation of the particle radius, σ_R (solid circles) as a function of time for *random ensembles* at five nucleation densities. These BD simulation results represent the mean of three simulations, each of which involved a different silver plating solution and different positions for nuclei on the surface. The standard deviation of the current and σ_R are indicated by the error bars ($\pm 1\sigma$) which are shown at five times. Data sets are labeled with the nucleation density in units of 10^9 cm^{-2} (From Ref. 6). (Reproduced with permission from J. Phys. Chem. B, 1999, 103, 7643. Copyright 1999, Am. Chem. Soc.)

time is in excellent agreement with the predictions of Eq. (3). At experiment times, $t < t_c$, the excursion of the depletion layer is small compared with the nearest-neighbor spacing and silver particles grow independently of one another. The total current for the ensemble of silver particles is therefore equal to the current for a single, isolated particle multiplied by the number of particles, N, growing on the surface (23):

$$i(t) = \frac{\pi F N (2 D_{Ag^+} C^*_{Ag^+})^{3/2} (M_{Ag} t)^{1/2}}{\rho_{Ag}^{1/2}} \quad (4)$$

Equation (4) predicts that the current in this time domain increases and is proportional to $t^{1/2}$. At times $t > t_c$, depletion layers at adjacent particles merge and an approximately planar diffusion layer blankets the entire geometric surface area of the electrode. The current in this "coupled" time regime is exactly the same as that at a planar electrode having the same geometric area, A, and is given by the familiar Cottrell equation (27):

$$i(t) = \frac{nFAD_{Ag^+}^{1/3} C^*_{Ag^-}}{(\pi t)^{1/2}} \quad (5)$$

Equation (5) predicts that the current decays with time and is proportional to $t^{-1/2}$. The presence of a peak in the current-time plot is therefore required to connect the increasing current at short times with the decaying current which is observed at long times.

Also plotted in Fig. 4 is the standard deviation of the particle diameter, σ_{dia}, as a function of deposition time. Three distinct domains can be distinguished in the σ_{dia}-versus-time transients: Initially, σ_{dia} rapidly increases and peaks at ≈ 100 μs; at intermediate times (100 μs $< t < t_c$), σ_{dia} decreases nearly until t_c; finally for $t > t_c$, σ_{dia} increases approximately linearly ("divergent growth"). The origin of this behavior for σ_{dia} has been discussed in detail (6). Briefly, the initial rapid increase of σ_{dia} derives from the stochastic nature of deposition at short times; it is unrelated to the arrangement of particles on the surface (6). In this time regime, σ_{dia} is proportional to $N_{dep}^{-1/6}$ (where N_{dep} is the mean aggregation number for particles on the surface) (6). In the intermediate time domain, the distribution narrowing that is observed mimics the behavior of colloidal particles growing in solution. In other words, distribution narrowing in this time domain is a consequence of the fact that $r(t) = kt^{1/2}$ (16,17). Divergent growth at longer times is caused by "interparticle coupling": the flux experienced by a particular particle is a function of its proximity to other particles on the surface. If a particle is located in proximity to one or more neighboring particles, then its growth rate is retarded relative to particles that are better isolated on the surface. This phenomenon can be viewed from another perspective: Following t_c and the transition to a planar depletion layer, the flux of metal ions per unit area on the surface is spatially uniform. Nuclei are randomly located on the surface, however, and the nucleation density is locally variable. Densely nucleated areas can therefore be expected to grow more slowly than regions of the same size (and sharing the same planar flux) but encompassing a smaller number of nanoparticles. Interparticle coupling should lead to a recognizable correlation between the mean radius of nearest neighbors on the surface and the distance separating them, and a statistical analysis of the data confirms (6) that this correlation exists.

It is important to appreciate that the interparticle coupling seen in Brownian dynamics simulations like those of Fig. 4 can account for all of the size dispersion observed in experiments involving the growth of silver or platinum nanoparticles for relatively short time (e.g., 10 ms). For example, the slope of the σ_{dia}-versus-time plot for $N = 5 \times 10^9$ cm^{-2} (Fig. 4, top) is approximately 1.0 Å ms^{-1}. If this linear increase can be extrapolated to 10 ms, a σ_{dia} of 10 Å is predicted. As shown in Fig. 2, the experimentally observed value (for platinum particles at a nucleation density of 2×10^9 cm^{-2}) is 9 Å.

Once the mechanism by which size dispersion develops in a growing ensemble of particles is understood, the growth conditions for metal nanoparticles can be engineered to yield improved size monodispersity. We address this issue next.

IV. THREE NEW ROUTES TO SIZE MONODISPERSE METAL NANOPARTICLES

If interparticle coupling is the primary mechanism by which size dispersion develops in metal nanoparticle ensembles, how might it be defeated in order to improve the size monodispersity of our metal particles? A clear prediction of the BD simulation results is that particle size monodispersity can be improved by locating nucleation sites in a two-dimensional geometric array. In such an array, the number of nearest neighbors and the interparticle distances are the same for every particle in the array. Thus, although interparticle coupling still occurs, it does not introduce flux inhomogeneities from particle to particle in the array. BD simulations (6) confirm that extremely narrow particle size distributions are attainable for the electrochemical growth of particles in such arrays. Unfortunately, the creation of a regular array of nuclei requires the imposition of a periodic array of nucleation sites on the surface, for example, using lithography. Can interparticle coupling be eliminated without resorting to such an array?

In principle, the answer to this question is yes. Possible remedies include the following: First, lower the nucleation density. Fig. 3 reveals that at sufficiently small values of the nucleation density, no interparticle coupling will occur. Provided nucleation is temporally discrete, even randomly nucleated particles at such nucleation densities are expected to exhibit convergent growth. Second, grow slow. If the growth rate is decreased from diffusion control, the concentration at the surface of the metal particle will be nonzero, and the steady-state radius of the diffusion layer will be reduced in size. In fact, the steady-state radius of the depletion layer is inversely proportional to the quantity $C^* - C(r_0,\infty)$:

$$C(r,\infty) = C^* - \frac{[C^* - C(r_0,\infty)]r_0}{r} \qquad (6)$$

where r_0 is the radius of the metal particle, $C(r_0,\infty)$ is the steady-state concentration of metal ion at the surface of the particle, and r is the distance measured from the center of the metal particle. Finally, move the particles on the surface during growth. If this can be accomplished, then the growth of particles on a surface can be made to resemble the growth of colloidal particles in solution. Because particles in motion will not persist in proximity to another particle during growth, flux inhomogeneities between particles are temporally averaged and the average growth rate for particles becomes more similar.

A. Pulsed Electrodeposition

We have developed three experimental methods for preparing metal particle dispersions exhibiting excellent size monodispersity. The first of these involves the application of a train of negative-going voltage pulses approximately 10 ms in duration separated by much longer (e.g., 1 s) rest periods at open circuit (Scheme 1a). The objective of pulsed electrodeposition is to reduce the average growth rate of particles which, as already indicated, is expected to reduce the interactions between them.

To date, this pulsed electrodeposition strategy has been applied to the growth of cadmium nanoparticles (25). The success of this approach has been assessed by examining the optical properties of cadmium sulfide nanoparticles synthesized from these cadmium precursor particles (12,13). Since the conversion of cadmium nanoparticles to cadmium sulfide nanoparticles occurs on a particle-by-particle basis, the optical properties of CdS particles prepared by this electrochemical/chemical route (26) provide an indirect indication of the size dispersion of the parent cadmium nanoparticles. To be more specific, the photoluminescence (PL) spectrum for ≈100-Å-diameter CdS "quantum dots" is inhomogeneously broadened by the size dispersion of these particles: Particles at the small end of the distribution emit light which is bluer than particles at the large end of the size distribution. This size dependence of the emission energy is a characteristic of the quantum confinement of excitons in these diminutive particles (see, for example, Refs. 27–29). Consequently, improved size monodispersity for these CdS nanoparticles will translate into a reduced PL emission linewidth.

Typical PL spectra for CdS nanoparticles prepared using a single cadmium plating pulse, and particles prepared using multiple plating pulses are compared

Electrodeposition of Metal Nanoparticles

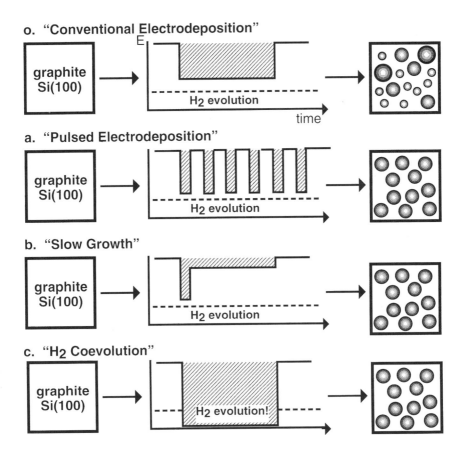

Scheme 1. Three experimental methods for electrodepositing metal nano- and microparticle dispersions exhibiting excellent size monodispersity.

in Fig. 5 (25). The PL emission lines of the spectra shown are blue-shifted from the CdS single crystal (spectrum "sc") by ≈130 meV, indicating that the mean diameters of the CdS cores are ≈42 Å. However, the spectrum for CdS particle prepared using a single cadmium plating pulse (labeled "sp") is 125 meV in width, whereas the spectrum for CdS particles prepared using a sequence of 10-ms voltage pulses (labeled "x") exhibits a linewidth of just 18 meV. The narrowing of the PL emission line is directly attributable to the narrowing of the particle size distribution for the CdS particles. Even narrower PL emission lines—down to 15 meV—were obtained using this approach (25). Although we estimate that approximately 300,000 CdS particles are present within the detection volume of our spectrometer in these experiments, the "multipulse" PL emission lines are nearly

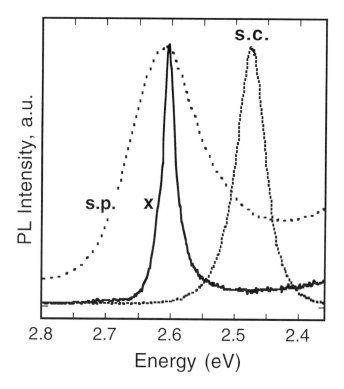

Fig. 5. Photoluminescence spectra of CdS/S NCs on the graphite (0001) surface at 20 K using $h\nu_{ex} = 3.53$ eV. The PL spectra for CdS/S NCs prepared from Cd° NCs deposited using a single, 100-ms plating pulse (spectrum "s.p."), multipulse CdS/S NCs deposited using a sequence of 10 × 10 ms Cd° plating pulses (spectrum "x"), and a macroscopic (0001) oriented CdS single crystal (spectrum "s.c.") are compared (From Ref. 25). (Reproduced with permission from J. Phys. Chem. B, 1999, 103, 5750. Copyright 1999, Am. Chem. Soc.)

as narrow as the PL emission spectrum of *single* CdS particles (30,31). Unfortunately, a direct measurement of the particle size distribution for CdS nanoparticles is not possible (using TEM, for example) because these particles are covered with an amorphous sulfur shell (12). The fundamental validity of the pulsed electrodeposition strategy has nevertheless been demonstrated.

B. Electrochemical Slow Growth

In principle, growing nanoparticles at a reduced deposition overpotential (Scheme 1b) has much the same effect as applying multiple plating pulses. In practice, there are two beneficial effects of this strategy: A lower nucleation density is obtained, and the diffusive coupling between particles is reduced. These two effects combine to produce metal nanoparticle dispersions exhibiting excellent particle size monodispersity.

Typical scanning electron micrographs (SEMs) of silver nano- and microcrystals deposited from acetonitrile solutions on graphite are shown in Fig. 6 (32). The particle areal density in these experiments varies from 5×10^5 cm^{-2} to 1.5×10^7 cm^{-2}—two orders of magnitude smaller than in experiments involving deposition at large overpotentials as described in Sec. II. The SEM images of the silver particles produced using this technique also reveal facets that are nearly as large as the particles themselves. This observation suggests that these particles are single crystals.

Particle size histograms (32) (Fig. 7) reveal that the absolute standard deviation of the particle diameter, $\sigma_{dia.}$, remains nearly constant at 100 nm as the mean particle diameter, $\langle dia.\rangle$, increases from 240 nm to 2.0 μm. This behavior is consistent with kinetic, not diffusional, control of the growth rate (18). At the upper end of the size range explored in this work, the relative standard deviation ($RSD_{dia.} = \langle dia.\rangle/\sigma_{dia}$) is 6–12%, which is truly state-of-the-art.

Since by design the growth rate is dramatically reduced in these experiments, the deposition durations required to obtain micron-scale silver crystals can be 100 s, two to three orders of magnitude longer than the plating durations employed in the experiments described in Sec. II. However, the nucleation densities of 10^5–10^7 cm^{-2} are reduced by approximately the same factor. Consequently, as shown in Fig. 3, these parameters (corresponding to region 2) border on the uncoupled growth regime. We therefore hypothesize that the growth of individual silver particles at low growth rates is uncoupled.

There is compelling experimental support for this hypothesis (32): The narrow particle size distributions seen in Fig. 7 are only obtained for the growth of particles at low overpotentials of -70 mV versus Ag$^+$/Ag^0. At higher overpotentials of -400 to -500 mV, much broader size distributions are reproducibly obtained. As in the analysis of simulation data (6), we can ask whether a correlation exists between the mean radius of two nearest neighbors on the surface, $\langle r \rangle$, and the distance separating them, d. If the growth of neighboring particles on the surface is coupled, then the plot of $\langle r \rangle$ versus d should have a positive slope. A statistical analysis of SEM images data for "fast" and "slow" growth experiments reveals that evidence of this positive correlation exists for fast growth experiments only. The relevant data are summarized in Fig. 8. Here is plotted the slope obtained for a plot of $\langle r \rangle$ versus d as a function of the linear regression coefficient, R, for this correlation. Data for fast and slow growth experiments are compiled in this plot.

Fig. 8 shows that the slope for fast growth experiments (i.e., those conducted using a plating potential of -500 mV versus Ag$^+$/Ag^0) is reproducibly positive where the slope in slow growth experiments is near zero, and sometimes positive and sometimes negative. Moreover, the $\langle r \rangle$ versus d data are more strongly correlated for fast growth experiments (i.e., R is reproducibly larger). In fact, it is obvious from this plot that no overlap exists between fast growth and

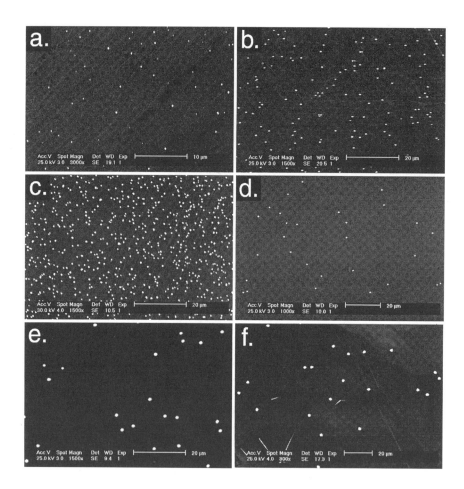

Fig. 6. Scanning electron micrographs of silver nano- and microcrystals electrodeposited using the slow growth technique. For each sample shown silver particles were deposited using a double-pulse experiment in which a 5.0-ms nucleation pulse was followed by a longer growth pulse having an amplitude of −70 mV vs. $E_{eq \cdot Ag+/Ag°}$. The plating solution in each case was 1.0 mM $AgClO_4$ in 0.1 M $LiClO_4$ acetonitrile. The deposition durations employed for the six samples shown were (from top to bottom) 1.0, 5.0, 0.64, 30, 30, 120 s (From Ref. 32). (Reproduced with permission from J. Phys. Chem. B, 2000, 104, 9131. Copyright 2000, Am. Chem. Soc.)

slow growth experiments with respect to the values of m and R that are measured for the $\langle r \rangle$ versus d correlation. The interpretation of this statistical analysis of the SEM data is clear: Interparticle diffusional coupling exists in the −500-mV experiments, but not in the −70-mV experiments which were responsible for the narrow size dispersions in Fig. 7.

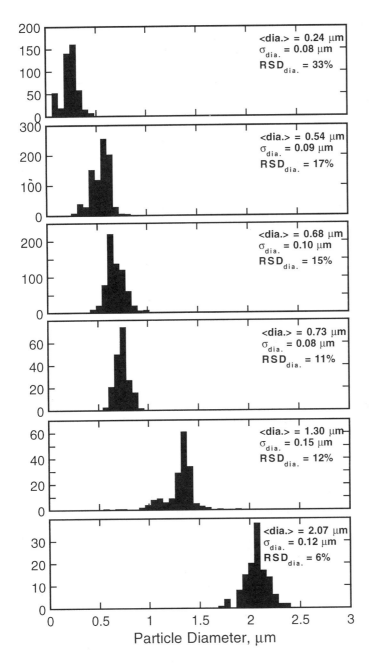

Fig. 7. Particle size histograms for the same silver crystals shown in Fig. 6 (From Ref. 32). (Reproduced with permission from J. Phys. Chem. B, 2000, 104, 9131. Copyright 2000, Am. Chem. Soc.)

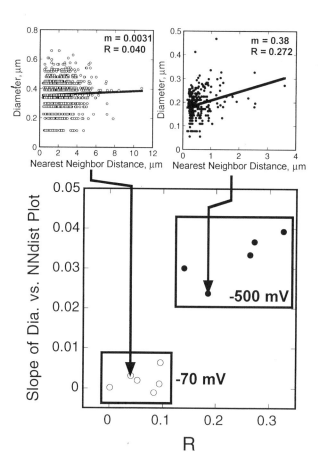

Fig. 8. (**Top**) Plots of the mean diameter of two nearest neighbors as a function of the distance between them for electrodeposition experiments conducted at −70 mV vs. (open circles; top left) and −500 mV (solid circles; top right). (**Bottom**) Plot of the slopes for these correlation as a function of the correlation coefficient, R, derived from a least-squares analysis of the data. (From Ref. 32) (Reproduced with permission from J. Phys. Chem. B, 2000, 104, 9131. Copyright 2000, Am. Chem. Soc.)

C. H$_2$ Coevolution

A radically different approach involves the application of an extremely negative voltage during metal plating (Scheme 1c). This voltage must be so negative that it causes the decomposition of solvent to occur in parallel with the plating of metal onto the substrate surface. In the case of aqueous solutions, an applied potential of −1.5 to −2.0 V versus the normal hydrogen electrode is sufficient to cause the rapid reduction of protons to H_2 and the evolution of hydrogen gas bubbles at the cathode. At these potentials virtually any transition metal can be electroplated as well.

We have applied this strategy initially to the growth of nickel nano- and microparticles on graphite surfaces (33). The beneficial effect of "H_2 coevolution" is demonstrated by the experiments in Fig. 9. Here are shown representative SEM images of nickel particles obtained at three deposition potentials: -1.2 V versus the saturated mercurous sulfate reference electrode (MSE), -1.6 V and -2.0 V. At all three of these potentials, the electrodeposition of nickel is diffusion controlled ($E_{eq.Ni^{2+}/Ni°} = -0.8$ V versus MSE), but the onset for H_2 evolution occurs at approximately -1.2 V versus MSE. Therefore, as the plating potential is reduced from -1.2 to -2.0 V, the rate of H_2 evolution dramatically increases. The SEMs in Fig. 9 show that this increased H_2 coevolution rate is associated with a pronounced narrowing of the particle size distribution for nickel nanoparticles. At -2.0 V, values for $RSD_{dia.}$ of 10–15% are routinely obtainable for nickel particles having a mean diameter between 50 and 600 nm (33). In contrast to the silver particles in Fig. 6, the nickel particles prepared using H_2 coevolution are *nanocrystalline* (33). Specifically, electron diffraction analysis and dark-field TEM imaging data demonstrate that the mean size of nickel grains in these particles is approximately 2.0 nm.

Why does H_2 coevolution work? Although more experimental data is needed, we believe two factors combine to enable the preparation of size monodisperse nickel microparticles by this unusual method. First, the generation of H_2 bubbles at the electrode surface stirs the plating solution in the immediate vicinity of the growing nickel particles (these particles actually catalyze the proton reduction reaction). This stirring has the potential to directly "erase" the depletion layer that develops at each growing nickel particle, thereby eliminating or reducing deleterious interparticle coupling. Although statistical data analogous to that of Fig. 9 has not yet been generated for nickel particles prepared using H_2 coevolution, it is absolutely clear form the available SEM data that no correlation of the particle diameter with interparticle distance exists in these experiments. These images show many instances of dimers and trimers of nickel particles in which two or three particles are disposed in direct contact with one another. The mean diameter of these particle aggregates is identical to that of other particles located in relative isolation on the surface.

There is also evidence from the SEM data for the movement of nickel particles across the graphite surface during growth. This movement is apparently a consequence of forces imparted to nickel particles by the formation and relttease of hydrogen bubbles from their surfaces. If particle motion is actually occurring (more experimental data is needed to confirm this fact), it is likely to have a beneficial effect on the size dispersion of particles for the reasons already indicated at the beginning of this section.

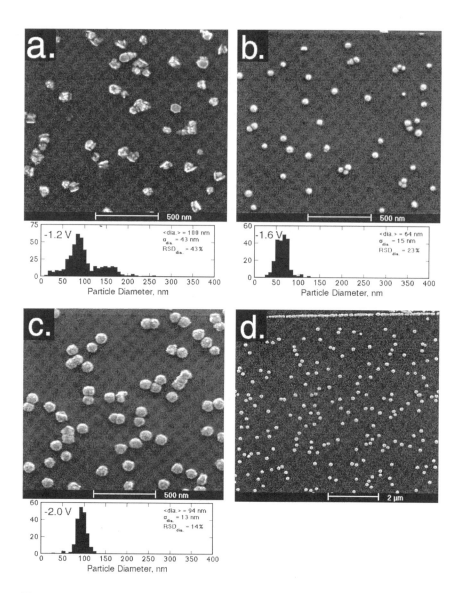

Fig. 9. Scanning electron microscope (SEM) images of HOPG electrode surfaces following the electrodeposition of nickel at various potentials for various deposition times: (a) −1.2 V vs. MSE × 25 min; (b) −1.6V × 2.8 s; (c, d) −2.0 V × 1.5 s. The nickel plating solution in all three experiments was N_2 sparged aqueous 10 mM Ni(NO$_3$)$_2$*6H$_2$O, 1.0 M NH$_4$Cl, 1.0 M NaCl, pH of 8.3 (Reproduced with permission from Ref. 33).

V. SUMMARY

The electrodeposition of metal nano- and microparticles on graphite electrode surfaces has been discussed in this monograph. The main conclusions of the experimental work carried out are the following:

1. The electrodeposition of metals on the low-energy surfaces of graphite or H-Si(100) occurs via a Volmer-Weber mechanism (immediate formation of three-dimensional metal "islands" or particles). Thus, dispersions of nanoscopic metal particles are obtained upon the deposition of submonolayer quantities of metals onto these surfaces.
2. The nucleation of metal on these surfaces is instantaneous, not progressive. Thus, an essential prerequisite for the growth of size monodisperse metal particles is satisfied.
3. For electrodeposition experiments conducted at large deposition overpotentials (e.g., ≈ -500 mV versus $E_{eq.,M^{n+}/M^{o}}$), the size dispersion of metal particles increases as a function of the deposition time. This "coarsening" of the particle distribution is opposite to what is normally seen in the growth of colloidal suspensions of metal particles.
4. The origins of size dispersion in the electrodeposition of metal particles can be elucidated using Brownian dynamics simulations of particle growth. The coarsening seen experimentally is reproduced in simulations involving randomly nucleation particles; an analysis of the BD simulation results reveals that interparticle diffusional coupling provides the most important mechanism by which particle dispersion develops in these metal particle dispersions.
5. Three experimental methods have been described which are capable of generating metal particles that are very narrowly dispersed in diameter: pulsed electrodeposition (25), slow growth (32), and H_2 coevolution (33). The success of all three methods can be traced to the elimination of interparticle diffusional coupling.

With the development of these new experimental methods, the size-selective electrodeposition of metal nanoparticle dispersions is, for all intents and purposes, a solved problem. From our perspective the new frontier in nanometer-scale electrodeposition is the imposition of short-range order on the electrodeposition of two or more different materials on an electrode surface. Can "functional ensembles" of nanoparticles be assembled on an electrode surface? Suitably engineered ensembles have the potential to function as transistors, light-emitting diodes, optical switches, and amperometric electrochemical sensors. The development of methods for assembling nanometer-scale particle ensembles represents a challenge that is likely to occupy electrochemists for some time to some.

ACKNOWLEDGMENTS

Presently, research in the author's laboratory is funded by the National Science Foundation (#DMR-9876479) and the Petroleum Research Fund of the American Chemical Society (#33751-AC5). The author also gratefully acknowledges an A. P. Sloan Foundation Fellowship and a Camille Dreyfus Teacher-Scholar award. Finally, the work described here would have been much more expensive without the generosity of Dr. Art Moore of Advanced Ceramics, who has provided my group with graphite for the past nine years.

REFERENCES

1. J-M Leger, C Lamy. Ber. Bunsenges, Phys. Chem. 94:1021, 1990.
2. S Mukerjee, S Srinivasan. J. Electroanal. Chem. 357:201, 1993.
3. A Capon, R Parsons. J. Electroanal. Chem. 45:205, 1973.
4. J Clavilier, R Parsons, R Durand, C Lamy, JM Leger. J. Electroanal. Chem. 124:321, 1981.
5. H Dahms, JOM Bockris. J. Electrochem. Soc. 111:728, 1964.
6. JL Fransaer, RM Penner. J. Phys. Chem. B 103:7643, 1999.
7. A Zangwill. Physics at Surfaces; Cambridge University Press, Cambridge, 1988.
8. M Volmer, A Weber. Z. Phys. Chem. 119:277, 1926.
9. JV Zoval, RM Stiger, PR Biernacki, RM Penner. J. Phys. Chem. 100:837, 1996.
10. JV Zoval, J Lee, S Gorer, RM Penner. J. Phys. Chem. 102:1166, 1998.
11. GS Hsiao, MG Anderson, S Gorer, D Harris, RM Penner. J. Am. Chem. Soc. 119:1439, 1997.
12. S Gorer, JA Ganske, JC Hemminger, RM Penner. J. Am. Chem. Soc. 120:9584, 1998.
13. M Anderson, S Gorer, RM Penner. J. Phys. Chem. 101:5895, 1997.
14. RM Nyffenegger, B Craft, M Shaaban, S Gorer, RM Penner. Chem. Mat. 10:1120, 1998.
15. R Stiger, B Craft, RM Penner. Langmuir 15:790, 1999.
16. H Reiss. J. Chem. Phys. 19:482, 1954.
17. VK LaMer, RH Dinegar. J. Am. Chem. Soc. 72:4847, 1950.
18. T Sugimoto. Adv. Coll. Interface Sci. 28:65, 1987.
19. JTG Overbeek. Adv. Coll. Int. Sci. 15:251, 1982.
20. TT Ngo, RS Williams. Appl. Phys. Lett. 66:1906, 1995.
21. J Crank. The Mathematics of Diffusion, 2nd ed.; Oxford University Press., New York, 1975.
22. B Scharifker, G Hills. J. Electroanal. Chem. 130:81, 1981.
23. PA Bobbert, MM Wind, J Vlieger. Physica A 146:69, 1987.
24. AJ Bard, LR Faulkner. Electrochemical Methods: Fundamentals and Applications Wiley, New York, 1980.
25. S Gorer, RM Penner. J. Phys. Chem. B 103:5750, 1999.

26. RM Penner. Acc. Chem. Res. 33:78, 2000.
27. N Herron, Y Wang, H Eckert. J. Am. Chem. Soc. 112:1322, 1990.
28. A Henglein. Chem. Rev. 89:1861, 1989.
29. MG Bawendi, ML Steigerwald, LE Brus. Ann. Rev. Phys. Chem. 41:477, 1990.
30. J Tittel, W Göhde, F Koberling, T Basché, A Kornowski, H Weller, A Eychmüller. J. Phys. Chem. B 101:3013, 1997.
31. SA Blanton, A Dehestani, PC Lin, P Guyot-Sionnest. Chem. Phys. Lett. 229:317, 1994.
32. H Liu, RM Penner. J. Phys. Chem. B 104:9131, 2000.
33. MP Zach, RM Penner. Adv. Mat.: 12:878, 2000.
34. G Erley, S Gorer, RM Penner. Appl. Phys. Lett., 1998.

11
Synthesis, Characterization, and Applications of Dendrimer-Encapsulated Metal and Semiconductor Nanoparticles

**Richard M. Crooks, Victor Chechik*, Buford I. Lemon III[†], Li Sun,
Lee K. Yeung[‡], and Mingqi Zhao[§]**
Texas A&M University, College Station, Texas

I. OVERVIEW

In this chapter we discuss a novel template-based strategy for preparing metal and semiconductor nanoparticles (Fig. 1). This approach is unique in that discrete polymers, known as dendrimers (1–5), are used as the templates rather than the more well-established approaches of using less-well-defined random polymers (6–15) or monolithic metal, ceramic, or polymers as templates (16–20). As we will show, dendrimers are particularly well suited for hosting metal particles for the following reasons: (a) they can act as both template and stabilizer for the nanoparticle; (b) they act as selective gates that control access of small molecules to the encapsulated nanoparticle after it is synthesized; (c) the terminal groups can be tailored to enhance solubility and as handles for facilitating surface immobilization.

---

[*] *Current Affiliation:* University of York, Heslington, York, U.K.
[†] *Current Affiliation:* Dow Chemical Co., Midland, Michigan
[‡] *Current Affiliation:* Dow Chemical Co., Freeport, Texas
[§] *Current Affiliation:* ACLARA Bio Sciences, Inc., Mountain View, California

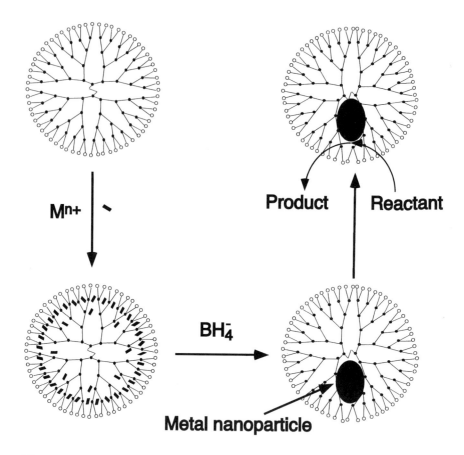

Fig. 1. Schematic of the synthesis of metal nanoparticles within dendrimer templates. The composites are prepared by mixing of the dendrimer and metal ion, and subsequent chemical reduction. These materials can be immobilized on electrode surfaces where they serve as electrocatalysts or dissolved in essentially any solvent (after appropriate end-group functionalization) as homogeneous catalysts for hydrogenation and other reactions. Reprinted with permission of Springer–Verlag, (Ref. 47). Copyright 2000 Springer–Verlag.

II. BACKGROUND

A. A Brief Retrospective on the Use of Polymers other than Dendrimers as Templates for Nanoparticle Growth

Materials *other* than dendrimers have been used for nanoparticle templating in the past. Polymers with metal-ion affinities can be used to sequester metal ions into localized domains that are then reduced or reacted to form metallic (or semiconducting) nanostructures (6). Block copolymers often serve this purpose, forming inverse micelle-like domains. A wide range of metal particles may be formed

Semiconductor Nanoparticles

within such polymer templates, including Cu, Ag, Au, Pt, Pd, Rh, and others (7, 8) The polymer template can serve to control particle size and to passivate the surface of the nanoparticles (9,10). Semiconductor nanoparticles (e.g., CdS and CdSe) have also been prepared within polymeric templates (11,12), and it has been shown that the polymer can be used to tune the size, and therefore the optical and electronic properties, of the embedded semiconductors (13). When semiconductor nanoparticles are grown or embedded within a semiconducting polymer matrix, electroluminescent films with desirable characteristics (low turn-on voltages, high stabilities, and tunable emission) can be obtained (14,15).

B. A Brief Retrospective on the Use of Metal, Ceramic, and Polymer Monoliths as Nanoparticle Templates

In addition to bulk-phase polymers of various sorts, nanoparticles have also been prepared within nanoporous monolithic templates. For example, the well-defined pores in alumina or polymeric filtration membranes can be used to define the geometrical and chemical properties of metal, semiconductor, and polymeric nanomaterials (16,20). In many cases the template can be removed chemically or thermally, leaving behind the naked nanomaterial (17–20). The obvious advantages to this technique are that highly monodisperse particles with a variety of shapes and aspect ratios (from spherical particles to nanorods) can be prepared. After release from the template, nanoparticles prepared in this way can be utilized in other applications that do not necessarily involve the original polymer support (19). A comprehensive review of nanoscopic materials synthesized by this approach is described in Chapter 7.

C. Dendrimers

1. Introduction to Dendrimers

Dendrimers are outstanding candidates for template synthesis of nanoparticles because of their regular structure and chemical versatility. Dendrimers have three basic anatomical features: a core, repetitive branch units, and terminal functional groups (1–5). The physical and chemical properties of dendrimers depend strongly on the chemical structure of all three components as well as on the overall size and dimensionality of the dendrimer. For example, larger dendrimers are more-or-less spherical in shape and contain interior void spaces, whereas lower-generation materials are flat and open. Also, terminal groups largely, but not solely, determine the solubility and adsorption properties of dendrimers.

2. Synthesis of Dendrimers

The two families of dendrimers used in the studies reported here are commercially available (21) and synthesized by a strategy known as the divergent method.

An excellent introduction to the basic principles of dendrimer synthesis (both the divergent and convergent approaches) is given in Ref. 3, and therefore only a brief introduction to the more relevant divergent method is provided here. In the divergent method, growth of the dendrimer is outward from the core to the dendrimer surface. This method of synthesis generally involves serial repetition of two chemical reactions and appropriate purification steps. For example, the generation 0 dendrimer (G0) is formed after the first cycle of reactions on the dendritic core. The generation, and thus the diameter of a dendrimer, increases more-or-less linearly with the number of growth cycles. The number of surface functional groups increases exponentially with each ensuing cycle, and, because two or three monomers are usually added to each branch point in the reaction cycle, the maximum size or generation of a dendrimer is governed by steric crowding at the periphery. The structures of G1 PAMAM and PPI dendrimers are shown in Fig. 2.

3. Chemical and Physical Properties of Dendrimers

Table 1 provides some general information about the evolution of size and molecular conformation as a function of generation for PAMAM and PPI dendrimers (22). It is important to recognize that the data in this table are for ideal-structure dendrimers, while in practice PAMAM and PPI dendrimers contain a statistical distribution of defects (3,4). The diameter of PAMAM dendrimers increases by roughly 1 nm per generation, while the molecular weight and number of functional groups increase exponentially. The surface density of dendrimer terminal groups, normalized to the expanding surface area, also increases nonlinearly. Simulation results (23) show that up to G2, PAMAM dendrimers have an expanded or "open" configuration, but as the dendrimer grows in size crowding of the surface functional groups causes the dendrimer to adopt a spherical or globular structure. Perhaps it is helpful to think of the structure of G4 PAMAM as resembling that of a wet sponge and of G8 as having a somewhat hard surface like that of a beach ball. That is, the interior of high-generation dendrimers is rather hollow, while their exteriors are far more crowded. Both of these factors figure prominently in the work described in this chapter.

As a consequence of their three-dimensional structure and multiple internal and external functional groups, higher-generation dendrimers are able to act as hosts for a range of ions and molecules. Endoreception occurs when substrates penetrate interstices present between densely packed surface groups and are incorporated into the interior cavities. Exoreception occurs when such species interact strongly with functional groups on the dendrimer surface. To prepare dendrimer-encapsulated metal and semiconductor nanoparticles, we rely on endoreception to bind the metal ions of choice to the dendrimer interior prior to chemical reduction (Fig. 1). The exoreceptors are useful for attaching dendrimers to surfaces, other polymers, and a broad range of discrete molecular compounds or metal ions.

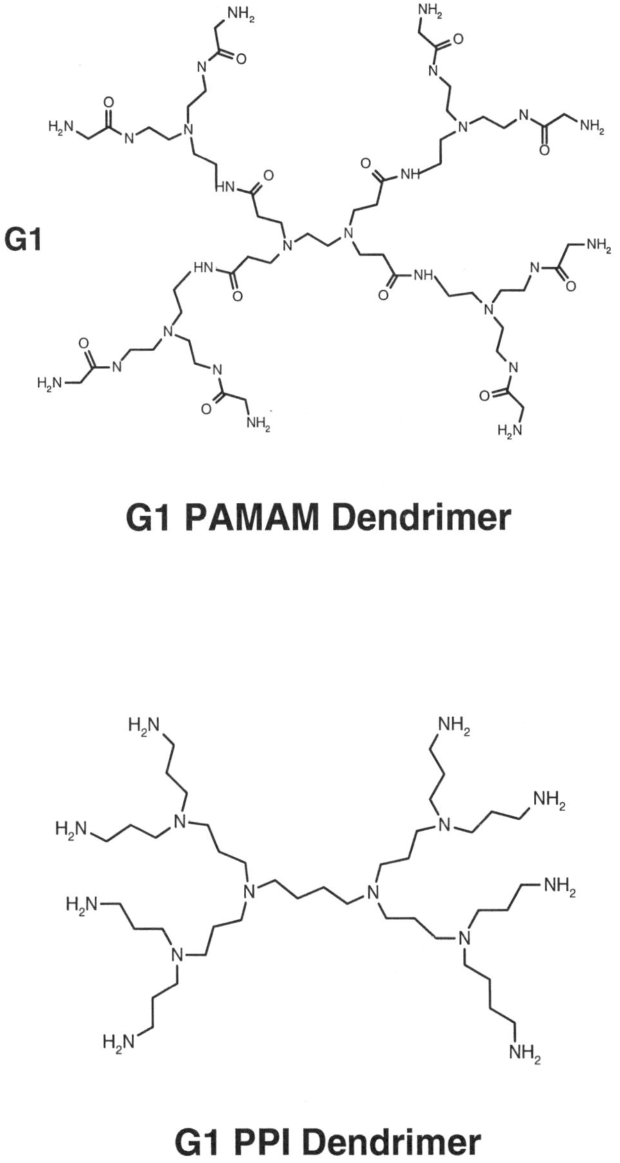

G1 PAMAM Dendrimer

G1 PPI Dendrimer

Fig. 2. Chemical structures of generation 1 (G1) poly(amidoamine) (PAMAM) and poly(propylene imine) (PPI) dendrimers.

Table 1. Physical Characteristics of PAMAM and PPI Dendrimers

Generation	Surface groups	Tertiary amines	Molecular weight[a]		Diameter[b], nm	
			PAMAM	PPI[c]	PAMAM	PPI[c]
0	4	2	517	317	1.5	0.9
1	8	6	1,430	773	2.2	1.4
2	16	14	3,256	1,687	2.9	1.9
3	32	32	6,909	3,514	3.6	2.4
4	64	62	14,215	7,168	4.5	2.8
5	128	126	28,826	14,476	5.4	—
6	256	254	58,048	29,093	6.7	—
7	512	510	116,493	58,326	8.1	—
8	1024	1022	233.383	116,792	9.7	—
9	2048	2046	467,162	235,494	11.4	—
10	4096	4094	934,720	469,359	13.5	—

[a] Molecular weight is based on defect-free, ideal-structure dendrimers.
[b] For PAMAM dendrimers the molecular dimensions were determined by size-exclusion chromatography, and the dimensions of PPI dendrimers were determined by SANs; data for the high-generation PPI dendrimers are not available.
[c] We have used the generational nomenclature typical for PAMAM dendrimers throughout this chapter. In the scientific literature the PPI family of dendrimers is incremented by one. That is, what we call a G4 PPI dendrimer (having 64 end groups) is often referred to as G5. Reprinted with permission of Springer–Verlag, (Ref. 47). Copyright 2000 Springer–Verlag.

PAMAM dendrimers are large (G4 is 4.5 nm in diameter) and have a hydrophilic interior and exterior; accordingly, they are soluble in many convenient solvents (water, alcohols, and some polar organic solvents). Importantly, the interior void spaces are large enough to accommodate nanoscopic guests, such as metal clusters, and are sufficiently monodisperse in size so as to ensure fairly uniform particle size and shape. As we will show, the space between the terminal groups can act as size-dependent gates between the dendrimer exterior and interior. For example, the exterior of G8 can distinguish linear and branched hydrocarbons (vide infra), which is useful for size-selective catalysis at encapsulated metal particles.

As shown in Table 1, the diameter of the amine-terminated G4 PPI dendrimers is 2.8 nm, so it is considerably smaller than the equivalent G4 PAMAM (4.5 nm) (22). Like the PAMAM dendrimers, the PPI dendrimers have interior tertiary amine groups that may interact with guest molecules and ions, but in contrast they do not contain amide groups. As a consequence, PPI dendrimers are stable at very high temperatures (the onset of weight loss for G4 PPI is 470°C) (24), which is a critical factor for some applications, including catalysis. In contrast, PAMAM dendrimers undergo retro-Michael addition at temperatures higher than about 100°C (25). This can be considered an advantage in some instances; for example, one could imagine templating a nanoparticle within a PAMAM dendrimer and then releasing it cleanly by thermal decomposition of the host dendrimer. Commercially available PPI dendrimers are terminated in primary amines, and they are soluble in water, short-chain alcohols, DMF, and dichloromethane. Simple amidation chemistry can be used to functionalize the end groups of either dendrimer family and thereby control solubility.

4. Dendrimers as Host Molecules

Dendrimer interior functional groups and cavities retain guest molecules selectively, depending on the nature of the guest and the dendritic endoreceptors, the cavity size, and the structure and chemical composition of the terminal groups. The driving force for guest encapsulation within dendrimers can be based on electrostatic interactions, complexation reactions, steric confinement, various types of weaker forces (van der Waals, hydrogen bonding, the hydrophobic force, etc.), and combinations thereof. Many examples of dendrimer-based host-guest chemistry have been reported (3–5). Meijer and co-workers were the first to demonstrate physical encapsulation and release of guest molecules from a "dendritic box" (26,27). As discussed in the next section, dendrimer endoreceptors can also be used to sequester metal ions within dendrimers. It is this property that makes it possible to prepare nanocomposite materials consisting of a dendritic shell (the host) and a metal or semiconductor nanoparticle (the guest).

D. Metal and Semiconductor Nanoparticles

As discussed in other chapters in this volume, small clusters of metals (28) and semiconductors (29) are interesting because of their unique mechanical, electromagnetic, and chemical properties. Of particular interest to us are transition-metal nanoclusters, which are useful for applications in catalysis and electrocatalysis (30–34). There are two main challenges in this area of catalysis. The first is the development of methods for stabilizing the nanoclusters by eliminating aggregation without blocking most of the active sites on the cluster surfaces or otherwise reducing catalytic efficiency. The second key challenge involves controlling cluster size, size distribution, and perhaps even particle shape. Because dendrimers can act as both "nanoreactors" for preparing nanoparticles and nanoporous stabilizers for preventing aggregation, we reasoned that they would be useful for addressing these two issues.

E. Metals Contained within Dendrimers

There are two classes of metal-containing dendrimers: those that contain metal ions and those that contain metal or semiconductor nanoparticles. The focus of this chapter is on the former, but an understanding of these materials requires some discussion of the metal-ion-containing precursors from which they are prepared (2–5,35–75).

There are three general categories of metal-ion-containing dendrimers. The first is composed of dendrimers that use metal ions as an integral part of their chemical structure. This includes, for example, dendrimers having an organometallic core and dendrimers that use metal ligation to assemble branches into the complete dendrimer. The second class consists of dendrimers that have peripheral groups that are good ligands for metal ions. The third group contains internal ligands that bind to nonstructural metal ions. Only dendrimers associated with nonstructural metal ions are relevant to this chapter.

Dendrimers complexed to exterior metal ions do not take full advantage of the many unique structural features of dendrimers, such as the hollow interior, the unique chemical properties of the interior microenvironment, the terminal-group-tunable solubility, or the nanofiltering capability of the dendrimer branches (vide infra). Moreover, exterior metal ions can lead to dendrimer cross-linking, agglomeration, and even precipitation. However, PAMAM and PPI dendrimers have functional groups within their interior that are also able to bind metal ions. Specifically, PAMAM dendrimer interiors contain tertiary and secondary (amide) amines, while PPI dendrimers lack amide groups (Fig. 2).

We are especially interested in trapping metal ions *exclusively* within the interiors of unmodified, commercially available PAMAM dendrimers, because such composites are easy to prepare and retain the desirable structural properties of the uncomplexed dendrimers. It is possible to prevent metal-ion complexation to amine-terminated PAMAM dendrimers by either selective protonation of the primary amines [for PAMAM dendrimers, the surface primary amines (pK_a = 9.5) are more basic than the interior tertiary amines (pK_a = 5.5)] (47,53,57) or by functionalization with noncomplexing terminal groups. The latter approach eliminates the restrictive pH window necessitated by selective protonation and generally results in more easily interpretable results (2,76). Accordingly, most of our work has focused on hydroxyl-terminated PAMAM dendrimers (Gn-OH). Indeed, we have shown that many metal ions, including Cu^{2+}, Pd^{2+}, Pt^{2+}, Ni^{2+}, Au and Ru^{3+}, sorb into Gn-OH interiors over a broad range of pH via complexation with interior tertiary amines (57,58,77–79).

III. SYNTHESIS AND CHARACTERIZATION OF DENDRIMER-ENCAPSULATED METAL NANOPARTICLES IN PAMAM AND PPI DENDRIMERS

This section briefly describes dendrimer-encapsulated metal nanoparticles, a new family of composite materials first described by us in 1998 (57), and their applications to catalysis.

A. Intradendrimer Complexes between PAMAM Dendrimers and Metal Ions

The first step in the preparation of dendrimer-encapsulated metal and semiconductor particles involves complexation of metal ions within the dendrimer interior (Fig. 1). Because the size and composition of the sequestered nanoparticles depend on this step, it is worth considering it in some detail.

1. Intradendrimer Complexes between PAMAM Dendrimers and Cu^{2+}

The first studies of dendrimer-encapsulated metal nanoparticles focused on Cu (57). This is because Cu^{2+} complexes with PAMAM and PPI dendrimers are very well behaved and have easily interpretable UV-vis and EPR spectra. In the absence of dendrimer and in aqueous solutions Cu^{2+} exists primarily as $[Cu(H_2O)_6]^{2+}$, which gives rise to a broad, weak absorption band centered at 810 nm (Fig. 3a). This corresponds to the well-known d-d transition for Cu^{2+} in a tetragonally distorted octahedral or square-planar ligand field.

In the presence of G4-OH, λ_{max} for the d-d transition shifts to 605 nm ($\varepsilon \sim$ 100 M^{-1} cm^{-1}). In addition, a strong ligand-to-metal-charge-transfer (LMCT) transition centered at 300 nm ($\varepsilon \sim$ 4000 M^{-1} cm^{-1}) emerges. The complexation interaction between dendrimers and Cu^{2+} is strong: the d-d transition band and the LMCT transition do not decrease significantly even after 36 hr of dialysis against pure water. These data show that Cu^{2+} partitions into the dendrimer from the aqueous phase and remains there.

To learn more about the Cu^{2+} ligand field, we quantitatively assessed the number of Cu^{2+} ions extracted into each dendrimer by spectrophotometric titration. Spectra of a 0.05 mM G4-OH solution containing different amounts of Cu^{2+} are given in Fig. 3b. The absorbance at 605 nm increases with the ratio of $[Cu^{2+}]/[G4\text{-}OH]$, but only slowly when the ratio is larger than 16. The titration results are given in the inset of Fig. 3b, where absorbance at the peak maximum of 605 nm is plotted against the number of Cu^{2+} ions per dendrimer. We estimated the titration endpoint by extrapolating the two linear regions of the curve, and this treatment indicates that each G4-OH dendrimer can strongly sorb up to 16 Cu^{2+} ions. Because a G4-OH dendrimer contains 62 interior tertiary amines and Cu^{2+} is tetravalent, it is tempting to conclude that each Cu^{2+} is coordinated to about four amine groups. However, EPR and ENDOR data (74) indicate that most of the ions bind to the outermost 16 pairs of tertiary amine groups, and CPK models reveal that the dendrimer structure is not well configured for complexation between the innermost amines and Cu^{2+}. Thus, on average, each Cu^{2+} is coordinated to two amine groups, and the remaining positions of the ligand field are occupied by more weakly binding ligands such as amide groups or water.

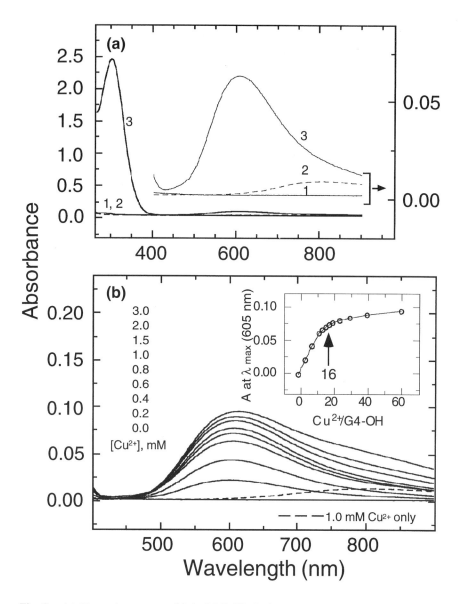

Fig. 3. (a) Absorption spectra of 0.6 mM $CuSO_4$ in the presence (spectrum #3) and in the absence (spectrum 2) of 0.05 mM G4-OH. The absorption spectrum of 0.05 mM G4-OH vs. water is also shown (spectrum #1). (b) Absorption spectra as a function of the Cu^{2+}/G4-OH ratio. The inset is a spectrophotometric titration plot showing absorbance at the peak maximum of 605 nm as a function of number of Cu^{2+} ions per G4-OH. Reprinted with permission of Springer-Verlag (Ref. 47). Copyright 2000 Springer-Verlag.

We also investigated the effect of dendrimer generation on the maximum number of Cu^{2+} ions that can bind within dendrimers. Figure 4a shows absorption spectra of 0.05 mM Gn-OH (n = 2, 4, and 6) in the presence of a fixed concentration of $CuSO_4$ in the Cu^{2+} d-d transition region. For G2-OH, there is an absorption shoulder at 605 nm and a band centered at 810 nm, which indicates only partial complexation of Cu^{2+}. For G4-OH the band at 605 nm becomes more pronounced, and it is the only absorption feature for G6-OH. This behavior is due to the increasing intradendrimer concentration of Cu^{2+} as a function of increasing generation, which is a consequence of the exponential increase in the number of tertiary amines (Table 1). The endpoints of the spectrophotometric titration curves for G2-OH and G6-OH (Fig. 4b and 4c) indicate strong binding of 4 and 64 Cu^{2+} ions, respectively. A similar titration was carried out for G3-OH and it was found to tightly bind up to 8 Cu^{2+} ions. Interestingly, G2-OH, G3-OH, and G6-OH contain 4, 8, and 64 pairs of tertiary amines, respectively, in their outermost generational shell, and therefore these titration results are fully consistent with the one-Cu^{2+}-per-two-outermost-tertiary-amines model proposed for G4-OH. Indeed, Fig. 4d shows that there is a linear relationship between the number of Cu^{2+} ions complexed within Gn-OH and the number of tertiary amine groups within Gn-OH. Results obtained from MALDI-TOF mass spectrometry confirm these findings (73).

In addition to hydroxyl-terminated PAMAM dendrimers, we also investigated the binding ability of amine-terminated G4-NH_2. The results are consistent with the model proposed for G4-OH, but the situation is somewhat complex and beyond the scope of this discussion. Details can be found elsewhere (47).

2. Intradendrimer Complexes between PAMAM Dendrimers and Metal Ions other than Cu^{2+}

Using chemistry similar to that just discussed for Cu^{2+}, we have shown that many other transition-metal ions, including Pd^{2+}, Pt^{2+}, Ni^{2+}, and Ru^{3+}, can be extracted into dendrimer interiors (58,59,77). For example, a strong absorption peak at 250 nm (ε = 8000 M^{-1} cm^{-1}) arising from a ligand-to-metal charge-transfer (LMCT) transition indicates that $PtCl_4^{2-}$ is sorbed within Gn-OH dendrimers. The spectroscopic data also indicate that the nature of the interaction between the dendrimer and Cu or Pt ions is quite different. As discussed earlier, Cu^{2+} interacts with particular tertiary amine groups by complexation, but $PtCl_4^{2-}$ undergoes a slow ligand-exchange reaction, which is consistent with previous observations for other Pt^{2+} complexes (80). The absorbance at 250 nm is proportional to the number of $PtCl_4^{2-}$ ions in the dendrimer over the range 0–60 [G4-OH(Pt^{2+})$_n$, n = 0–60], which indicates that it is possible to control the G4-OH/Pt^{2+} ratio. Control experiments confirm that the Pt^{2+} ions are inside the dendrimer rather than complexed to the exterior hydroxyl groups (47,77).

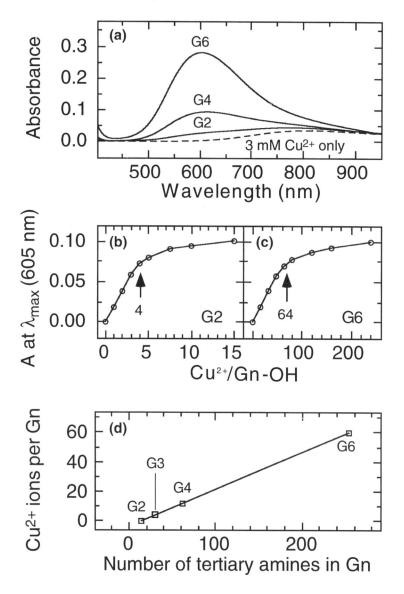

Fig. 4. (a) The effect of dendrimer size on the absorbance of 3.0 mM Cu^{2+} + 0.05 mM Gn-OH solutions. (b,c) Spectrophotometric titration plots of the absorbance at the peak maximum of 605 nm as a function of the number of Cu^{2+} ions per G2-OH or G6-OH. The initial concentration of G2-OH and G6-OH was 0.2 or 0.0125 mM, respectively. (d) The relationship between the number of Cu^{2+} ions complexed within Gn-OH and the number of tertiary amine groups within Gn-OH. Reprinted with permission of Springer-Verlag (Ref. 47). Copyright 2000 Springer-Verlag.

B. Synthesis and Characterization of Dendrimer-Encapsulated Metal Nanoparticles

In this section two methods used to prepare dendrimer-encapsulated metal nanoclusters are discussed: direct reduction of dendrimer-encapsulated metal ions and displacement of less-noble-metal clusters with more noble elements.

1. Direct Reduction of Dendrimer/Metal-Ion Composites

Chemical reduction of Cu^{2+}-loaded G4-OH dendrimers (G4-OH/Cu^{2+}) with excess $NaBH_4$ results in formation of intradendrimer Cu clusters (Fig. 1). Evidence for this comes from the immediate change in solution color from blue to golden brown; that is, the absorbance bands originally present at 605 nm and 300 nm disappear and are replaced with a monotonically increasing spectrum of nearly exponential slope toward shorter wavelengths (Fig. 5). This behavior results from the appearance of a new interband transition corresponding to formation of intradendrimer Cu clusters. The measured onset of this transitions at 590 nm agrees with the reported value (81), and the nearly exponential shape is characteristic of a bandlike electronic structure, strongly suggesting that the reduced Cu does not exist as isolated atoms but rather as clusters (82). The absence of an absorption peak arising from Mie plasmon resonance (around 570 nm) (83) indicates that the Cu clusters are smaller than the Mie-onset particle diameter of about 4 nm (83–85). Plasmon resonance cannot be detected for very small metal clusters, because the peak is flattened due to the large imaginary dielectric constant of such materials (82). The presence of metal clusters is also supported by loss of signal in the EPR spectrum (86) following reduction of the dendrimer Cu^{2+} composite. Transmission electron microscopy (TEM) results also indicate the presence of intradendrimer Cu clusters after reduction. Micrographs of Cu clusters within G4-OH reveal particles having a diameter less than 1.8 nm,* much smaller than the 4.5-nm diameter of G4-OH (87–89).

Intradendrimer Cu clusters are extremely stable despite their small size, which provides additional strong evidence that the clusters reside within the dendrimer interior. Clusters formed in the presence of G4-OH or G6-OH dendrimers and with a Cu^{2+} loading less than the maximum threshold values were found to be stable (no observable agglomeration or precipitation) for at least one week in an oxygen-free solution. However, in air-saturated solutions the clusters revert to intradendrimer Cu^{2+} ions overnight. In contrast, when excess Cu^{2+} is added to a dendrimer solution, Cu^{2+} is present inside the dendrimer and as hydrated ions in solution. After reduction, the excess Cu^{2+} forms a dark precipitate within a few hours, but the remaining transparent solution yields the same absorption spectrum as one

---

*The value of 1.8 nm represents an upper limit on the cluster size. The actual particles certainly have a critical dimension of less than 1 nm (based on CPK models), which is below our TEM resolution.

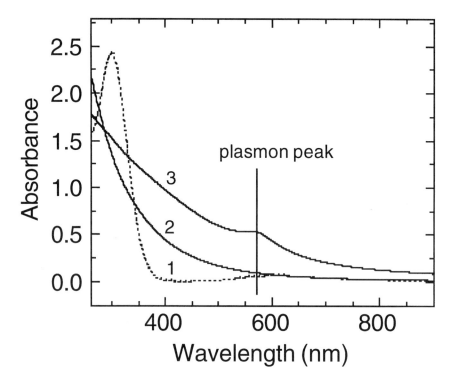

Fig. 5. Absorption spectra of a solution containing 0.6 mM $CuSO_4$ and 0.05 mM G4-OH before (dashed line, spectrum 1) and after (solid line, spectrum 2) reduction with a fivefold molar excess of $NaBH_4$. spectrum 3 was obtained under the same conditions as those for spectrum 2 except 0.05 mM G4-NH_2 was used in place of G4-OH. Reprinted with permission of the American Chemical Society (Ref. 57). Copyright 1998 the American Chemical Society.

prepared with a stoichiometric amount of Cu^{2+}. TEM images of the particles in these solutions reveal two size regimes: the first is 9 ± 4 nm in average diameter and is responsible for the dark precipitate; the second, which is estimated to have an upper limit of 1.8 nm in diameter,* corresponds to intradendrimer clusters.

The ability to prepare well-defined intradendrimer metal nanoclusters depends strongly on the chemical composition of the dendrimer. For example, when G4-NH_2, rather than the just-described hydroxyl-terminated dendrimers, is used as the template, a maximum of 36 Cu^{2+} ions is sorbed; most of these bind to the terminal primary amine groups. Reduction of a solution containing 0.6 mM $CuSO_4$ and 0.05 mM G4-NH_2 results in a clearly observable plasmon resonance band at 570 nm (Fig. 5) (82–84), indicating that the Cu clusters prepared in this way are larger than 4 nm in diameter. This larger size is a consequence of agglomeration of Cu particles adsorbed to the unprotected dendrimer exterior (64,65).

The approach for preparing dendrimer-encapsulated Pt metal particles is similar to that used for preparation of the Cu composites: chemical reduction of an aqueous solution of G4-OH(Pt^{2+})$_n$ yields dendrimer-encapsulated Pt nanoparticles [G4-OH(Pt$_n$)]. A spectrum of G4-OH(Pt$_{60}$) is shown in Fig. 6a; it displays a much higher absorbance than G4-OH(Pt^{2+})$_{60}$ throughout the wavelength range displayed. This change results from the interband transition of the encapsulated zero-valent Pt metal particles.

Spectra of G4-OH(Pt)$_n$, $n = 12, 40, 60$, obtained between 280 and 700 nm and normalized to $A = 1$ at $\lambda = 450$ nm, are shown in Fig. 6b; all of these spectra display the interband transition of Pt nanoparticles. Control experiments clearly demonstrate that the Pt clusters are sequestered within the G4-OH dendrimer. For example, BH$_4^-$ reduction of G4-NH$_2$(Pt^{2+})$_n$, which exist as cross-linked emulsions, results in immediate precipitation of large Pt clusters. In contrast, Gn-OH-encapsulated particles do not agglomerate for up to 150 days, and they redissolve in solvent after repeated solvation/drying cycles.

The absorbance intensity of the encapsulated Pt nanoparticles is related to the particle size. A plot of log A versus log λ provides qualitative information about particle size: the negative slopes are known to decrease with increasing particle size. For aqueous solutions of G4-OH(Pt$_{12}$), G4-OH(Pt$_{40}$), and G4-OH(Pt$_{60}$), the slopes are -2.7, -2.2, and -1.9, respectively (Fig. 6b, inset). These results confirm that the size of the intradendrimer particles increases with increasing Pt^{2+} loading.

High-resolution transmission electron microscopy (HRTEM) images (Fig. 7) clearly show that dendrimer-encapsulated particles are nearly monodisperse and that their shape is roughly spherical. For G4-OH(Pt$_{40}$) and G4-OH(Pt$_{60}$) particles, the metal-particle diameters are 1.4 ± 0.2 and 1.6 ± 0.2 nm, which are slightly larger than the theoretical values of 1.1 and 1.2 nm, respectively, calculated by assuming that particles are contained within the smallest sphere circumscribing a fcc Pt crystal. When prepared in aqueous solution, Pt nanoparticles usually have irregular shapes and a large size distribution. The observation of very small, predominantly spherical particles in this study is a consequence of the dendrimer cavity, that is, the template in which they are prepared. Note that when the dendrimers are loaded with metal ions at maximum capacity (middle frame of Fig. 7), the resulting nanoparticles are more monodisperse than when a lower Pt^{2+} loading is used (top frame). This is an expected statistical consequence of the substoichiometric loading. X-ray energy dispersive spectroscopy (EDS) and X-ray photoelectron spectroscopy (XPS) analyses were also carried out, and they unambiguously identify the particle composition as zero-valent Pt (77,90). Interestingly, the XPS data also indicate the presence of Cl prior to reduction, but it is not detectable after reduction. This finding is consistent with the change in valency from the chloride-containing G4-OH(Pt^{2+})$_n$ complex to the zero-valent metal.

Results similar to those discussed for dendrimer-encapsulated Cu and Pt also obtain for Pd, Ru, and Ni nanoclusters. An example of 40-atom Pd nanoclusters confined within G4-OH is shown in the bottom micrograph of Fig. 7.

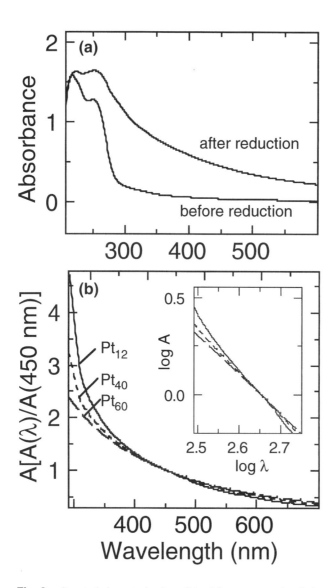

Fig. 6. Spectral characterization of dendrimer-encapsulated clusters containing different numbers and types of atoms. (a) Absorption spectra of solutions containing 0.05 mM G4-OH(Pt^{2+})$_{60}$ before and after reduction. (b) UV-vis spectra of solutions containing G4-OH(Pt$_{12}$) (solid line), G4-OH(Pt$_{40}$) (short dashes), and G4-OH(Pt$_{60}$) (long dashes) normalized to $A = 1$ at $\lambda = 450$ nm. Logarithmic plots of these data, shown in the inset, demonstrate that larger particles result in less negative slopes. Reprinted with permission of Wiley-VCH (Ref. 77). Copyright 1999 Wiley-VCH.

Semiconductor Nanoparticles

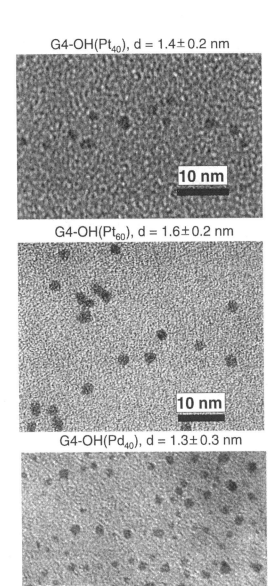

Fig. 7. HRTEM images of G4-OH(Pt$_{40}$), G4-OH(Pt$_{60}$), and G4-OH(Pd$_{40}$), which illustrate the size and shape distribution of the encapsulated metal nanoparticles. Reprinted with permission of Wiley-VCH (Ref. 58). Copyright 1999 Wiley-VCH.

2. Intradendrimer Metal Displacement Reactions

In the previous section it was shown that dendrimer-encapsulated metal nanoclusters can be prepared by direct reduction if the corresponding metal ions can be extracted into the dendrimer interior. However, this approach is not suitable for encapsulation of Ag particles, because the equilibrium between PAMAM dendrimers and Ag^+ ions does not strongly favor the intradendrimer complex.

Although it is not possible to prepare Ag particles inside Gn-OH by direct reduction of interior ions, stable, dendrimer-encapsulated Ag particles can be prepared by a metal exchange reaction. In this approach, dendrimer-encapsulated Cu nanoclusters are prepared as described in a previous section (57), and then upon exposure to Ag^+ the Cu particles oxidize to Cu^{2+} ions, which stay entrapped within the dendrimer at pH values larger than 5.5, and Ag^+ is reduced to yield a dendrimer-encapsulated Ag nanoparticle (Fig. 8).

Figure 9 shows UV-vis spectra of an aqueous G6-OH(Cu^{2+})$_{55}$ solution before and after reduction. At pH 7.5 a strong LMCT band is evident at 300 nm ($\varepsilon \sim 4000$ M^{-1} cm^{-1}), which (as discussed earlier) is not present in the absence of the dendrimer or Cu^{2+} (90). As signaled by the loss of the band at 300 nm, chemical reduction of G6-OH(Cu^{2+})$_{55}$ with excess of BH_4^- results in formation of intradendrimer Cu nanoclusters (Fig. 9a). The postreduction spectrum is dominated by a monotonic and nearly exponential increase toward shorter wavelengths, which results from the interband transition of intradendrimer Cu clusters (57).

Ag^+ is a stronger oxidizing agent than Cu^{2+}, and therefore Eq. (1) leads to conversion of a dendrimer-encapsulated Cu nanoparticle to Ag.*

$$Cu + 2Ag^+ = Cu^{2+} + 2Ag \tag{1}$$

Figure 9b shows the spectroscopic evolution of such a reaction. Spectrum 1 is the UV-vis spectrum of G6-OH(Cu$_{55}$), which is the same as spectrum 2 in Fig. 9a. When Ag^+ is added to a G6-OH(Cu$_{55}$) solution at pH 3.0, a new absorbance band centered at 400 nm appears (spectrum 2), which corresponds to the plasmon resonance of Ag particles (82). When the pH of the solution is adjusted to 7.5 (spectrum 3), the Ag plasmon peak does not change much, but a new peak at 300 nm ($\varepsilon \sim 4000$ M^{-1} cm^{-1}) appears. This is the G6-OH(Cu^{2+})$_{55}$ LMCT band discussed earlier, and it indicates that when the interior tertiary amines are deprotonated Cu^{2+} resides within the dendrimer after intradendrimer oxidation of Cu. Thus, both the metallic Ag nanoparticles and the Cu^{2+} ions generated by the displacement reaction are present within the dendrimers simultaneously. Such versatile and well-defined microenvironments are unique and should find a variety of

---

* Because the dendrimer-encapsulated nanoparticles are so small, they do not have the properties of bulk metals. Therefore, it is not possible to calculate the reduction potential for the exchange reactions from tabulated literature data.

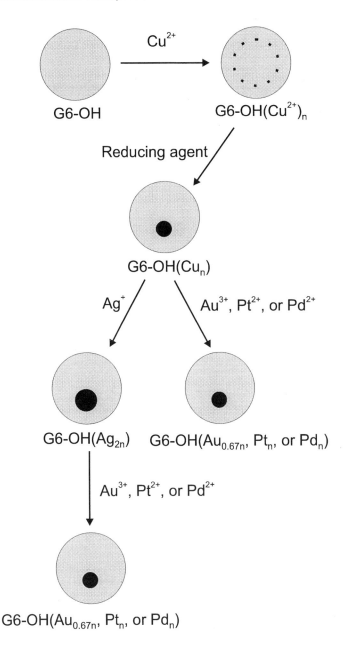

Fig. 8. Schematic of the method used to prepare dendrimer-encapsulated Ag, Au, Pd, and Pt nanoclusters by primary and secondary displacement reactions using G6-OH(Cu_n) or G6-OH(Ag_{2n}) as starting materials. Reprinted with permission of the American Chemical Society (Ref. 59). Copyright 1999 American Chemical Society.

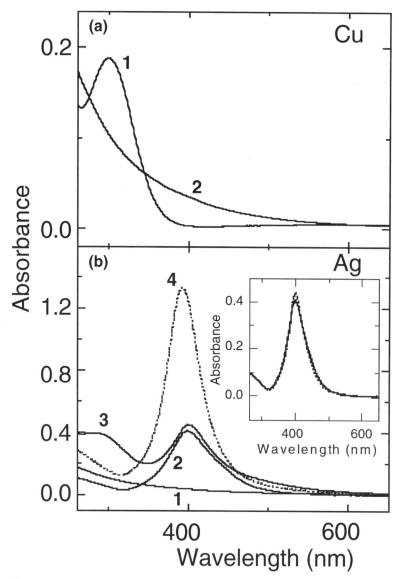

Fig. 9. (a) Absorption spectra of aqueous 0.55 mM $Cu(NO_3)_2$ + 0.01 mM G6-OH before (spectrum 1) and after (spectrum 2) reduction. (b) The absorption spectra of 0.01 mM G6-OH(Cu_{55}) + 1.1 mM Ag^+ (spectrum 2, pH 3.0 and spectrum 3, pH 7.5). Spectrum 1 is the same as spectrum 2 in (a). Spectrum 4 corresponds to the direct reduction of 1.1 mM Ag^+ + 0.01 mM G6-OH 1 hr after the addition of BH_4. The inset shows spectra 10 min (solid line) and 18 hr (dashed line) after displacement. Reprinted with permission of the American Chemical Society (Ref. 59). Copyright 1999 American Chemical Society.

fundamental and applied applications, especially in the area of catalysis. The Ag particles synthesized by this displacement reaction are very stable; for example, spectra taken 10 min and 18 hr after reduction are nearly identical (Fig. 9b, inset) and no precipitation is observed even after storage in air for more than two months. XPS and TEM analysis confirm the presence of dendrimer-encapsulated Ag metal particles formed by exchange. Of course, this displacement method can be used to prepare other types of noble-metal particles (Fig. 8), such as Au, Pt, and Pd, because the standard potentials ($E°$) of the corresponding half-reactions are more positive than for Cu^{2+}/Cu. Finally, Ag nanoparticles synthesized by primary displacement of dendrimer-encapsulated Cu nanoparticles can themselves be displaced to yield Au, Pt, or Pd nanoparticles by secondary displacement reactions (Fig. 8).

C. Dendrimer-Encapsulated Bimetallic Nanoclusters

Bimetallic metal particles are important materials because their characteristics, especially their catalytic properties, are often quite different from those of pure metal particles. Dendrimer-encapsulated bimetallic clusters can be synthesized by any of three methods (Fig. 10): (1) partial displacement of the dendrimer-encapsulated cluster, (2) simultaneous co-complexation of two different metal ions followed by reduction, or (3) sequential loading and reduction of two different metal ions.

Preparation of mixed metal intradendrimer clusters by partial displacement is a straightforward extension of the complete displacement approach for forming single-metal clusters described in the previous section. If less than a stoichiometric amount of Ag^+, Au^{3+}, Pd^{2+}, or Pt^{2+} is added to a G6-OH(Cu_{55}) solution, or if less than a stoichiometric amount of Au^{3+}, Pd^{2+}, or Pt^{2+} is added to G6-OH(Ag_{110}) solution, it is possible to form Ag/Cu, Au/Cu (Au/Ag), Pd/Cu (Pd/Ag), and Pt/Cu (Pt/Ag) bimetallic clusters inside dendrimers.

Dendrimer-encapsulated bimetallic clusters can also be prepared by simultaneous co-complexation of two different metal ions, followed by a single reduction step. For example, the absorption spectrum of a solution containing G6-OH, $PtCl_4^{2-}$, and $PdCl_4^{2-}$ is essentially the sum of the spectra of a solution containing G6-OH + $PtCl_4^{2-}$ and a second solution containing G6-OH + $PdCl_4^{2-}$, which strongly suggests co-complexation of Pt^{2+} and Pd^{2+} within individual dendrimers. After reduction of these co-complexed materials, a new interband transition, which has an intensity different from that of either a pure Pt or a pure Pd cluster, is observed.

The sequential loading method is also effective for preparing bimetallic clusters. For example, dendrimer-encapsulated Pt/Pd clusters are synthesized as follows. First, a solution containing G6-OH(Pt_{55}) is prepared by using the direct reduction approach already described. Next, K_2PdCl_4 is added to this solution to

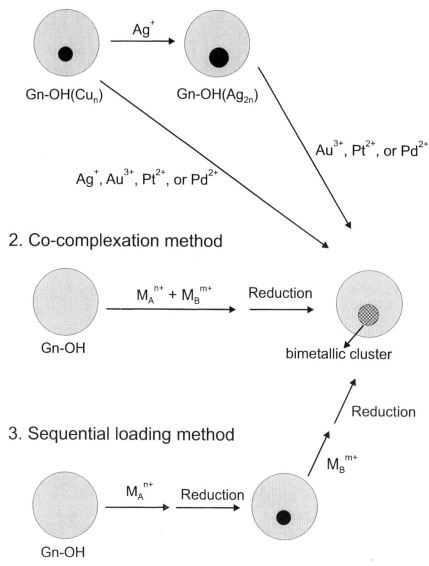

Fig. 10. Schematic of the preparation of dendrimer-encapsulated bimetallic clusters by displacement reaction, co-complexation, and sequential loading. Reprinted with permission of Springer–Verlag (Ref. 47). Copyright 2000 Springer–Verlag.

form a mixed metal-ion intradendrimer composite: G6-OH[(Pt_{55}) + (Pd^{2+})$_{55}$]. That is, Pd^{2+} partitions into the dendrimer and complexes with tertiary amine sites vacated by Pt^{2+} after the first reduction step. The existence of the mixed valent composite can be confirmed by the UV-vis spectrum, which is nearly the sum of the two individual components. However, after reduction of G6-OH[(Pt_{55}) + (Pd^{2+})$_{55}$] the intense LMCT band arising from Pd^{2+} complexed to interior tertiary amines essentially disappears and the interband transition due to the formation of the bimetal clusters appears. After the first reduction step, HRTEM indicates that the Pt nanoclusters in G6-OH(Pt_{55}) are 1.4 ± 0.2 nm in diameter. However, after addition of Pd^{2+} and subsequent reduction the particles grow to 3.0 ± 1.0 nm. The relatively wide distribution in particle size may be due to interdendrimer transfer of Pd atoms during the second reduction process.

Presumably these three methods for preparing bimetallic, dendrimer-encapsulated nanoparticles can be extended to trimetallics, bi and trimetallics having unique structures (such as core/shell materials), and interesting combinations of two (or more) zero-valent metals plus intradendrimer ions. However, analysis of such materials awaits more sophisticated analytical methods than are currently at our disposal.

IV. CATALYSIS USING TRANSITION-METAL NANOPARTICLES

Transition-metal nanoparticles are important in the field of catalysis (30,91–94). Synthetic routes to metal nanoparticles include evaporation and condensation, and chemical or electrochemical reduction of metal salts in the presence of stabilizers (91,92,95–97). The purpose of the stabilizers, which include polymers, ligands, and surfactants, is to control particle size and prevent agglomeration. However, stabilizers also passivate cluster surfaces. For some applications, including catalysis, it is desirable to prepare small, stable, but not fully passivated, particles so that substrates can access the encapsulated clusters. Another promising method for preparing clusters and colloids involves the use of templates, such as reverse micelles (98,99) and porous membranes (93,100,101). However, even this approach results in at least partial passivation and mass-transfer limitations unless the template is removed. Unfortunately, removal of the template may result in slow agglomeration of the naked particles. By using dendrimers as both template and stabilizer, we achieve control over particle size and particle stability while allowing substrates to penetrate the dendrimer interior and access the cluster surface (at this time the extent of mass-transfer resistance suffered by the substrate is unknown). To the best of our knowledge this advantageous set of properties is unique.

A. Dendrimer-Encapsulated Pt Nanoparticles as Heterogeneous Electrocatalysts for O_2 Reduction

Pt is the most effective practical catalyst known for O_2 reduction, which is an important electrode reaction in fuel cells (32,102). To achieve efficient utilization of Pt in such applications, a large amount of work has been devoted to developing methods for fabricating very small, stable, O_2-accessible clusters. Because the dendrimer-encapsulated metal nanoparticles seemed likely to exhibit these properties, we thought it reasonable to test them as heterogeneous O_2 reduction electrocatalysts. Such applications require immobilization of the dendrimer composites on a conducting substrate in such a way that electrons can transfer from the substrate to the metal nanoparticle (77). Methodologies for attaching dendrimers to surfaces by electrostatic and covalent means, and by coordination, have been discussed in the literature (37,89,103–107). We have generally found it most convenient to prepare surface-confined, dendrimer-encapsulated catalysts by first depositing the nanoclusters within the dendrimer and then affixing the dendrimers to a gold electrode by dipping the it into the dendrimer solution (77,89,107). Even dendrimers terminated in hydroxyl groups adhere to gold, presumably via interactions with internal tertiary amine groups that are able to configure themselves in a way that puts them in intimate contact with the gold surface (77,89).

Cyclic voltammograms (CVs) obtained from Au electrodes coated with a monolayer of either G4-OH or G4-OH(Pt_{60}) reveal this catalytic effect (77). Specifically, in the presence of O_2 a G4-OH–modified electrode yields a relatively small current having a peak potential (E_p) of 2150 mV. However, when a G4-OH(Pt_{60})–modified electrode is examined in the same solution, a much larger current is observed and the peak potential shifts positive, indicating a substantial catalytic effect. In the absence of O_2, only a small background current is observed, confirming that the process giving rise to the peak is O_2 reduction. Importantly, results such as these conclusively demonstrate that the surface of at least some of the dendrimer-encapsulated Pt nanoparticles are accessible to reactants in the solution and can exchange electrons with the underlying electrode surface.

B. Homogeneous Catalysis in Water Using Dendrimer-Encapsulated Metal Nanoparticles

Ligand- or polymer-stabilized colloidal noble metals have been used for many years as catalysts for the hydrogenation of unsaturated organic molecules (108–111). Additionally, there is a special interest in developing "green" methodologies for catalyzing organic reactions in aqueous solutions (108). Accordingly, we investigated the homogeneous catalytic hydrogenation of alkenes in aqueous solutions using dendrimer-encapsulated nanoparticles (58). The hydrogenation activities for dendrimer-encapsulated Pd nanoparticles for a simple, unbranched

Semiconductor Nanoparticles

alkene (allyl alcohol) and an electron-deficient, branched alkene (*N*-isopropyl acrylamide) in water are given in Fig. 11. Similar data have been obtained for dendrimer-encapsulated Pt nanoparticles and the results are analogous. GC and NMR confirmed the product of these hydrogenation reactions. G4-OH(Pd$_{40}$) shows a high catalytic activity for the hydrogenation of both alkenes. For example, the turnover frequencies (TOFs, mol of H$_2$/mol of metal atoms-h) for G4-OH (Pd$_{40}$) for hydrogenation of *N*-isopropyl acrylamide and for allyl alcohol hydrogenation compare favorably to water-soluble polymer-bound Rh(I) catalysts (109,110) and are comparable to PVP-stabilized colloidal Pd dispersions in water (111). The key result is that substrates (the alkenes and hydrogen in this case) can permeate dendrimers, encounter the nanoparticle therein, and undergo intradendrimer chemical transformation.

Importantly, the hydrogenation reaction rate can be controlled by using dendrimers of different generations. This key finding is a consequence of the fact that dendrimer porosity is a function of generation: higher-generation materials are more sterically crowded on their exterior and thus less porous and less likely to admit substrates to interior metal nanoparticles than the lower generations (58,103). That is, the dendrimer acts as a selective nanoscopic filter that controls the catalytic activity of the composite. For example, the TOFs for G6-OH(Pd$_{40}$) and G8-OH(Pd$_{40}$) are only 10% and 5%, respectively, that of G4-OH(Pd$_{40}$) for *N*-isopropyl acrylamide. However, when the same materials are used to reduce the linear alkene, a much smaller decrease in activity is noted. This key finding shows that it is possible to control reaction rates and do selective catalysis by adjusting the "mesh" of the dendrimer "nanofilter." In this case the high-generation dendrimers selectively excluded the branched alkene, but the linear molecule was able to reptate through the dense G6 and G8 exteriors and encounter the catalyst.

C. Homogeneous Catalysis in Organic Solvents Using Dendrimer-Encapsulated Metal Nanoparticles

From an environmental and economic perspective it would be highly desirable to carry out all catalytic reactions in water. Unfortunately, many substrates and products are water insoluble, and therefore many commercially important catalytic reactions are run in organic solvents. Accordingly, we also tested the activity of dendrimer-encapsulated Pd nanoparticles for their hydrogenation activity in organic phases. As mentioned previously, the solubility of dendrimers can be controlled by functionalization of the terminal groups. For example, alkyl functionalization renders the otherwise water-soluble PAMAM dendrimers soluble in organic solvents. Importantly, functional groups need not be added to the dendrimer by covalent grafting, although this is also a viable strategy (vide infra) (4). A simpler and more versatile approach that is suitable for many applications takes

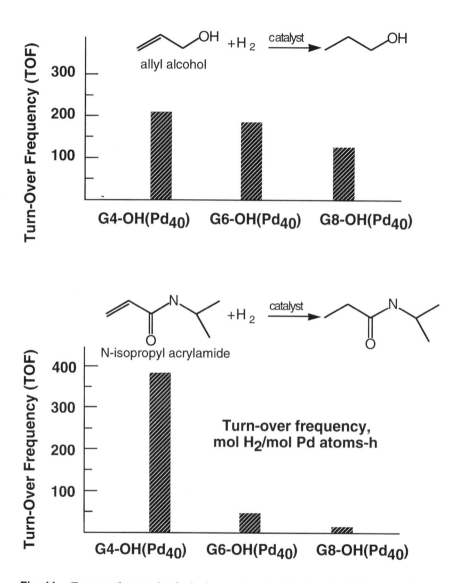

Fig. 11. Turnover frequencies for hydrogenation of allyl alcohol and N-isopropyl acrylamide obtained in water using dendrimer-encapsulated Pd nanoparticles of constant size. The hydroxyl-terminated PAMAM dendrimer generation varies from G4 to G8. Reprinted with permission of Springer–Verlag (Ref. 47). Copyright 2000 Springer–Verlag.

advantage of the acid-base interaction between fatty acids and dendrimer terminal amine groups (53).

Using an approach similar to that discussed previously, Pd nanoparticles were prepared within amine-terminated PAMAM dendrimers. To prevent coordination of Pd^{2+} to the primary amine groups of the dendrimers, the solution pH was adjusted to ~2, which preferentially protonates the exterior amines to a greater extent than the interior tertiary amines. Accordingly, Pd^{2+} binds preferentially to the interior tertiary amines, and upon reduction Pd particles form only within the dendrimer interior. $G4-NH_2$ dendrimer-encapsulated nanoparticles can then be quantitatively transported from an aqueous phase into toluene by addition of 10–20% of dodecanoic acid to the organic phase (53). This transition is readily visualized by the color change: the brown aqueous solution of Pd nanoparticles becomes clear after addition of the acid, while the toluene layer turns brown. Our studies have shown that this is a consequence of the formation of monodisperse inverted micelles templated by the dendrimer. These interesting hybrid materials have a catalytic activity for hydrogenation similar to that observed for the same reactions run in aqueous solutions (vide supra).

D. Homogeneous Catalysis in Fluorous Solvents Using Dendrimer-Encapsulated Metal Nanoparticles

Reactions in biphasic fluorous/organic systems were suggested by Horvath and co-workers in 1994 (112) to facilitate recovery and recycling of soluble catalysts. Such systems consist of organic and fluorous layers. The catalyst is selectively soluble in the fluorous phase, while the reactants are preferentially soluble in the organic solvent. Stirring and/or heating of the mixture leads to formation of a fine emulsion and partial homogenization (with some solvents complete homogenization is obtained at elevated temperatures) and the catalytic reaction proceeds at the interface between the two liquids. When the reaction is over, the liquid phases are allowed to settle, the product is isolated from the organic phase, and the catalyst-containing fluorous layer is recycled. Such easy separation and recycling are particularly attractive in terms of "green chemistry," and a number of fluorous phase-soluble catalysts have been reported in the literature, including some based on metal complexes (113,114). We recently described two new approaches for using dendrimer-encapsulated metal particles to perform biphasic catalysis. The first is hydrogenation catalysis using PAMAM dendrimers rendered soluble in the fluorous phase by electrostatic attachment of perfluoroether groups (78). The second demonstrates the use of perfluoroether groups covalently linked to the exterior of PPI dendrimers to carry out a Heck reaction (75). In both cases the fluorous-soluble, dendrimer-based catalyst was recyclable. Moreover, the reactions resulted in selectivities and products that clearly reflected the unique microenvironment of the dendrimers. Note also that the PPI dendrimers are stable at quite high temperatures (in contrast to PAMAM dendrimers, which tend to decompose rather quickly above 100°C). Finally, the successful Heck chemistry (75) conclusively proves that it is possible to catalyze synthetically useful carbon-carbon bond-forming reactions between two reactants, which have significant size and mass, within a dendrimer interior.

E. Homogeneous Catalysis in Supercritical CO_2 Using Dendrimer-Encapsulated Metal Particles

As mentioned earlier, there are good reasons to search for reaction conditions that eliminate the need for organic solvents. The use of liquid or supercritical (SC) CO_2 addresses some of these issues, including catalyst recovery, reduced toxicity, and simpler product recovery (115). Until recently, however, the use of SC CO_2 had been limited to organometallic Pd complexes functionalized with perfluorinated ligands (116–118), due to the limited solubility of metal colloids in CO_2, and often required the use of water as a cosolvent (119). The work described here shows that dendrimers can be used to solubilize Pd nanoclusters in liquid and SC CO_2. This new finding opens the door to the combined benefits of a catalyst that promotes Heck couplings but without the need for toxic ligands or solvents.

As discussed earlier, perfluorinated polyether "ponytails" can be covalently grafted onto dendrimers, and DeSimone et al. recently showed that such materials are soluble in liquid CO_2 (120). Preliminary results from a study of catalytic activation of the heterocoupling between arylhalides and alkenes using ponytail-functionalized dendrimer-encapsulated Pd nanoparticles have been promising. For example, the classic Pd-catalyzed Heck coupling between arylhalides and methacrylate yields predominately (>97%) the *trans*-cinnimaldehyde product (121). On the other hand, the CO_2-soluble dendrimer nanocomposite exclusively catalyzes the production of the highly unfavored 2-phenyl-acrylic acid methyl ester isomer at 5000 psi and 75°C (122).

V. DENDRIMER-ENCAPSULATED SEMICONDUCTOR NANOPARTICLES

Up to this point we have focused on the intradendrimer synthesis of metal nanoparticles, but in principle any type of particle can be prepared inside a dendrimer template if a means can be found to first sequester the components and then chemically transform them into the desired product. We recently demonstrated the versatility of the dendrimer template approach by preparing dendrimer-encapsulated semiconductor nanoparticles (123,124). Specifically, we have shown that it is possible to control the size, and thus the photoluminescent properties, of encapsulated CdS quantum dots (QDs) (125).

Semiconductor Nanoparticles

Efforts to prepare composites of dendrimers and technologically important QDs (126–129) have so far yielded primarily *inter*dendrimer composites. Such materials have been shown to be agglomerates in which multiple dendrimers stabilize relatively large CdS QDs. Results from these important studies suggested to us that it might be feasible to prepare *intra*dendrimer QD nanocomposites equivalent to the dendrimer-encapsulated metals described early in this chapter.

Figure 12 shows absorption and luminescence spectra of an alcoholic solution of G4 PAMAM dendrimers in which ultrasmall (<2 nm) CdS QDs have been grown. The extreme blue shift of both the absorbance and emission spectra (from that of bulk CdS), as well as the narrow linewidths, point to the existence of intradendrimer CdS (125). Evidence for intradendrimer templating comes from generation-dependent studies in which CdS QDs prepared in the presence of the higher-generation dendrimers G6 (spectrum b) and G8 (spectrum c) yield larger particles that have red-shifted absorbance and emission

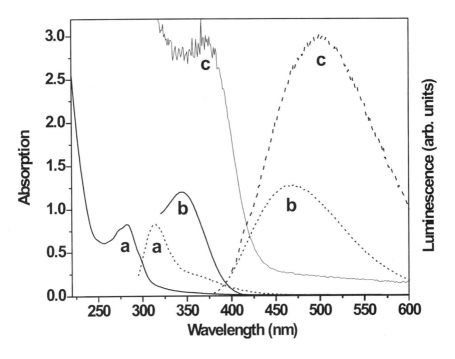

Fig. 12. Absorption (solid lines) and luminescence (dotted lines) spectra of dendrimer-encapsulated CdS nanoparticles: (a) G4-OH; (b) G6-OH; (c) G8-OH. Reprinted with permission of Springer–Verlag (Ref. 47). Copyright 2000 Springer–Verlag.

spectra. HRTEM data confirm these generation-dependent increases in size of the QDs. As for dendrimer-encapsulated metal nanoparticles, the size of the CdS nanoparticles is related to the number of tertiary amine groups in the dendrimer's outer shell.

VI. SUMMARY AND CONCLUSIONS

This chapter has comprehensively reviewed new composite materials called dendrimer-encapsulated nanoparticles. We first reported these interesting inorganic/organic hybrid materials in 1998 (57), and within a matter of months our results were confirmed by others (60). Between the time of the first report and the present, we have shown that dendrimers can template a vast array of metal nanoparticles, including Cu, Pt, Pd, Ru, Ag, Ni, and Au. Metal ions that bind directly to ligands on the interior of the dendrimer are prepared by direct reduction of the composite, but it is also possible to prepare intradendrimer metal nanoparticles from nonbinding ions by displacing a less noble metal such as Cu. We have also shown that bimetallic, and presumably multimetallic, dendrimer-encapsulated metal nanoparticles can be prepared and that even semiconductor quantum dots are accessible.

There are a number of desirable aspects of this synthetic approach and of the resulting materials. First, because the dendrimer can act as a template for preparing the particles, and because dendrimers are commercially available in different sizes (generations), nearly monodisperse particles ranging in size from <1 nm to perhaps 5 or 6 nm in diameter can be prepared. Second, once synthesized, the nanoparticles are stabilized by the dendrimer host. Therefore, the particles do not agglomerate even after repeated drying and resolution cycles. Third, the dendrimer acts as a nanofilter that selectively allows particular substrates to encounter the encapsulated particle. As we showed, this attribute allows the composites to perform size-selective catalysis. Moreover, because the particles are contained within the dendrimer primarily by steric considerations (e.g., in contrast to only ligand stabilization), the nanoparticle surface is sufficiently accessible that catalytic reactions proceed at reasonable rates. Fourth, because the particle is confined within the dendrimer, functional groups on the dendrimer surface can be used to control solubility, as a synthetic handle for stabilization of the nanocomposite in polymer coatings, and as a means for enabling direct surface immobilization of dendrimers on metal, semiconducting, and insulating surfaces. Thus, the approach described here for preparing dendrimer-encapsulated nanoparticles takes advantage of each of the unique aspects of dendrimer structure: the chemistry of the terminal groups, the generation-dependent size, the three-dimensional structure, the low-density region near the core, and the endoreceptors present within the dendrimer interior.

Semiconductor Nanoparticles

These materials are likely to find three major uses in the years ahead. First, interactions between metalated dendrimers, particularly those containing semiconductor quantum dots, and biologically relevant materials will be important. In this application the dendrimer can act as a site-specific shuttle for any number of guests (DNA, drugs, contrast agents, and the like), while the encapsulated semiconductor acts as a highly stable, biocompatible luminescent tag for the composite. Second, the application of dendrimers to catalysis is likely to be of great importance for reasons discussed in the preceding paragraph. For example, the unique microenvironment of the dendrimer interior can be used to house one or more catalysts or cocatalysts, while the exterior functional groups can be used to control solubility (even in fluorous and supercritical fluids), enhance catalyst recovery, and selectively gate access of substrates into the dendrimer interior. Moreover, intriguing recent results suggest that it may be possible to control the dendrimer interior to selectively favor particular products, perhaps even particular enantiomers. Finally, the ability to precisely control the size (and perhaps shape) of metal and semiconductor particles opens the door to a vast range of fundamental studies relevant to the fields of electronics, photonics, catalysis, and magnetism.

The preparation of dendrimer-encapsulated nanoparticles is generally rather simple from an experimental viewpoint. For example, the dendrimers are all commercially available from either Dendritech or DSM, end-group modifications normally involve a single synthetic step, and separations are generally carried out by dialysis. Such simplicity should be a strong enticement to workers in the fields of catalysis, biotechnology, medicine, electronics, photonics, physics, materials science, and engineering to (a) begin to use the types of materials described herein for technological applications and fundamental discoveries, (b) invent new, related materials, and (c) begin to use these dendrimer composites as components in the next generation of molecular machines and devices.

ACKNOWLEDGMENTS

We acknowledge our co-workers past and present that contributed to the results described in this chapter. Professor Victoria J. DeRose and Dr. Tomasz Wasowicz designed and executed EPR and ENDOR experiments that support the conclusions described in this chapter. Likewise, Professor David H. Russell and Mr. Li Zhou collaborated with us on MALDI-TOF MS experiments that confirmed the presence of Cu ions within the dendrimers. Professors Keith P. Johnston and C. Ted Lee at the University of Texas in Austin worked with us on the catalysis experiments in CO_2. Mr. Mark Kaiser (formerly of Dendritech, Inc.), with whom we have had the pleasure of collaborating for many years, provided invaluable technical advice and most of the PAMAM dendrimers used in our early studies.

Finally, we would also like to express our gratitude to the Office of Naval Research, the National Science Foundation, and the Robert A. Welch Foundation for providing financial support of our work. Additional support was provided via subcontract from Sandia National laboratories, which is supported by the U.S. Department of Energy (Contract DE-AC04-94AL8500). Sandia is a multiprogram laboratory operated by the Sandia Corporation, a Lockheed-Martin company, for the U.S. Department of Energy. M. Z. acknowledges fellowship support from The Electrochemical Society, Inc., the Eastman Chemical Company, and Phillips Petroleum. We are also indebted to Ms. Mary Sacquety for her careful and diligent editing and formatting of this manuscript.

REFERENCES

1. E Buhleier, W Wehner, F Vögtle. Synthesis:155, 1978.
2. F Vögtle. Topics in Current Chemistry. Berlin: Springer, 1998.
3. F Zeng, SC Zimmerman. Chem Rev 97:1681, 1997.
4. AW Bosman, HM Janssen, EW Meijer. Chem Rev 99:1665, 1999.
5. M Fischer, F Vögtle. Angew Chem Int Ed 38:884, 1999.
6. S Forster, M Antonietti. Adv Mater 10:195, 1998.
7. RT Clay, RE Cohen. New J Chem 22:745, 1998.
8. ABR Mayer, JE Mark. Coll Poly Sci 275:333, 1997.
9. ABR Mayer. Mater Sci Eng C C6:155, 1998.
10. T Hashimoto, M Harada, N Sakamoto. Macromolecules 32:6867, 1999.
11. M Moffitt, H Vali, A Eisenberg. Chem Mater 10:1021, 1998.
12. A Weitz, J Worrall, F Wudl. Adv Mater 12:106, 2000.
13. T Vossmeyer, L Katsikas, M Girsig, IG Popovic, K Diesner, A Chemseddine, A Eychmuller, H Weller. J Phys Chem 98:7665, 1994.
14. AV Firth, DJ Cole-Hamilton, JW Allen. Appl Phys Lett 75:3120, 1999.
15. MC Schlamp, X Peng, AP Alivisatos. J Appl Phys 82:5837, 1997.
16. RM Penner, CR Martin. J Electrochem Soc 133:2206, 1986.
17. JC Hulteen, CR Martin. J Mater Chem 7:1075, 1997.
18. KB Shelimov, M Moskovits. Chem Mater 12:250, 2000.
19. SM Marinakos, JP Novak, LC Brousseau, AB House, EM Edeki, JC Feldhaus, DL Feldheim. J Am Chem Soc 121:8518, 1999.
20. D Routkevitch, T Bigioni, M Moskovits, JM Xu. J Phys Chem 100:14037, 1996.
21. PAMAM dendrimers are available from Dendritech, Inc. (Midland, MI) or the Aldrich Chemical Co. (Milwaukee, WI). PPI dendrimers may be purchased from DSM in the Netherlands.
22. Technical data from DSM Fine Chemicals (The Netherlands) and Dendritech, Inc (Midland, MI, USA).
23. AM Naylor, IWA Goddard, GE Kiefer, DA Tomalia. J Am Chem Soc 11:2339, 1989.
24. EMM De Brabander-van den Berg, A Mijenhuis, J Mure M, Keulen, R Reintjens, F Vandenbooren, B Bosman, D de Raat, T Frijins, S van den Wal, M Castelijns, J Put, EW Meijer. Macromol Symp 77:51, 1994.

25. A Miedaner, CJ Curtis, RM Barkley, DL DuBois. Inorg Chem 33:5482, 1994.
26. JFGA Jansen, EW Meijer, EMM de Brabander-van den Berg. J Am Chem Soc 117:4417, 1995.
27. JFGA Jansen, EMM de Brabander-van den Berg, EW Meijer. Science 266:1226, 1994.
28. G Schmid. Clusters and Colloids. Weinheim: VCH, 1994.
29. PV Kamat. Prog Inorg Chem 44:273, 1997.
30. GJK Acres, GA Hards. Phil Trans R Soc Lond A 354:1671, 1996.
31. P Stonehart. In: JAG Drake, ed. Electrochemistry and Clean Energy. Cambridge: The Royal Society of Chemistry, 1994.
32. A Hamnett. Phil Trans R Soc Lond A 354:1653, 1996.
33. ITE Fonseca. In: Sequeira CAC, ed. Chemistry and Energy I. New York:Elsevier, 1991.
34. AJ McEvoy, M Grätzel. In: CAC Sequeira, ed. Chemistry and Energy I. New York:Elsevier, 1991.
35. GR Newkome, CN Moorefield, F Vögtle. Dendritic Molecules. Weinheim: VCH, 1996.
36. H Frey, C Lach, K Lorenz. Adv Mater 10:279, 1998.
37. F Gröhn, G Kim, BJ Bauer, EJ Amis Macro–Molecules 34:2179, 2001.
38. RJ Puddephatt. J Chem Soc Chem Commun:1055, 1998.
39. MA Hearshaw, JR Moss. J Chem Soc Chem Commun:1, 1999.
40. EC Constable. J Chem Soc Chem Commun:1073, 1997.
41. V Balzani, S Campagna, G Denti, A Juris, S Serroni, M Venturi. Acc Chem Res 31:26, 1998.
42. MHP Rietveld, DM Grove, G van Koten. New J Chem 21:751, 1997.
43. M Enomoto, TJ Aida. Am Chem Soc 121:874, 1999.
44. PJ Dandliker, F Diederich, M Gross, M Knobler, A Louati, EM Sanford. Angew Chem Int Ed Engl 33:1739, 1994.
45. KW Pollak, JW Leon, JMJ Fréchet, M Maskus, HD Abruña. Chem Mater 10:30, 1998.
46. CB Gorman, JC Smith, MW Hager, BL Parkhurst, H Sierzputowska-Gracz, CA Haney. J Am Chem Soc 121:9958, 1999.
47. RM Crooks, BI Lemon III, LK Yeung, M Zhao. In: F Vögtle, ed. Topics in Current Chemistry. Heidelberg: Springer, 2000, p. 81–135.
48. GR Newkome, F Cardullo, EC Constable, CN Moorefield, AMWC Thompson. J Chem Soc, Chem Commun:925, 1993.
49. K Takada, DJ Díaz, HD Abruña, I Cuadrado, C Casado, B Alonso, M Morán, J Losada. J Am Chem Soc 119:10763, 1997.
50. D Seyferth, T Kugita, AL Rheingold, GPA Yap. Organometallics 14:5362, 1995.
51. JW Kriesel, S König, MA Freitas, AG Marshall, JA Leary, TD Tilley. J Am Chem Soc 120:12207, 1998.
52. MF Ottaviani, E Cossu, NJ Turro, DA Tomalia. J Am Chem Soc 117:4387, 1995.
53. V Chechik, M Zhao, RM Crooks. J Am Chem Soc 121:4910, 1999.
54. DA Tomalia, HD Durst. Top Curr Chem 165:193, 1993.
55. MF Ottaviani, C Turro, NJ Turro, SH Bossmann, DA Tomalia. J Phys Chem 100:13667, 1996.

56. S Watanabe, SL Regen. J Am Chem Soc 116:8855, 1994.
57. M Zhao, L Sun, RM Crooks. J Am Chem Soc 120:4877, 1998.
58. M Zhao, RM Crooks. Angew Chem Int Ed Engl 38:364, 1999.
59. M Zhao, RM Crooks. Chem Mater 11:3379, 1999.
60. L Balogh, DA Tomalia. J Am Chem Soc 120:7355, 1998.
61. MF Ottaviani, ND Ghatlia, SH Bossmann, JK Barton, H Durr, NJ Turro. J Am Chem Soc 114:8946, 1992.
62. MF Ottaviani, S Bossmann, NJ Turro, DA Tomalia. J Am Chem Soc 116:661, 1994.
63. MF Ottaviani, F Montalti, M Romanelli, NJ Turro, DA Tomalia. J Phys Chem 100:11033, 1996.
64. MF Ottaviani, F Montalti, NJ Turro, DA Tomalia. J Phys Chem B 101:158, 1997.
65. AW Bosman, APHJ Schenning, RAJ Janssen, EW Meijer. Chem Ber/Recueil 130:725, 1997.
66. K Takada, GD Storrier, M Morán, HD Abruña. Langmuir 15:7333, 1999.
67. DJ Díaz, GD Storrier, S Bernhard, K Takada, HD Abruña. Langmuir 15:7351, 1999.
68. K Vassilev, WT Fored. J Polym Sci A37:2727, 1999.
69. SC Bourque, F Maltais, WJ Xiao, O Tardiff, H Alper, P Arya, LE Manzer. J Am Chem Soc 121:3035, 1999.
70. T Muto, K Hanabusa, H Shirai. Macromolecular Rapid Comm 20:98, 1999.
71. RJMK Gebbink, AW Bosman, MC Feiters, EW Meijer, RJM Nolte. Chem Eur J 5:65, 1999.
72. MC Moreno-Bondi, G Orellana, NJ Turro, DA Tomalia. Macromolecules 23:910, 1990.
73. L Zhou, DH Russel, M Zhao, RM Crooks. Macromolecules 34: 3567, 2001.
74. T Wascowicz, M Zhao, RM Crooks, VJ DeRose. Unpublished results.
75. LK Yeung, RM Crooks. Nano Lett 1:14, 2001.
76. DA Skoog, DM West, FJ Holler. Analytical Chemistry, 6 ed. Harcourt Brace, Philadephia, 1994.
77. M Zhao, RM Crooks. Adv Mater 11:217, 1999.
78. V Chechik, RM Crooks. J Am Chem Soc 122:1243, 2000.
79. F Gröhn, BJ Bauer, Y Akpalu, CL Jackson, EJ Amis. Macromolecules 33:6042, 2000.
80. F Fanizzi, FP Intini, L Maresca, G Natile. J Chem Soc Dalton Trans:199, 1990.
81. CY Fong, ML Cohen, RRL Zucca, J Stokes, YR Shen. Phys Rev Lett 25:1486, 1970.
82. U Kreibig, M Vollmer. Optical Properties of Metal Clusters. Berlin: Springer, 1995.
83. H Abe, K-P Charle, B Tesche, W Schulze. Chem Phys 68:137, 1982.
84. AC Curtis, DG Duff, PP Edwards, DA Jefferson, BFG Johnson, AI Firkland, ASA Wallace. Angew Chem Int Ed 27:1530, 1988.
85. I Lisiecki, MP Pileni. J Am Chem Soc 115:3887, 1993.
86. KJ Klabunde. Free Atoms, Clusters, and Nanoscale Particles. San Diego: Academic Press, 1994.
87. Dendritech, Technology Review, 1995.

88. CL Jackson, HD Chanzy, FP Booy, DA Tomalia, EJ Amis. Proc Am Chem Soc 77:222, 1997.
89. H Tokuhisa, M Zhao, LA Baker, VT Phan, DL Dermody, ME Garcia, RF Peez, RM Crooks, TM Mayer. J Am Chem Soc 120:4492, 1998.
90. M Gerloch, EC Constable. Transition Metal Chemistry: The Valence Shell in d-Block Chemistry. Weinheim: VCH, 1994.
91. LN Lewis. Chem Rev 93:2693, 1993.
92. JS Bradley. In: G Schmid, ed. Clusters and Colloids. Weinheim: VCH, 1994.
93. GL Che, BB Lakshmi, ER Fisher, CR Martin. Nature 393:346, 1998.
94. E Reddington, A Sapienza, B Gurau, R Viswanathan, S Sarangapani, ES Smotkin, TE Mallouk. Science 280:1735, 1998.
95. G Schmid. Chem Rev 92:1709, 1992.
96. JD Aiken III, Y Lin, RGA Finke. J Mol Catal A Chem 114:29, 1996.
97. MT Reetz, W Helbig. J Am Chem Soc 116:7401, 1994.
98. C Petit, P Lixon, M Pileni. J. Phys. Chem. 97:12974, 1993.
99. CB Murray, DJ Norris, MG Bawendi. J Am Chem Soc 115:8706, 1993.
100. CR Martin. Science 266:1961, 1994.
101. Y Zhang, N Raman, JK Bailey, CJ Brinker, RM Crooks. J Phys Chem 96:9098, 1992.
102. BP Sullivan. Platinum Metals Rev 33:2, 1989.
103. M Wells, RM Crooks. J Am Chem Soc 118:3988, 1996.
104. H Tokuhisa, RM Crooks. Langmuir 13:5608, 1997.
105. VV Tsukruk, F Rinderspacher, VN Bliznyuk. Langmuir 13:2171, 1997.
106. VV Tsukruk. Adv Mater 10:253, 1998.
107. M Zhao, H Tokuhisa, RM Crooks. Angew Chem Int Ed 36:2596, 1997.
108. WA Herrmann, CW Kohlpaintner. Angew Chem Int Ed 32:1524, 1993.
109. F Joó, L Somsák, MT Beck. J Mol Catal 24:71, 1984.
110. DE Bergbreiter, Y-S Liu. Tetrahedron Lett 38:7843, 1997.
111. L Nádasdi, F Joó, I Horváth, L Vígh. Appl Catal A 162:57, 1997.
112. IT Horvath, J Rabai. Science 266:72, 1994.
113. B Cornils. Angew Chem Int Ed Engl 36:2057, 1997.
114. DP Curran. Angew Chem Int Ed 37:1174, 1998.
115. G Kaupp. Angew Chem Int Ed Engl 33:1452, 1994.
116. DK Morita, DR Pesiri, SA David, WH Glaze, W Tumas. Chem Commun 13:1397, 1998.
117. MA Carroll, AB Holmes. Chem Commun 13:1395, 1998.
118. BM Bhanage, Y Ikushima, M Shirai, M Arai. Tetrahedron Lett 40:6427, 1999.
119. M Ji, XY Chen, CM Wai, JL Fulton. J Am Chem Soc 121:2631, 1999.
120. AI Cooper, JD Londono, G Wignall, JB McClain, ET Samulski, JS Lin, A Dobrynin, M Rubinstein, ALC Burke, JMJ Fréchet, JM DeSimone. Nature 389:368, 1997.
121. RF Heck, JP Nolley. J Org Chem 37:2320, 1972.
122. LK Yeung, CJ Lee, Jr., KP Johnston, RM Crooks. Chem. Comm., submitted.
123. PV Kamat. Chem Rev 93:267, 1993.
124. BI Lemon, RM Crooks. J. Am. Chem. Soc. 122: 12886, 2000.
125. T Vossmeyer, L Katsikas, M Giersig, IG Popovic, K Diesner, A Chemseddine, A Eychmuller, H Weller. J Phys Chem 98:7665, 1994.

126. K Sooklal, LH Hanus, HJ Ploehn, CJ Murphy. Adv Mater 10:1083, 1998.
127. JR Lackowicz, I Gryczynski, Z Gryczynski, CJ Murphy. J Phys Chem B 103:7613, 1999.
128. LH Hanus, K Sooklal, CJ Murphy, HJ Ploehn. Langmuir: ASAP Article, 2000.
129. J Huang, K Sooklal, CJ Murphy, HJ Ploehn. Chem Mater 11:3595, 1999.

12
The Electrochemistry of Monolayer Protected Au Clusters

David E. Cliffel, Jocelyn F. Hicks, Allen C. Templeton, and Royce W. Murray
University of North Carolina, Chapel Hill, North Carolina

I. INTRODUCTION

The advent of improved synthetic methods for metal nanoparticles, bolstered by a report by Schiffrin (1), has generated a widespread research effort on the chemical and electronic nature of metallic nanoparticles. There are tremendous opportunities in the nanoparticle world for chemists to contribute understanding and diversity to this dimension of matter bridging small molecules and bulk materials. Schriffrin and co-workers' contribution was to combine classic two-phase colloid synthesis with metal-alkanethiolate self-assembled monolayer chemistry to produce nanoparticles much smaller (sub-4-nm) than traditional colloidal materials (1). We subsequently named these nanoparticles "monolayer protected clusters" (MPCs). From the viewpoint of a chemist, the great importance of MPCs, in addition to their small dimension, is that, owing to the protecting monolayer, they are stable as dry chemicals. MPCs resist aggregation of the metal cores when dried to a solvent-free state and can be repeatedly isolated and redissolved. This crucial feature allows subsequent monolayer chemical derivatization reactions and characterization steps, which we have reported (2–5) upon in a project aimed at developing a diverse chemistry of these large, polyfunctional molecules.

MPCs have proven to have a rich electrochemistry and electron transfer chemistry. Research efforts have focused on three aspects: (a) electronic charging of the electrical double layer of the MPC core dissolved in electrolyte solutions, (b) electronic conductivity of dry (e.g., solid-state) films of MPCs, and (c) the electrochemistry of redox active groupings attached to the protecting monolayer.

The chapter discusses some results in these areas. The results are oriented substantially toward MPCs with Au cores and organothiolate monolayers. Core charging will be emphasized since it comprises the strongest departure from previous electrochemical and electron transfer literature.

II. MPC QUANTIZED DOUBLE-LAYER CHARGING

Single electron transfer (SET) properties of nanoparticle structures have been known for some time (6–8), often under the experimental heading of "Coulomb staircase" behavior. A Coulomb staircase experiment is classically (6) conducted at reduced temperature, on single nanoparticles, using some kind of microjunction or scanning tunneling microscopy (STM) form of contact. Most are *chemically* poorly defined, often little more than double-junction grain boundaries between isolated fragments on a surface in which the junctions exhibited small capacitances. Recent reports (9–10) have employed better-defined and more manipulable single nanoparticles.

Our group recognized (11–14) that freely diffusing alkanethiolate-MPCs in electrolyte solutions should have well-defined electrical double-layer capacitances, which might be small enough to produce Coulomb staircase–like properties when charged by a working electrode. This indeed proved to be the case; the first results (11) were termed "ensemble Coulomb staircase charging" by analogy to previous Coulomb staircase experiments (6–10). Because the double-layer property is actually the more fundamental aspect of the phenomenon, and since double-layer concepts are well established in electrolyte solution chemistry, we later (13–15) adopted the phrase "quantized double layer charging" (QDL) to describe the SET charging of MPC nanoparticles.

When an MPC is diffusionally transported to a potentiostated working electrode/electrolyte interface, electron transfers occur to equilibrate the Fermi level of the MPC metallic core with that of the working electrode. The resulting current flow can be measured to follow this process. The electronic charging (i.e., gain/loss of electrons) of the MPC core leads to formation of an ionic space-charge layer, or electrochemical double layer, around the MPC. For sufficiently small MPCs the associated electrical capacitance (C_{CLU}) per MPC is tiny—a subattofarad (aF) number—so a SET causes a substantial and measurable change in the MPC potential, $\Delta V = [e/C_{CLU}]$, where e is the electronic charge. Since $\Delta V > k_B T$ (where k_B is the Boltzmann constant, 1.38×10^{-23} J/K, and T is absolute temperature), *sequential* SET reactions between the working electrode and MPCs diffusing to it occur at potential intervals (ΔV) that are well separated (i.e., resolvable) on the working electrode potential axis. We have come to recognize the resulting succession of resolved and evenly spaced electrochemical waves, an example of which is shown in Fig. 1, as the *signature* of electrochemical QDL processes.

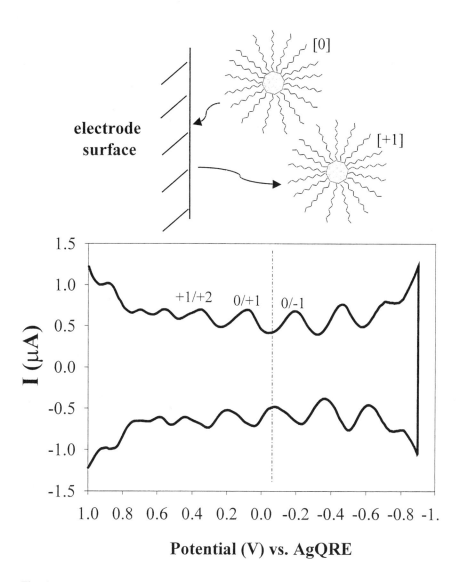

Fig. 1. Differential pulse voltammogram (DPV) showing QDL charging events for a ca. 0.1 mM fractionated C6 MPC in dichloromethane (DCM) at a 1.6-mm-diameter Au working electrode (0.05 M Bu_4NClO_4, potential versus Ag QRE reference with Pt coil counter-electrode). All charging events shown are above background. (From Ref. 15.) The background current is in part due to the non-monodisperse portion of the MPC sample. (Reproduced with permission from Anal. Chem., 1999, 71, 3703-3711. Copyright 1999, Am. Chem. Soc.)

The fractionation (16–17) of MPCs so as to produce narrow dispersities of core size was a crucial step in observing MPC double-layer charging, since QDL peaks are unresolved in a mixture of MPCs with sufficiently diverse core sizes. Owing to the dependence of C_{CLU} on core size, a mixture can simply produce a continuum of charging current. The charging current arising from a solution of polydisperse MPCs is, while featureless, distinct from that at a nanoscopic working electrode in that it is demonstrably (2,18–20) diffusion controlled by the MPCs. The average cluster capacitance can be estimated from the potential dependence of the diffusion-controlled MPC charging current in rotating-disk (18,19) and microelectrode voltammetry (2,20). For nanoparticles of large size (or C_{CLU}), the separation of SET peaks is unresolved whether they are monodisperse or not because $\Delta V < k_B T$. QDL is not seen on ordinary working electrodes, again for the same reason; their capacitance is too large and thus the separation between consecutive SET peaks is too small. In an elegant and pioneering experiment (21) with a nanoscopic working electrode, it was shown that SET steps can also be observed with a sufficiently small working electrode.

The regular spacing of the current peaks in Fig. 1 reflects the basically electrostatic nature of the MPC double-layer charging process (13) and is consistent with a metal-like core characteristic. If, on the other hand, the MPC metal core contains a sufficiently *small* number of Au (or other element) atoms, it will develop molecule-like properties. These properties would include a HOMO-LUMO gap (i.e., band gap) that can be detected optically and electrochemically. The electrochemical band gap is manifested (12) as a large separation between the initial oxidation and reduction, relative to the potential at which the MPC is uncharged (e.g., MPC potential-of-zero change or E_{PZC}, which has been estimated as ca. -0.2 V versus Ag/AgCl). This chapter discusses mainly MPCs (sizewise) near or above this molecular boundary, since there is more electrochemical information available for metal-like core MPCs.

The nature of a SET reaction, whether of a redox molecule, or QDL of a metal-like MPC core, or of a small molecule-like MPC core, specifies the electrochemical potential at which it occurs. If the reaction is kinetically fast, then it can be expected to follow classical principles governing electron transfer reactions of diffusing species (Nernst and Fick equations). That is, the QDL process will give rise (11–14) to mass-transport-controlled voltammetric current peaks with properties analogous to those of traditional redox reactions, such as the oxidation of ferrocene. The QDL formal potentials are referenced to the MPC's E_{PZC}.

A. QDL Charging Theory

This section summarizes a thermodynamic description of the potentials at which MPC charging occurs; the theory (13) is an extension of that developed by Weaver (22) in the context of fullerene electrochemistry (which represents, by analogy, molecule-like MPC behavior). Specifically the section explains, for a solution of monodisperse MPCs, the placement of the charging peaks on the working electrode potential axis.

The formal potential of the MPC (E_P) current peak, relative to its E_{PZC}, is

$$E_P = E_{PZC} + \frac{(z \pm \frac{1}{2})e}{C_{CLU}} \qquad (1)$$

where z is the (signed) number of electronic charges on the particle, and the (integral) capacitance of individual MPCs is C_{CLU}. Assuming spherical MPCs (the cores are actually thought to be truncated octahedra), C_{CLU} can be modeled as a capacitance of concentric spheres (with the monolayer being the insulating dielectric and the core and electrolyte solution the conductors):

$$C_{CLU} = A_{CLU}\frac{\varepsilon\varepsilon_0}{r}\frac{r+d}{d} = 4\pi\varepsilon\varepsilon_0\frac{r}{d}(r+d) \qquad (2)$$

where r is the monolayer thickness and d is core radius. This equation predicts that when r and d are comparable, C_{CLU} increases with increasing core radius, and for monolayers with similar dielectric property, decreases with increasing monolayer thickness. The limits of behavior are when $r < d$ where $C_{CLU} = \varepsilon\varepsilon_0 A_{CLU} r$ (e.g., a naked nanoparticle in a thick dielectric medium) and when $r > d$ where $C_{CLU} = \varepsilon\varepsilon_0 A_{CLU}$ (akin to a monolayer on a large flat surface). The MPCs that we have studied have dimensions intermediate between these limits. This vastly oversimplified equation is a surprisingly good predictor (13,15) of cluster capacitance.

The work W required to charge initially uncharged MPCs by z electrons to a potential E_P with a working electrode at potential E_{APP} is (at equilibrium)

$$W = \int_0^z e(E_{APP} - E_P)\,dz = ez\left[E_{APP} - \left(E_{PZC} + \frac{ze}{2C_{CLU}}\right)\right] \qquad (3)$$

At the electrode/solution interface, the corresponding average Boltzmann population of MPCs having z charges, relative to the population of uncharged MPCs (N_0) is

$$\frac{N_z}{N_0} = \exp\left(\frac{W}{k_B T}\right) = \exp\left\{\frac{ze}{k_B T}\left[E_{APP} - \left(E_{PZC} - \frac{ze}{2C_{CLU}}\right)\right]\right\} \qquad (4)$$

The ratio (α_z) of the populations of *adjacent* charged states (differing by one electron in charge), N_z/N_{z-1}, follows from Eq. (4) as

$$\alpha_z = \frac{N_z}{N_{z-1}} = \exp\left\{\frac{e}{k_B T}\left[E_{APP} - E_{PZC} - \frac{(z-\frac{1}{2})e}{C_{CLU}}\right]\right\} \qquad (5)$$

Eq. (5) is Nernstian in form. Thus an equimolar solution mixture of MPC nanoparticles of charges z and $z - 1$ comprises, in a formal sense, a 1:1 mixed valent "redox couple" solution, and has a formal potential, $E^\circ_{z,z-1}$ which is

$$E^\circ_{z,z-1} = E_{\text{PZC}} + \frac{(z - \frac{1}{2})e}{C_{\text{CLU}}} \tag{6}$$

Which is equivalent to EQ. (1). The nanoparticle double-layer charging behavior can be formally regarded as a multivalent redox system with equally spaced formal potentials. This condition requires that C_{CLU} be independent of the charge state of the MPCs, which for metal-like MPCs seems usually to be the case. When the MPC cores are small enough to be molecule-like, the above relations are still followed, but the SET potentials are not evenly spaced, due to the HOMO-LUMO gaps characteristic of molecular species.

When C_{CLU} is independent of z, Eq. (6) predicts that a plot of peak potential ($E_{z,z-1}$) versus MPC (relative) charge state (z) is linear with slope e/C_{CLU}. It is not necessary to know the absolute charge state of the MPC to evaluate C_{CLU}; knowing which charging steps correspond to $z = \pm 1/0$, etc., requires knowing E_{PZC}. We have estimated E_{PZC} by using AC impedance measurements of butanethiolate MPC solutions at a gold electrode (12).

Observing current peaks for quantized DL charging requires that the MPCs in the solution have reasonably uniform values of C_{CLU}; unresolved overlapping of differently spaced peaks would otherwise lead to a featureless charging background. A uniform C_{CLU} translates to uniform MPC core radius and monolayer thickness and composition. As-prepared MPCs from the Brust reaction (1) are normally somewhat polydisperse in core size (seen in transmission electron microscopy, TEM), and fractionation is required to procure monodisperse MPC samples. Fractionation has been described for C4 through C16 MPCs and confirmed by various analytical techniques that image the MPC core (15–17).

Fractionations to date have, however, generally been imperfect in the sense that a spread, albeit narrowed, of core sizes remain. To assess the impact on QDL charging voltammetry, of single and of bi-Gaussian distribution, the following equations, respectively, have been derived (13):

$$f(r) = \frac{1}{\sigma\sqrt{\pi}} \exp\left[-\frac{(r - r_0)^2}{\sigma^2}\right] \tag{7a}$$

$$f(r) = \frac{1}{\sqrt{\pi}}\left\{\frac{x}{\sigma_1}\exp\left[-\frac{(r - r_{01})^2}{\sigma_1^2}\right] + \frac{1-x}{\sigma_2}\exp\left[-\frac{(r - r_{02})^2}{\sigma_2^2}\right]\right\} \tag{7b}$$

where r_0 is mean core radius and σ its standard deviation, subscripts 1 and 2 denote the populations with r_{01} and r_{02} mean core radii, and x is the mole fraction of the r_{01} population. These distributions can be used to simulate the expected electrochemical response of an MPC solution, such as for steady-state microelectrode voltammetry, the limiting current equation for which is

$$i = 4nr_{EL}FDC^* \tag{8}$$

where n is the number of electrons transferred per MPC, r_{EL} the microelectrode radius, F Faraday's constant, and D and C^* the MPC diffusion coefficient and PC bulk concentration, respectively. The diffusion coefficient is size dependent and can be estimated (23) by the soft-sphere version of the Einstein-Stokes equation:

$$D = \frac{k_B T}{6\pi\eta r_H} \tag{9}$$

where η is solvent viscosity and r_H is the hydrodynamic radius. Combining the above equations gives

$$i_{NORM} = \int_{r_{low}}^{r_{high}} f(r) \left(\sum_{-N}^{0} \frac{-1}{1+\alpha_z} + \sum_{1}^{N} \frac{\alpha_z}{1+\alpha_z} \right) \frac{dr}{r+d} \tag{10}$$

where i_{NORM} is normalized current [$i_{NORM} = 6i\pi\eta/4nr_{EL}FC^*k_BT$], and r_{low} and r_{high} are, respectively, the lower and upper limits of the core sizes (as observed experimentally). In Eq. (10), the first and second summation terms represent current contributions from adding electrons to or removing electrons from the MPC in the charging step, respectively.

Equations (9) and (10) predict (13,23) that QDL charging peaks are best defined when there is a single monodisperse population of core sizes in a sample, and that with increasing dispersity in C_{CLU} the charging peaks become indistinct especially for peaks remote from the E_{PZC}.

B. Effect of Core Size on QDL Charging

Equations (1) and (2) predict the dependency of the spacing of QDL charging peaks on the working electrode potential axis. The lengths of the vertical lines in Fig. 2A show (13) the predicted variation of spacing of the $z = +1/0$ and $0/-1$ peaks (based on $\Delta V = e/C_{CLU}$ and centered about E_{PZC}) for a series of MPC core sizes (and C6 monolayers). The ● and ○ points are experimental results for the same series of core sizes. The $z = +1/0$ and $z = 0/-1$ peaks represent, formally, the solution analogs of ionization potential (IP) and electron affinity (EA), respectively. The two smaller core sizes (8 and 14 kDa MPCs, core radius < 0.7 nm) show a spacing between the $z = +1/0$ and $z = 0/-1$ peaks that is larger than that anticipated by Eq. (1) and (2) from the spacings of the three larger core sizes. This was interpreted (12) as the emergence of a HOMO-LUMO gap, which is also seen in the near-IR spectroscopy of these smaller and molecule-like clusters. Figure 2B presents the data as voltage spacing between the $z = +1/0$ and $z = 0/-1$ peaks, which shows the metal-to-molecule transition more dramatically and suggests that it begins at a cutoff core size of approximately 22 kD.

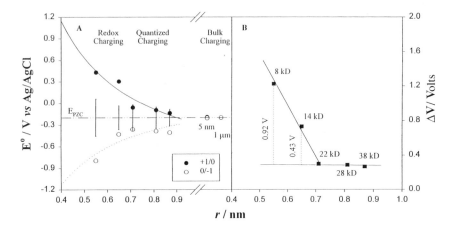

Fig. 2. (A) Formal potential of and (B) peak spacing between the $z = \pm 1/0$ peaks of hexanethiolate Au MPCs with various core sizes. (●) $z = +1/0$ and (○) $z = 0/-1$ (Data points of $r = 0.65$ nm are for C4 Au MPC, 22 kD). The lines (−) and (·) are only eye guides. Vertical bars in part A are the theoretical predictions of ΔV for the C6 Au MPCs based on their C_{CLU} estimated from Eq. (1). The values of ΔV for illustrative larger particle dimensions (the "bulk" domain) correspond to calculations from Eq. (2) using $\varepsilon = 3$, $d = 0.77$ nm, and $r = 5$ nm (a colloidal particle) and $r = 5$ μm (a small microelectrode disk), giving 13 mV and 0.37 μV, respectively. (From Ref. 13.) (Reproduced with permission from J. Phys. Chem B., 1998, 102, 9898-9907. Copyright 1998, Am. Chem. Soc.)

Figure 2A shows that the metal-like quantized double-layer capacitance charging will be observed only over a modest range of core size dimensions. Figure 2A shows predictions of ΔV for a larger metal cluster ($r = 5$ nm) and for a small microelectrode. The important point is that even for rather small cluster dimensions (e.g., 5 nm) the predicted ΔV values shrink to values below (room temperature) $k_B T$ and become unobservable as in the more conventional domain of double-layer charging phenomena. Figure 2A divides the core size regions into "redox" (molecule-like), "quantized" (metal-like), and bulk (metal) domains. The exact dimensions of the domains depend, of course, on monolayer dimensions and dielectric (and associated C_{CLU}) and, potentially, the core metal as well.

C. Evaluating C_{CLU} from Voltammetry of Polydisperse MPCs.

Average cluster capacitance can be evaluated electrochemically from the I-E slopes of the transport-controlled currents seen in MPC solutions with steady-state forms of voltammetry. This has been most successfully done for MPCs that additionally bear redox groupings at the monolayer chain termini. The pre- and/or the postredox wave I-E slopes have been analyzed in rotated-disk and microelectrode voltammetry (2,18–20). For a microelectrode the MPC double-layer capacitance per unit Au core surface area A_{CLU} is related to the slope of the current-potential curve ($\Delta i/\Delta E$) between potentials E_A and E_B by

$$\frac{i_A - i_B}{E_A - E_B} = 4nrDC\frac{\overline{\sigma_A} - \overline{\sigma_B}}{E_A - E_B} \tag{11}$$

$$C_{DL} = \frac{\overline{\sigma_A} - \overline{\sigma_B}}{(E_A - E_B)NA_{CLU}} \tag{12}$$

where C_{DL} is the apparent double layer capacitance of the working electrode, σ_A and σ_B are the charge per mole of MPC at E_A and E_B, N is Avogadro's number, D is the diffusion coefficient, and C is the concentration of the MPC (20).

D. Electrolyzing MPCs

Observing the manifestation of MPC core double-layer capacitances in the preceding way suggests that solutions of even polydisperse MPCs might be usefully charged to various potentials. The charge could then be used to drive electron transfer reactions, as is done with ordinary electrodes. After determining the average cluster capacitance, the number of electrons that an MPC solution electrolyzed at potential E_P can deliver in a reaction with a substance having a redox formal potential that is the same as E_{PZC} can be estimated by Eq. (1). If the reaction substrate has a different formal potential, then the number of available electrons is estimated by replacing E_{PZC} with that potential in Eq. (1).

This section summarizes results for the storage and use of electrochemical energy as MPC core charge in a chemical experiment. Solutions of charged MPCs were obtained by electrolyzing MPC solutions at fixed potentials (24). Generally the charged MPCs were used in as-prepared solutions, but the charged MPCs can also be isolated in dried form and redissolved with minimal loss of charge. Figure 3 shows a potentiometric titration of an ethyl ferrocene solution with a solution of positively charged MPCs. The end point of the titration (the upward break in potential) corresponds to the amount of redox charge in the MPC solution that was consumed by the ethyl ferrocene, which was quantitatively equal to the initial amount of ethyl ferrocene. This example shows that MPC charge can be used to drive a chemical reaction in a stoichiometrically quantitative manner.

Charged MPCs also undergo electron transfer reactions among themselves. This was demonstrated (24) by mixing solutions of differently charged MPCs and observing the resulting equilibrium potentials. The equilibrium potential should be describable from the Nernst equation, taking into account the stoichiometry of the mixed solutions and the mixed valent solution resulting from electron transfer reactions between "oxidizing" and "reducing" MPCs. This relation is, of course,

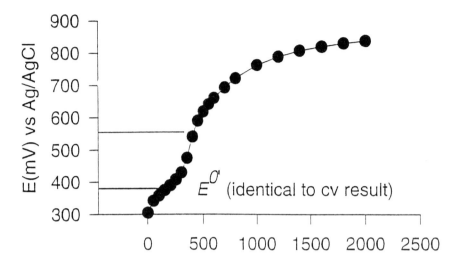

Fig. 3. Titration of 5 mL of 76 μM ethylferrocene in 0.1 M Bu$_4$NPF$_6$/2:1 toluene/CH$_3$CN by incremental addition of the indicated number of μL to a 0.3 mM MPC solution previously charged to +0.92 V. (From Ref. 24.)

$$E = E^{o\prime}_{MPC} + 0.059 \log\left\{\frac{N_z}{N_{z-1}}\right\} \quad (13)$$

It was also noted that mixing solutions of differently charged MPCs (prepared batchwise ahead of time and stored dry) is a convenient preparative route to solutions having a desired state of MPC charge. It was possible to easily prepare MPC solutions with a desired potential ±10 mV (24,28), and a higher level of accuracy is probably attainable.

E. Effect of Chain Length on QDL Charging

Equation (2) indicates that cluster capacitance increases with core radius r and decreases with monolayer chain length d; varying either will cause an observable change in cluster capacitance.

It is important to note the assumptions inherent in Eq. (2). First, the MPC core is considered to be not spherical but a truncated octahedron (16). Second, an effective dielectric constant of $\varepsilon = 3.0$ is used in the comparisons to data; actual values for, say, pentane and pentanethiol monomers would be 1.8 and 4.55, respectively. However, the used dielectric constant is consistent with self-assembled monolayer results (25). Third, we assumed that the dielectric thickness, d, is defined by fully extended alkanethiolate chains. Fourth, the monolayer/electrolyte solution interface is assumed to be sharp and spherical, and that the electrolyte solution is a good conductor over the entire sphere (i.e., no discreteness of ionic charge effects).

The assumption involving the monolayer chains was tested (15) in experiments like that in Fig. 4, where differential pulse voltammetry (DPV) is shown for solutions of fractionated C6, C8, and C12 MPCs in 2:1 toluene/CH_3CN. The QDL charging peaks are reasonably well defined, and the peak spacing (e.g., C_{CLU}) changes regularly with monolayer chain length, as anticipated. The ΔE_{PEAK} differences seen between the positive- and negative-going E_{DC} potential scans are attributed mainly to uncompensated resistance losses (iR_{UNC}). The C12 MPC has a peak spacing of 0.38 V and a smaller capacitance (0.40 aF) than the C6 sample ($\Delta V = 0.27$, $C_{CLU} = 0.72$ aF).

Plots of the Fig. 4 peak potentials [Eq. (1)] versus MPC charge state gave values of C_{CLU} that are presented, along with other analogous data, in Table 1 ("full" z-plot results). The right-hand column in Table 1 gives the ratio of these experimental results to values of C_{CLU} calculated from Eq. (2), using the assumptions noted above. The associated d values and core radii (r) values are given in the table. The ratio is near unity (generally within 10%). It is remarkable that over a fivefold variation in d and ca. twofold variation in C_{CLU}, such a simple, concentric sphere model [Eq. (2)] would predict experimental behavior so well. It is probable that further and more exacting measurements, and further chemical diversity of monolayer and electrolyte, will reveal its limitations. Acquiring such measurements will be strongly dependent on inventing procedures for more producing perfectly fractionated MPCs.

Equation (2) also predicts that C_{CLU} varies linearly with the monolayer dielectric constant. One would imagine that penetration of solvent (or electrolyte ions) into the monolayer would, by changing the effective monolayer dielectric constant, produce a solvent dependence of C_{CLU}. Experiments in Table 2 show, however, that cluster capacitance varies little over a threefold change in solvent dielectric constant. It follows that, at least for C6 alkanethiolate chain lengths, the thiolate monolayer properties dominate the dielectric constant and, in turn, the double-layer capacity and C_{CLU}.

Some DPV features are not exactly observant of the ideal Eq. (1). First, the spacing (ΔV) between DPV charging peaks of shorter-chain MPCs (Fig. 4A) decreases at the most positive potentials; i.e., the effective values of C_{CLU} increase for highly positively charged MPCs. The origin of this reproducible phenomenon in Fig. 4A is uncertain; one possibility is enhanced counterion penetration (e.g., specific adsorption) into the monolayer at high positive core potential. The decrease in ΔV at positive potentials is seen only for the shorter-chain MPCs, either because it does not occur for longer-chain MPCs or because they do not become as highly charged over the accessible potential range.

Exclusion of the two most positive data points in Eq. (1) plots lowers the average C_{CLU} result, Table 1, but in reference to the calculated C_{CLU} value this does not produce an obviously better agreement with the calculated C_{CLU}.

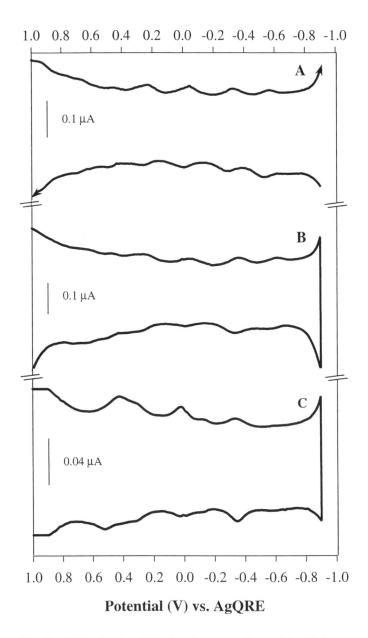

Fig. 4. DPVs showing QDL charging events for ca. 0.1 mM fractionated Au MPCs in toluene/CH$_3$CN at a 0.5-mm-diameter Au working electrode (0.05 M Bu$_4$NClO$_4$, potential versus Ag QRE reference with Pt coil counterelectrode). All charging events shown are above background: (A) C6 MPC, (B) C8 MPC, (C) C12 MPC. (From Ref. 15.) (Reproduced with permission from Anal. Chem., 1999, 71, 3703-3711. Copyright 1999, Am. Chem. Soc.)

Table 1. Capacitance as a Function of Monolayer Chain Length

Chain length (r,d^b) nm	Calc. C_{CLU}	Expt C_{CLU} Z-plots Full	$<+2^c$	Ratio[a] (expt./cal.)
In 2:1 Toluene/CH$_3$CN				
C4[d] (0.8[e],0.52)	0.69	0.59	0.66	0.9
C6[d] (0.8[e],0.77)	0.53	0.57	0.51	1.1
C6[d] (1.0[f],0.77)	0.77	0.72	0.63	1.1
C8[d] (0.7[f],1.02)	0.40	0.52		1.3
C10[d] (0.8[f],1.27)	0.44	0.47		1.1
C12[d] (0.8[e,f],1.52)	0.39	0.40		1.0
C16[d] (0.8[f],2.02[h])	0.36	0.39		1.1
In CH$_2$Cl$_2$				
C6[g] (0.8,0.77)	0.53	0.57	0.55	1.1
C6[d] (1.0[f],0.77)	0.77	0.70	0.62	1.1
C8[d] (0.7[f],1.02)	0.40	0.55		1.4
C10[d] (0.8[f],1.27)	0.44	0.53		1.2

[a] Ratio calculated using all charge states.
[b] Dimensions of the monolayer were obtained from molecular modeling data (Hyperchem).
[c] Capacitance calculated for charge states $<+2$.
[d] Core size selection achieved by fractional precipitation (Ref. 5).
[e] Average core size determined by LDI/MS.
[f] Average core size determined by TEM.
[g] Core size selection achieved by precipitation (Ref. 15).
[h] Value corrected (typo error) from original publication (15).
Source: Ref. 15.

Second, the definition of QDL peaks changes with solvent (15), being for example better in CH$_2$Cl$_2$ than in 2:1 toluene:CH$_3$CN. The peaks seen in CH$_2$Cl$_2$ are larger relative to background currents, and at negative potentials, in both solvents. The average FWHM of a DPV peak in a C6 sample in CH$_2$Cl$_2$ solvent can be as small as 115 mV but is usually larger. Ideally, the FWHM should be 90 mV for a reversible one-electron process (26). The difference could arise from slow-electron-transfer kinetics (unlikely for a tunneling barrier chain as short as C6) or from imperfect core size (or monolayer) monodispersity. The FWHM criterion is potentially a very sensitive measure of monodispersity.

The factors that determine the DPV peak resolution in these experiments are presently obscure. One factor, of course, is the level of monodispersity of core sizes (and thus C_{CLU} and ΔV) in the MPC sample. As noted above, simulations show (13) that in polydisperse samples the overlapping of QDL charging peaks progressively reduces peak definition for peaks more removed from the MPC

Table 2. C6 MPC Capacitance Measurements in Various Solvents[a]

Solvent	Dielectric constant (ε_s)	C_{CLU}, aF	Ratio[b]
2:1 Toluene/CH$_3$CN	14.1[c]	0.63[d]	1.0
Pyridine	12.4	0.67	0.9
3:1 Toluene/CH$_3$CN	11.5[e]	0.62	1.0
1,2-Dichloroethane	10.4	0.73	0.9
Dichloromethane	9.00	0.62	1.0
Tetrahydrofuran	7.58	0.66	0.9
Chlorobenzene	5.53	0.45[e]	1.3

[a] All measurements were performed at a 1.6-mm-diameter Pt working electrode (AgQRE reference and Pt coil counter) with 0.30 mM MPC solutions in each of the respective solvents (same sample was recovered and used for all experiments) containing 0.05 M Bu$_4$NClO$_4$ electrolyte.
[b] Ratio is capacitance relative to that measured in 2:1 toluene/CH$_3$CN.
[c] Estimated by fractional weighting (by volume) of data on pure solvents; ignores effect of specific solvation.
[d] Value differs from that reported in Table 1 for same solvent/electrolyte combination. The difference in capacitance may be attributed to variation in average core size and sample dispersity.
[e] DPV charging peaks were observed only at negative potentials for this sample, in contrast to the others.
Source: Ref. 15.

E_{PZC}. The difference in peak resolution at positive and negative potentials is thus likely due to imperfect monodispersity. Adsorption may also influence peak resolution. If some of the DPV response is due to adsorbed MPCs, that response may be solvent and/or potential dependent. A third factor could be specific adsorption of electrolyte counteranion at the monolayer/electrolyte interface, and the associated statistical distribution of MPCs with adsorbed counterions may influence the peak resolution seen at positive potentials. These factors will require further work to confirm their importance.

III. ELECTRON TRANSFER KINETICS FROM MPC SOLID-STATE CONDUCTIVITY

While electron transfer kinetics have not yet been measured for MPCs in solutions, the solid-state electronic conductivity of thin MPC films is directly related to the electron transfer kinetics between MPC clusters. At low applied voltages, thin films of MPCs exhibit ohmic behavior, and the conductivity of the film is strongly dependent on the chain length of the alkanethiol monolayer insulating the clusters from each other (27,28). The electron transfer coupling coefficient for tunneling, β, derived from the chain-length dependence was 0.8 Å^{-1} (28), which is comparable to other reports of β (ca. 1 Å^{-1}) for 2-D alkanethiolate monolayers (29,30).

At higher applied voltages, the current-potential curves rise exponentially as expected for an activated process (27). The electron transfer is best modeled as a bimolecular self-exchange of electrons hopping between neighboring clusters. Estimates of the electron transfer rate constants for self-exchange, k_{ex}, were 2×10^{10}, 7×10^8, and 2×10^8 M^{-1} s^{-1} for C8, C_{12}, and C_{16} alkanethiolate MPC clusters, respectively (27). A chain-length dependence of the electron transfer rate constant is, of course, expected at high, as well as low, applied voltages, given the role of electron tunneling in the conductivity. Using chemical charging of the MPCs to bias the initial rest potential of the cores (i.e., making "mixed valent," solid-state MPCs), it has been recently observed (28) that the conductivity is enhanced for mixtures of charge states and minimized when all the MPCs have the same charge state. This behavior is strongly supportive of a bimolecular electron transfer picture in the solid-state materials.

IV. ELECTROCHEMISTRY OF REDOX-FUNCTIONALIZED MPCS

The electrochemistry of MPCs functionalized with multiple copies of redox groups is also interesting since it provides a platform to examine MPC materials in the context of multielectron transfer electrocatalysis and biosensing applications, among others. We (2–5,31,32) and others (33–35) have pioneered methods, including ligand place-exchange and coupling (amide and ester) reactions, to prepare MPCs functionalized with multiple copies of redox and other interesting groups (for example, an MPC bearing 25 copies of a ferrocene group). Ligand place-exchange reactions are simple in that MPC and ω-functional thiol (bearing the group of interest) are simply codissolved and stirred for fixed periods to incorporate the ω-functionalized thiol into the cluster monolayer (2,5,31). A functional group of interest can alternatively be coupled via a reactive handle to pendant MPC reactive monolayer functionalities (alcohols, acids, etc.) by forming ester or amide bonds (3,4,32). Coupling reactions are often the preferred route since the synthesis of ω-functionalized thiol may not be facile.

Using place-exchange and/or coupling reactions, MPCs have been prepared to date bearing multiple copies of ferrocene (Fig. 5a) (2,18,19,31,36), anthraquinone (Fig. 5b) (2,3,20), phenothiazine (Fig. 5c) (4), and viologen (Fig. 5d) (32). While the study of redox-functionalized MPCs bearing such groups is still in its infancy, some of the significant questions to be addressed with these materials are: (a) What is the nature (kinetics, mechanism) of their polyelectron reactions at the electrode/solution interface? (b) What is the nature of the interactions of the multiple donor/acceptor sites with one another. (c) Can the donor/acceptor sites be spatially organized on an MPC periphery such that they interact with one another in a controlled manner? and (d) Can the redox-functionalized MPCs

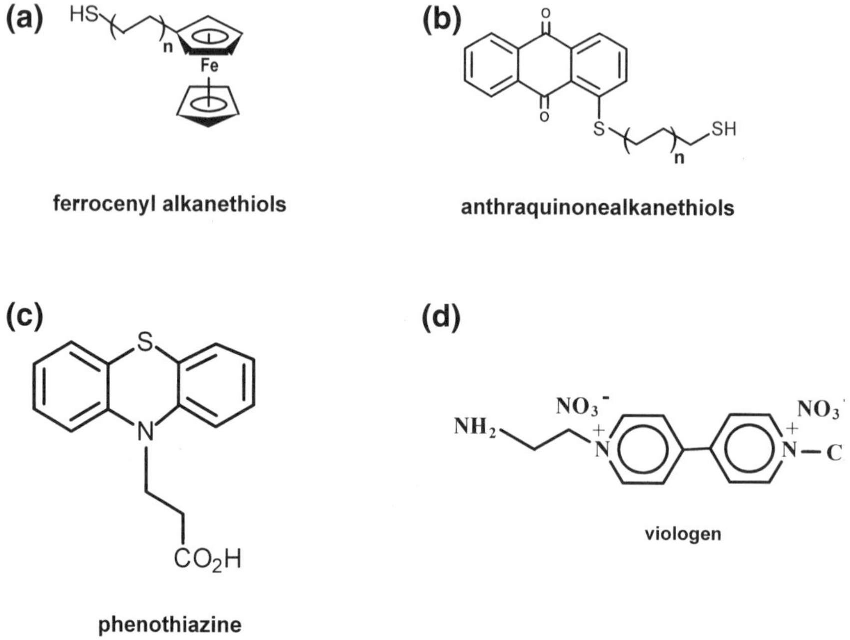

Fig. 5. Portfolio of redox groups used to functionalize MPCs to date (a) ferrocenealkanethiols, (b) anthraquinonealkanethiols, (c) phenothiazine derivative, and (d) viologen derivative.

perform mediated electrocatalysis with solution substrates, perhaps taking advantage of cooperative effects with additional appropriately positioned MPC functional groups? All of these questions are relevant to designing MPCs as conduits for truly nanoscale chemistry.

Our work on the small cadre of redox-functionalized MPCs (Fig. 5) has been directed at questions like the above. Figure 6a shows an illustrative cyclic voltammogram for Fc_9-MPC solution (31). The basic nature of the voltammetry of such a system depends on the strength of electronic coupling between each of the Fc sites and/or the underlying MPC core. Strong electronic coupling could potentially produce a concerted nine-electron Fc reaction; weak electron coupling, on the other hand, leads to a rapid sequence of nine one-electron transfer reactions. Thin-layer coulometry of phenothiazine-functionalized MPCs indicates all redox groups attached to clusters are electroactive (Fig. 6b inset) (4). Analysis of redox-functionalized MPC voltammetry, like that shown in Fig. 6a, b, indicates that the reaction occurs as a rapid series of diffusion-controlled, concerted one-electron transfers. The few known precedents of such unusual reactions include soluble redox polymers and redox-labeled dendrimers. The weak

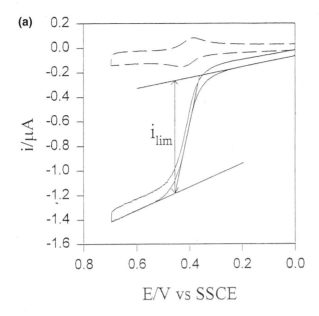

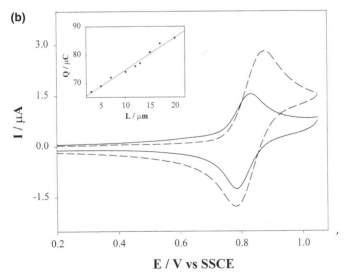

Fig. 6. (a) Cyclic voltammograms at 5 mV/s for the 1 μM 1:9.5 C8Fc/C8 MPC (avg. of 9 Fc/cluster) in 0.1 Bu$_4$NClO$_4$/CH$_2$Cl$_2$ at a 0.15 cm^2 stationary (- -) and rotated (—, 1600 rpm) glassy carbon electrode (from Ref. 31). (b) Cyclic voltammetry of 0.8 mM (in phenothiazine) cluster in 2:1 toluene:CH$_3$CN (v/v) at 100 mV/s. (Upper inset): Thin-layer coulometry charge Q vs. cell length ($r^2 = 0.98$), indicating that all phenothiazine groups are electroactive. (From Ref. 4)

electronic coupling of all the attached redox groups explored to date is not at all surprising given the length of ligand spacer employed. Shortening or altogether removing the connecting spacer would be expected to promote strong electronic coupling and, potentially, multielectron transfer reactions. MPCs bearing mixtures (11,20) of different redox groups (poly-*hetero*-functionalized MPCs), such as ferrocene and anthraquinone sites, exhibit voltammetry of the electron donor and acceptor groups occurring independently of each other, additional evidence for the weakly coupling natures of redox-functionalized MPCs with long alkane spacers.

The voltammetry of redox-functionalized MPCs includes currents for charging the double layer of nanoelectrode-like gold core. These currents are manifest in the pronounced slopes of the currents preceding ("prewave") and following ("postwave") the redox wave (Fig. 6a), which are observable in both macro- and microelectrode experiments. Simultaneous core charging and oxidation/reduction of attached redox groups tends to lead to underestimates in cluster hydrodynamic radii (r_H) and, in turn, diffusion coefficients obtained from electrochemical experiments. We have recently performed a comparison of diffusion coefficients obtained from electrochemical data and Taylor dispersion experiments that supports this notion (23).

The details of possible electronic interactions between core charging and oxidation/reduction of attached redox groups (especially for attached groups with very short or no spacers) is currently an unexplored topic.

Another goal was to explore the mechanistic pathway through which a diffusing redox-functionalized MPC becomes oxidized/reduced at the solution/electrode interface. At least three processes can be delineated by which multiple redox groups on MPCs may undergo reaction: rotational diffusion, electron hopping over the MPC surface, and/or electron tunneling down chains with transfer through the gold core. We have devised theoretical working models to estimated the rates of these processes. For the C8 alkanethiolate MPCs, rotational diffusion is predicted to be fastest, but not by a large margin (37). Further work is needed to validate the models and to establish how the individual reaction pathways can be detected and their rates systematically manipulated. The experimental approaches include varying the alkanethiolate linker chains, varying the number of redox groups/MPC, varying the rotational correlation time (through varying core size and solvent viscosity), and attaching the MPC to the electrode surface. In like manner, there is as yet no information on the electron transfer rates of redox-functionalized MPCs. One appealing approach is to employ a kinetically slower redox group than those studied thus far.

As mentioned briefly, multiple MPC electron donor and acceptor groups invite applications to mediated electrocatalysis. The basic prospects include concerted polyelectron exchanges with substrates, intramolecular electron transfer reactions during binding of MPCs to substrates (perhaps transiently), and ("cofactor") MPCs that are "multipurpose" by allowing both electron and proton transfers with a substrate in electron-proton coupled reactions. Our first probe of mediated electrocatalysis with MPCs was based on the reduction of *gem*-dinitrocyclohexane (DNC) by clusters bearing ca. 25 electrogenerated anthraquinone radical anion (AQ$^-$) sites (20,37). The two main findings of this study were that (a) the MPC AQ$^-$ site reactivities with the DNC are nearly the same as that of monomeric AQ$^-$ species, and (b) the observed electrocatalytic currents are enhanced, relative to freely diffusing AQ$^-$, due to the smaller diffusion coefficient of MPC AQ$^-$ and the ensuing compression of the electron transfer reaction layer (37).

Redox-functionalized MPCs have been found to adsorb readily to electrode surfaces, as evidenced from voltammetric data. Voltammetry data has been used to obtain adsorption amounts from the charges of surface-confined waves of ferrocene-functionalized MPCs. In a recent study, we used electrochemical quartz crystal microbalance (EQCM) experiments to detect and quantify the amount of adsorption of viologen-functionalized water-soluble MPCs (Fig. 7) (32). Chief findings from these studies were that water-soluble MPCs adsorb at single-monolayer coverages upon reduction and are, in turn, stripped upon being scanned positively and oxidized.

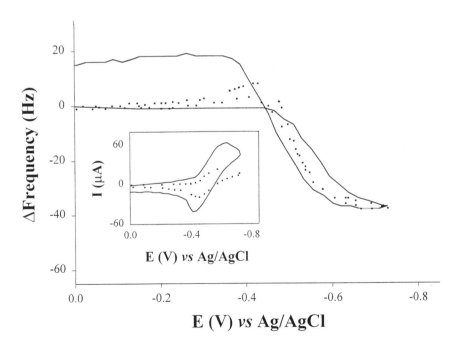

Fig. 7. EQCM crystal frequency change vs. potential at 10 mV/s (●●●) and 100 mV/s (—) in 2 mM (in viologen) cluster functionalized with viologen (36/cluster), in 0.1 M borate buffer (pH 9.2). Corresponding voltammograms for these experiments are shown as insets to the figure (From Ref. 32).

V. CONCLUSION

The electrochemistry of MPCs has revealed a great deal about the thermodynamics of quantized charging. Improving the synthetic routes to monodisperse clusters will enable electrochemical methods to elucidate the kinetics of electron transfers and increase our understanding of the thermodynamics of the quantized nature of electron transfer. In lieu of solution kinetics, the use of solid-state conductivity has provided information about the rates of electron transfer in solid films of alkanethiolate MPCs. Attaching electrochemically active tags to MPC clusters has yet to provide evidence of significant interaction between the tag and the core, nor has the ability to improve the rates of electron transfer at MPCs been successfully demonstrated. These challenges remain, but the rapidly improving synthetic methods of MPC formation may provide avenues to address them. The story of MPC electrochemistry has been exciting, but is hardly complete.

Added in proof: MPC–MPC electron hopping dynamics have been measured (38) in multilayer MPC films on electrodes (39).

ACKNOWLEDGMENT

This research was supported by grants from the Office of Naval Research, the National Science Foundation, and the Department of Energy, and by student fellowship support from the Lord Corporation, The Electrochemical Society, and the American Chemical Society (Perkin-Elmer Corporation).

REFERENCES

1. M Brust, M Walker, D Bethell, DJ Schiffrin, RJ Whyman. *J. Chem. Soc., Chem. Commun.* **1994**, 801–802.
2. RS Ingram, MJ Hostetler, RW Murray. *J. Am. Chem. Soc.* **1997**, *119*, 9175.
3. AC Templeton, MJ Hostetler, CT Kraft, RW Murray. *J. Am. Chem. Soc.* **1998**, *120*, 1906–1911.
4. AC Templeton, MJ Hostetler, EK Warmoth, S Chen, CM Hartshorn, VM Krishnamurthy, MDEF Forbes, RW Murray. *J. Am. Chem. Soc.* **1998**, *120*, 4845–4849.
5. MJ Hostetler, AC Templeton, RW Murray. *Langmuir* **1999**, *15*, 3782–3789.
6. R Wilkins, E Ben-Jacob, RC Jaklevic. *Phys. Rev. Lett.* **1989**, *63*, 801–804.
7. M Amman, R Wilkins, E Ben-Jacob, PD Maker, RC Jaklevic. *Phys. Rev. B* **1991**, *43*, 1146–1149.
8. AE Hanna, M Tinkham. *Phys. Rev. B* **1991**, *44*, 5919–5922.
9. RP Andres, T Bein, M Dorogi, S Feng, JI Henderson, CP Kubiak, W Mahoney, RG Osifchin, R Reifenberger. *Science* **1996**, *272*, 1323–1325.
10. JW Gerritsen, SE Shafranjuk, EJG Boon, G Schmid, H Vankempem. *Europhys. Lett.* **1996**, *33*, 279.
11. RS Ingram, MJ Hostetler, RW Murray, TG Schaaff, JT Khoury, RL Whetten, TP Bigioni, DK Guthrie, PN First. *J. Am. Chem. Soc.* **1997**, *119*, 9279–9280.

12. S Chen, RS Ingram, MJ Hostetler, JJ Pietron, RW Murray, TG Schaaff, JT Khoury, MM Alvarez, RL Whetten. *Science* **1998**, *280*, 2098–2101.
13. S Chen, RW Murray, SW Feldberg. *J. Phys. Chem. B* **1998**, *102*, 9898–9907.
14. S Chen, RW Murray. *Langmuir* **1999**, *3*, 682–689.
15. JF Hicks, AC Templeton, S Chen, KM Sheran, R Jasti, RW Murray, J Debord, TG Schaaff, RL Whetten. *Anal. Chem.* **1999**, *71*, 3703–3711.
16. RL Whetten, JT Khoury, MM Alvarez, S Murthy, I Vezmar, ZL Wang, PW Stephens, CL Cleveland, WD Luedtke, U Landman. *Adv. Mater.* **1996**, *8*, 428–433.
17. TG Schaaff, MN Shafigullin, JT Khoury, I Vesmar, RL Whetten, WG Cullen, PN First, C Gutièrrez-Wing, J Ascensio, MJ Yose-Yacamn. *J. Phys. Chem. B* **1997**, *101*, 7885–7891.
18. SJ Green, JJ Stokes, MJ Hostetler, JJ Pietron, RW Murray. *J. Phys. Chem. B* **1997**, *101*, 2663–2668.
19. SJ Green, JJ Pietron, JJ Stokes, MJ Hostetler, H Vu, WP Wuelfing, RW Murray. *Langmuir* **1998**, *101*, 2663.
20. RS Ingram, RW Murray. *Langmuir* **1998**, *14*, 4115–4121.
21. F-RF Fan, AJ Bard. *Science* **1997**, *277*, 1791–1793.
22. MJ Weaver, X Gao. *J. Phys. Chem.* **1993**, *97*, 332–338.
23. WP Wuelfing, AC Templeton, JF Hicks, RW Murray. *Anal. Chem.* **1999**, *71*, 4069–4074.
24. JJ Pietron, JF Hicks, RW Murray. *J Am. Chem. Soc.* **1999**, *121*, 5565–5570.
25. MD Porter, TB Bright, DL Allara, CED Chidsey. *J. Am. Chem. Soc.* **1987**, *109*, 3559–3568.
26. AJ Bard, LR Faulkner. *Electrochemical Methods: Fundamentals and Applications*; Wiley: New York, 1980.
27. RH Terrill, TA Postlethwaite, CH Chen, CD Poon, A Terzis, A Chen, JE Hutchison, MR Clark, G Wingall, JD Londono, R Superfine, M Falvo, CS Johnson Jr., ET Samulski, RW Murray. *J. Am. Chem. Soc.* **1995**, *117*, 12537–12547.
28. WP Wuelfing, SJ Green, JJ Pietron, DE Cliffel, RW Murray. Manuscript in preparation. *J. Am. Chem. Soc.* **2000**, *122*, 11465–11472.
29. SB Sachs, SP Dydek, RP Hsung, LR Sita, JF Smalley, Newton, SW Feldberg, CED Chidsey. *J. Am. Chem. Soc.* **1997**, *119*, 10563–10564.
30. JF Smalley, SW Feldberg, CED Chidsey, MR Linford, MD Newton, YP Liu. *J. Phys. Chem.* **1995**, *99*, 13141.
31. MJ Hostetler, SJ Green, JJ Stokes, RW Murray. *J. Am. Chem. Soc.* **1996**, *118*, 4212–4213.
32. AC Templeton, DE Cliffel, RW Murray. *J. Am. Chem. Soc.* **1999**, *121*, 7081–7089.
33. M Brust, J Fink, D Bethell, DJ Schiffrin, C Kiely. *J Chem. Soc., Chem. Comm.* **1995**, 1655–1656.
34. PA Buining, BM Humbel, AP Philipse, AJ Verkleij. *Langmuir* **1997**, *13*, 3921–2926.
35. KJ Watson, J Zhu, ST Nguyen, CA Mirkin. *J. Am. Chem. Soc.* **1999**, *121*, 462–463.
36. W-Y Lee, MJ Hostetler, RW Murray, M Majda. *Israel J. Chem.* **1997**, *37*, 213–223.
37. JJ Pietron, RW Murray. *J. Phys. Chem. B* **1999**, *103*, 4440–4446.
38. FP Zamborini, RW Murray. *J. Am. Chem. Soc.* **2000**, *122*, 4514–4515.
39. JF Hicks, FP Zamborini, AJ Osisek, RW Murray. *J. Am. Chem. Soc.* **2000**, *123*, in press.

13
Nanoparticle Electronic Devices: Challenges and Opportunities

Wyatt McConnell, Louis C. Brousseau III, A. Blaine House, Lisa B. Lowe, Robert C. Tenent, and Daniel L. Feldheim
North Carolina State University, Raleigh, North Carolina

I. INTRODUCTION

Recent advances in scanning probe microscopies, electron beam lithography, and other wet-chemical etching and patterning techniques have enabled the characterization of electron transport in individual nanoscale objects such as carbon tubes (1), alkanethiolates (2), metal and semiconductor nanoparticles (3), and atomic point contacts (4). These studies have revealed electronic behaviors unique to the nanometer-size regime, including resonant tunneling, quantum interference, resistance quantization, and single-electron tunneling. While there is still much debate over the ultimate utility of such behaviors in computing, certain niche of nanoscale devices have already been demonstrated in practice [e.g., electrometers (5) and chemical sensing (1,6)] or suggested in theory.

Electron transport in gold nanoparticles has been of interest to our group (6) and others recently (3). Gold is an appealing material for nanoscale electronics because gold particles may be synthesized with virtually any diameter from 0.8 nm to >200 nm, and highly size-monodisperse samples of gold particles may be isolated either from a 1-pot synthesis or by solvent extraction/fractionation (7). In addition, methods for modifying gold surfaces with alkylthiols, alkylselenides, phosphines, polymers, silicates, etc., are now well established (8); consequently a large number of size-equivalent nanoscale "electronic elements" with tailored solubilities and reactivities may be collected.

Of particular interest to our group is how surface chemistry affects electron transport in nanoscale objects. Our motive for addressing this question is that ultimately nanoparticle or nanotube electronics will rely heavily on molecules. For example, alkyldithiols may be used as interconnects to wire up particles to each other or to contact pads. Particle or nanotube surface coatings may also be used as molecular insulators, controlling electron tunneling rates or "crosstalk" between nanostructures (9). Finally, functional molecular monolayers or thin films may be useful as analyte recognition agents for the construction of nanoscale chemical sensors. A basic understanding of surface chemistry—nanoparticle electronic function relationships is thus an important step toward the implementation of many types of nanoscale electronic devices.

II. BACKGROUND

Electron transport through metal nanoparticles was described in Chapter 1 and has been reviewed extensively elsewhere. Briefly, because gold particles with diameters as small as ca. 2 nm behave as free-electron metals (e.g., contain a continuum of electronic states), a metal nanoparticle connected to two conductive leads through thin insulating tunnel junctions and driven by an ideal voltage source can be treated as a simple series RC circuit (10). Staircase-shaped I–V curves (the so-called Coulomb staircase) are then expected with charging potentials of

$$V = \frac{(Q_0 - 1/2)e}{C_2 + V_{\text{offset}}} \tag{1}$$

and current steps of

$$I = \frac{e}{2R_2C_T} \tag{2}$$

where Q_0 is the charge on the particle, C_2 and R_2 are the capacitance and resistance, respectively, of the most resistive particle-lead junction, C_T is the total particle capacitance, and V_{offset} accounts for any initial misalignment in tip-particle or particle-substrate Fermi levels and any charged impurities residing near the particle.

Equations (1)–(2) can be used to determine particle capacitance and junction resistances as a function of particle surface chemistry. For example, Kubiak measured single-electron tunneling through a gold cluster with an STM tip to determine the resistance of a few molecules assembled in the cluster/substrate junction (11). Similar measurements have also been reported recently by Mulvaney on SiO_2-coated gold particles (12), by Rao on poly(vinylpyrrolidone)-capped palladium and gold particles (13), and by Whetten on alkanethiol-coated gold nanoparticles (14). In Rao's work, particle charging energies followed a scaling law of the form $E = A + B/d$, where d is the diameter of the nanoparticle.

Measurements of single-particle $I–V$ behaviors have also been performed recently, using two-point conductance probes on planar Si substrates (15). Assembly of a single palladium particle in a ca. 4-nm gap between two electrodes was accomplished by electrostatic trapping (ET). In ET a voltage is applied between the two electrodes in the presence of a solution of nanoparticles. The electric field polarizes the metal particles and draws them into the gap. As soon as a single particle makes contact, the electrodes are shorted together and the field disappears. Thus, only one particle becomes trapped in the gap. $I–V$ measurements of the single particle subsequently revealed single-electron tunneling effects.

The measurements discussed above were all performed under DC bias conditions with a two-electrode configuration (source and drain). A third gate electrode can be employed to modulate the source-drain (S-D) single-electron currents. The electric field from a gate electrode causes shifts in the S-D current staircase along the potential axis (Fig. 1, top). At constant S-D bias, a gate bias can therefore be used to induce periodic oscillations of the S-D single-electron current (Fig. 1, bottom). Each period corresponds to a single electron induced on the particle. This behavior is in fact the principle behind the single-electron transistor electrometer and other devices which measure rapid charge fluctuations (5). Remarkably, single-electron electrometers are capable of detecting sub-single-electron amounts of charge placed on the gate electrode.

Charge placed on a gate electrode is one means of modulating single-electron currents in nanoparticles. However, other factors can influence electron transport energetics in nanoparticles, and thus the position of the current staircase along the voltage axis. Single charged impurities in the substrate can introduce shifts in nanoparticle $I–V$ curves (16). These shifts are highly undesirable in nanoscale computer circuitry, however, because it is unlikely that any two devices will be electronically equivalent.

Our initial hypothesis was that chemical transformations involving particle capping ligands could also alter electron transport through a nanostructure either by changing particle capacitance or via an electric field effect. Modulating nanoscale electronic behaviors chemically could ultimately obviate the need for gate electrodes, thereby alleviating inherent difficulties with attaching three contacts to a single nanoscale object (e.g., source, drain, gate), and could lead to sensitive nanoscale "chemical electrometers". In the following sections our initial results aimed at establishing basic capping ligand chemistry–nanoparticle electrical property relationships are described. In accord with our hypothesis, we have found that chemical transformations involving particle capping ligands can indeed influence electron transport through the nanoparticle significantly.

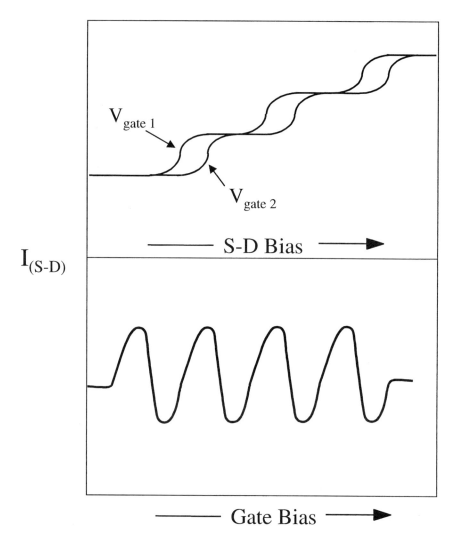

Fig. 1. Illustration of source-drain current vs. gate bias at constant source-drain bias (bottom) and for constant gate biases $V_{gate\ 1}$ and $V_{gate\ 2}$ (top).

III. ELECTRON TRANSPORT IN INDIVIDUAL LIGAND-MODIFIED GOLD PARTICLES

The effects of surface charge on single-electron tunneling were studied by synthesizing gold nanoparticles [ca. 5 nm in diameter using known protocols (7)] containing surface-bound ω-functionalized alkythiolates. The thiols were terminated with the pH-responsive ligands galvinol and NH_3 (Scheme 1). The particles were cast on hexanethiol-coated gold substrates and *I–V* curves recorded on individual particles with an STM tip in aqueous solution under controlled pH and ionic strength conditions.

I–V curves recorded at pH 5 for galvinol-gold across a wide bias range show several well-defined current steps indicative of single-electron tunneling (Fig. 2). These data are typical of the *I–V* curves acquired under all pH conditions, although peak-shaped steps were frequently observed at higher pH during negative-bias sweeps (vide infra).

The effects of increasing the solution pH on nanoparticle *I–V* behavior are shown in Fig. 3 (left). First, the entire staircase shifted to positive bias upon converting the neutral galvinol monolayer to the anionic galvinoxide species; i.e., each successive voltage plateau occurred at a more positive bias potential (Fig. 3).

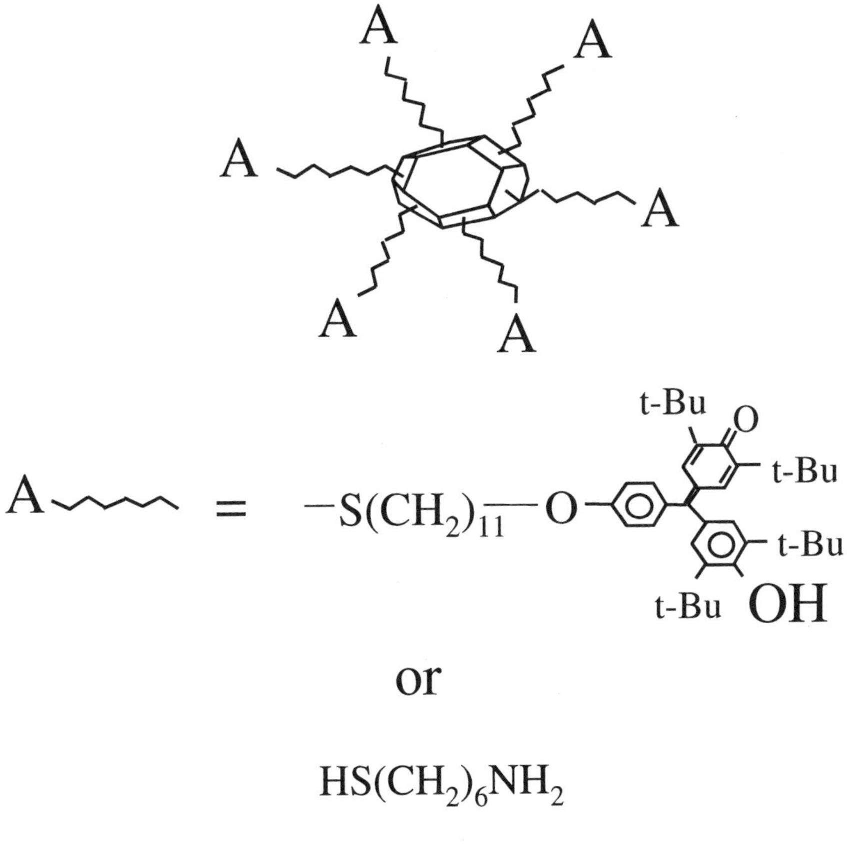

Scheme 1.

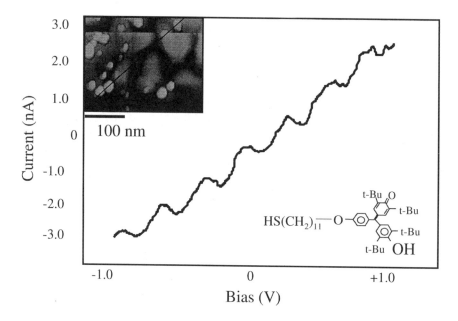

Fig. 2. Current vs. voltage for a single galvinol-coated gold particle acquired in aqueous solution at pH 5. Inset shows an STM image of the sample. Tip was coated with apiezon wax, gold substrate was insulated with hexanethiol.

The second effect observed upon monolayer charging was a decrease in the average potential plateau width (Table 1).

Similar behavior was observed in *I–V* curves recorded for mercaptohexylamine-modified gold particles, although peaked-shaped steps were now most often observed at low pH. Converting the neutral amine to the ammonium cation also caused shifts in the current staircase (Fig. 4). In contrast to galvinol-gold, however, the voltage plateau widths decreased upon lowering the pH (Table 1).

The contrasting behavior in voltage plateau widths versus pH for galvinol-gold and amino-gold particles provides evidence against systematic tip or substrate effects which could alter the current staircase independent of capping ligand chemistry (e.g., shifts in pzc, ionic strength effects). Further evidence against a tip or substrate effect was obtained through two control experiments: (a) particles containing pH-insensitive ligands (hexanethiol) were synthesized and *I–V* curves recorded versus pH (Fig. 5); and (b) *I–V* curves for galvinol-gold were recorded versus ionic strength, holding pH constant. In both control experiments relatively little change was observed in step potential (Table 1) or average step width.

The observations thus reported can be rationalized by considering the voltage offset term in Eq. (1) and the well-documented effects of monolayer charging

Nanoparticle Electronic Devices

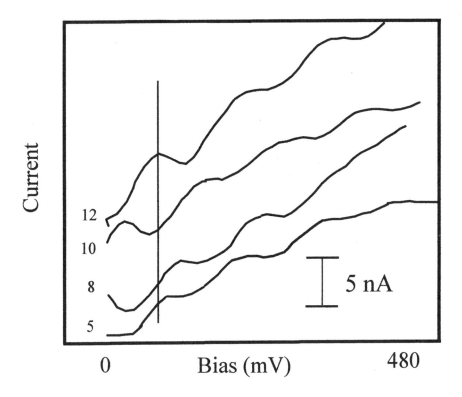

Fig. 3. Current vs. voltage curves for galvinol-coated gold particles acquired in aqueous solution at the pH indicated (curves have been offset for clarity). The vertical line marks the position of the second voltage plateau at pH 5.

Table 1. Single-Particle Capacitance as a Function of pH

Nanoparticle	Medium	Diameter (nm)	C(aF)
Gal-Au	H_2O (pH 5)	8	1.2 ± 0.12
	H_2O (pH 8)		1.8 ± 0.14
	H_2O (pH 12)		2.4 ± 0.20
Amino-Au	H_2O (pH 5)	5	0.70 ± 0.08
	H_2O (pH 8)		0.58 ± 0.011
	H_2O (pH 12)		0.54 ± 0.05
C_8-Au	H_2O (pH 5)	5	2.7 ± 0.30
	H_2O (pH 12)		2.5 ± 0.10

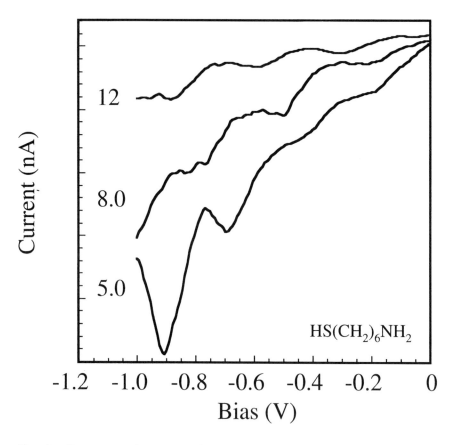

Fig. 4. Current vs. voltage curves for mercaptohexylamine-coated gold particles acquired in aqueous solution at the pH indicated.

on the capacitance of metal electrodes. Qualitatively, excess charge near the metal island of a single-electron transistor causes a pseudocapacitance which shifts the entire staircase to higher or lower bias, depending on the sign of the charge. These charges are often unwanted impurities in the substrate. In this work the "impurity" charge has been designed in as a functionalized organic capping ligand.

The decreases in voltage plateau widths upon converting the neutral galvinol and amine ligands to charged galvinoxide and ammonium ions quantitatively indicate an increase in particle capacitance (Table 1). For galvinol-coated gold particles, the increase in capacitance results from an increase in solution pH and corresponding conversion of the neutral galvinol to the anionic galvinoxide species. In amino-gold, decreasing solution pH charges the amine, causing the in-

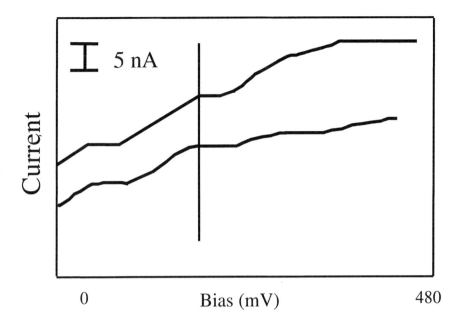

Fig. 5. Current vs. voltage curves for hexanethiol-coated gold nanoparticles acquired in aqueous solution at pH 5 and 12 (curves have been offset for clarity). The vertical line marks the position of the second voltage plateau at pH 5.

crease in particle capacitance. These observations are consistent with capacitance changes reported previously for similar monolayers assembled on macroscopic planar gold electrodes (17).

Finally, the observation of a peak at a bias potential where a voltage plateau is expected is a common feature of *I–V* curves recorded for single molecules and metal and semiconductor nanoparticles. This so-called negative differential resistance (NDR) is often attributed to resonant tunneling through quantum states in the molecule or particle, but can in fact occur for many reasons, including tunneling into surface traps or interaction with surface step edges (18). While we typically observe NDR when the particle capping ligand is charged (at high pH for galvinoxide and low pH for ammonium), at this time we are unsure of the origin of NDR in these systems.

We conclude from these data that single-electron tunneling energies in gold nanoparticles can be manipulated by chemical receptors (e.g., ω-substituted thiols) bound to the nanoparticle. Conversely, if environmental sensitivity is not a preferred property, alkylthiols appear to prevent an electronic response to certain environmental changes (e.g., ionic strength, pH).

IV. ELECTRON TRANSPORT IN "DOUBLE DOT" SYSTEMS

In addition to probing electron transport in gold nanoparticles bound to metal substrates, we have found that nanoparticles may be attached directly to a Pt/Ir STM tip via hexanedithiol linkers. $I-V$ curves recorded at a single location over a gold surface versus tip height confirm that SET behavior is observed with the particle-modified tip (Fig. 6). As predicted by SET theory, a current staircase evolves with increasing current step heights as the tip approaches the substrate. The scanning single-electron tunneling tip may then be rastered over the substrate while collecting $I-V$ curves (Fig. 7).

A scanning SET tip provides the ability to probe the sensitivity of single-electron tunneling to molecules or other nanostructures bound to a substrate. For example, the particle-modified tip may be placed over a gold particle bound to the surface to explore electron transport in coupled metal nanostructures. As the tip approaches the particle, single-electron tunneling through the coupled particle system elicits a very sharp Coulomb blockade response (zero current at zero-bias plateau; Fig. 8, curves 1–3). With decreasing particle-particle distance, however,

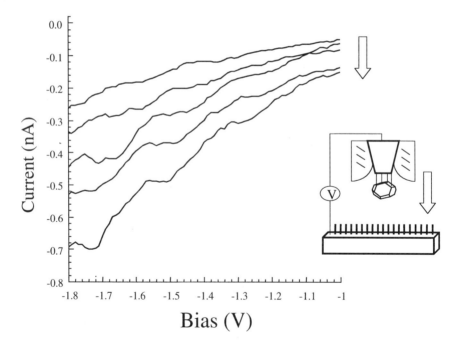

Fig. 6. Current vs. voltage curves for a gold particle-modified Pt/Ir tip at decreasing tip-substrate distances (particle diameter was 5 nm).

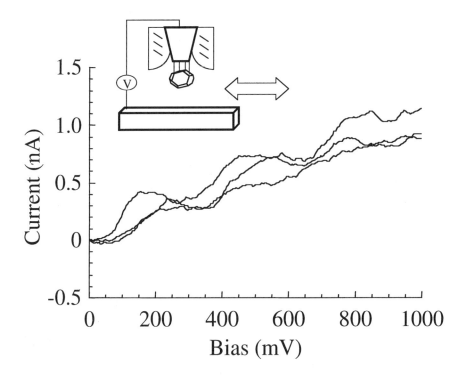

Fig. 7. Current vs. voltage curves for a gold particle-modified Pt/Ir tip at three different locations on a polycrystalline bare gold surface. Approximately 10% of $I-V$ curves acquired with particle-modified tips contain current steps on the polycrystalline surface.

electronic coupling between the two particles becomes too large to observe SET behavior at room temperature. The nonlinear $I-V$ behavior thus begins to appear ohmic again (curve 4).

V. SOLVENT EFFECTS ON NANOPARTICLE CHARGING ENERGIES

In addition to characterizing the effects of capping ligand charge on single-electron tunneling in individual gold particles, solvent effects on particle charging energies have recently been assessed by our group. In these experiments, solvent dielectric constant was manipulated through changes in toluene:acetonitrile composition. However, because apiezon wax or polyethylene tip insulators expand or dissolve in organic solvents, it was not possible to measure electron

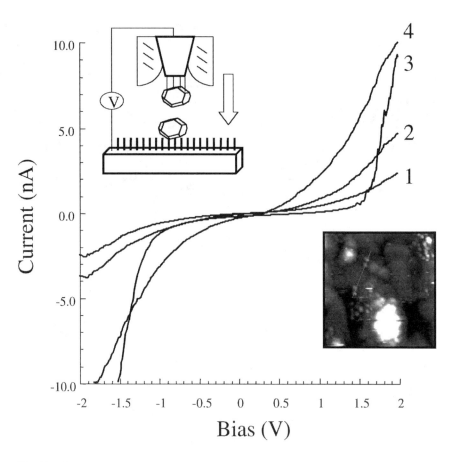

Fig. 8. Current vs. voltage curves for a gold particle-modified Pt/Ir tip at decreasing tip heights over a surface-bound gold particle (curve 4 is the closest tip-particle distance). Inset at lower right shows an STM image acquired with the tip. The line section revealed particle heights of ca. 5 nm.

transport in individual particles. To obviate the need for an insulated tip, we chose to perform differential pulse voltammetry (DPV) on solutions containing gold nanoparticles.

As originally reported by Murray and co-workers (19), size monodisperse fractions of alkanethiol-capped gold nanoclusters display ensemble-averaged single-electron charging peaks when subjected to DPV. Average particle charging energies in DPV are described essentially as given in Eq. (1), with V_{offset} equal to the particle pzc. As with single-electron tunneling in individual particles, charging energies measured in a DPV experiment could conceivably be influenced by changes in monolayer capacitance brought on by changes in solvent dielectric properties.

In their DPV studies, Murray and co-workers found that particle capacitance is independent of solvent composition for alkanethiols longer than C_6. (Shorter capping ligands were not investigated.) Similar results were obtained in our lab for octanethiol-capped gold nanoparticles. However, when triphenylphosphine ligands were bound to the surface, particle capacitance was found to depend on solvent composition (Table 2). Differential pulse voltammograms for triphenylphosphine-capped gold nanoclusters versus solvent composition are shown in Fig. 9. Note that the voltage between charging events increases with decreasing solvent dielectric constant. Large voltage windows between charging events signal a decrease in particle capacitance. The data in Fig. 9 thus suggest that solvent molecules can penetrate the triphenylphosphine monolayer and effect monolayer dielectric constant, but are prevented from entering the octanethiol capping layer.

VI. SUMMARY AND FUTURE WORK

Nanometer-sized metal and semiconductor particles are certain to be fundamental building blocks of advanced electronic and optoelectronic devices. Indeed, it has already been demonstrated that a single nanocrystal (15) or carbon tube (20) can serve as a transistor in which logic and memory functions are controlled by the flow of individual electrons. Such nanoscale electronic devices portend a revolution in device speed and circuit density.

Prospects for fabricating electronic devices out of single nanoparticles are thus realistic and exciting. However, enthusiasm for nanoscale electronic devices must be tempered by the fact that relatively little is understood about how electron transport in nanoparticles and nanotubes is affected by their chemical environment, including the environment created when these objects are integrated together.

This work represents our initial attempts at characterizing electron transport in ligand-modified metal nanoparticles and interacting nanoparticle dimers. Other key challenges for the future vis-à-vis nanoscale electronics are (a) establishing

Table 2. Solvent Dependence of Triphenylphosphine-Capped Gold Nanoparticles

Toluene:acetonitrile	$C(aF)$
2:1	0.315 ± 0.004
1:1	0.364 ± 0.005
1:2	0.383 ± 0.004
1:3	0.697 ± 0.019

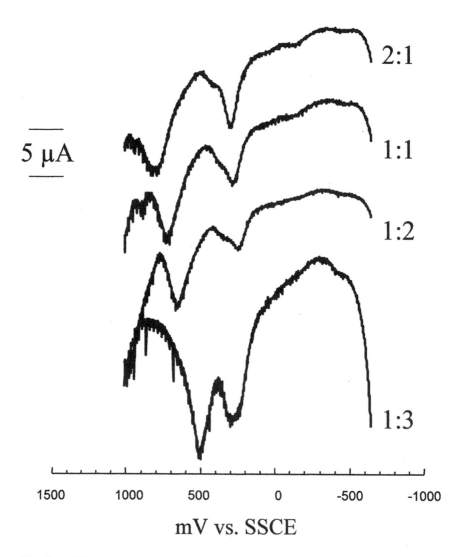

Fig. 9. Differential pulse voltammograms showing the negative-going sweep for solutions containing triphenylphosphine-capped gold nanoparticles. The solutions were toluene:acetonitrile at the indicated ratio and 0.1 M tetrabutylammonium hexafluorophosphate. Working electrode was 3-mm-diameter gold, counter was a Pt wire, and reference was Ag/AgCl. Scan rate = 10 mV/s.

electrical contacts to individual nanoparticles on planar substrates (e.g., Si) reliably and routinely; (b) assembling nanoparticles into symmetrically and spatially well-defined arrays in order to assess collective electronic and electromagnetic behaviors; and (c) identifying new applications for integrated nanoscale assemblies beyond the dream of the all-self-assembled terabit nanoscale computer.

ACKNOWLEDGMENTS

The authors thank The Arnold and Mabel Beckman Foundation, the National Science Foundation, and the Office of Naval Research for their support.

REFERENCES

1. PG Collins, K Bradley, M Ishigami, A. Zettl. Science 287:1801, 2000.
2. Y Xue, S Datta, S Hong, R Reifenberger, JI Henderson, CP Kubiak. Phys. Rev. B 59:R7852, 1999.
3. B Alperson, S Cohen, I Rubinstein, G Hodes. Phys. Rev. B 52:R17017, 1995.
4. CZ Li, HX He, A Bogozi, JS Bunch, NJ Tao. Appl. Phys. Lett. 76:1333, 2000.
5. MJ Yoo, TA Fulton, HF Hess, RL Willett, LN Dunkleberger, RJ Chichester, LN Pfeiffer, KW West. Science 276:579, 1997.
6. DL Feldheim, CD Keating. Chem. Soc. Rev. 27:1, 1998.
7. AC Templeton, WP Wuelfing, RW Murray. Acc. Chem. Res 33:27, 2000.
8. F Caruso. Adv. Mater., in press.
9. CB Gorman, JC Smith, MW Hager, BL Parkhurst, H Sierzputowska-Gracz, CA Haney. J. Am. Chem. Soc. 121:9958, 1999.
10. LP Kouwenhoven, PL McEuen. In: G Timp, ed. Nanotechnology. New York: Springer-Verlag, 1999, p. 471.
11. R Andres, T Bein, M Dorogi, S Feng, JI Henderson, CP Kubiak, W Mahoney, RG Osifchin, R Reifenberger. Science 272:1323, 1996.
12. S-T Yau, P Mulvaney, W Xu, GM Spinks. Phys. Rev. B 57:R15124, 1998.
13. PJ Thomas, GU Kulkarni, CNR Rao. Chem. Phys. Lett. 321:163, 2000.
14. RS Ingram, MJ Hostetler, RW Murray, TG Schaaff, JT Khoury, RL Whetten, TP Bigioni, DK Guthrie, PN First. J. Am. Chem. Soc. 119:9279, 1997.
15. A Bezryadin, C Dekker. Appl. Phys. Lett. 71:1273, 1997.
16. JGA Dubois, ENG Verheijen, JW Gerritsen, H van Kempen. Phys. Rev. B 48: 11260, 1993.
17. MD Porter, TB Bright, DL Allara, CED Chidsey. J. Am. Chem. Soc. 119:9279, 1997.
18. P Avouris, I-W Lyo. Science 264:942, 1994.
19. S Chen, RS Ingram, MJ Hostetler, JJ Pietron, RW Murray, TG Schaaff, JT Khoury, MM Alvarez, RL Whetten Science 280:2098, 1998.
20. MS Fuhrer, J Nygard, L Shih, M Forero, Y-G Yoon, MSC Mazzoni, HJ Choi, J Ihm, SG Louie, A Zettl, PL McEuen. Science 288:494, 2000.

Index

Aggregates 114–115, 148–151, 208
Alumina (see also Aluminum Oxide) 121, 263
Aluminum Oxide 121–123
Atomic Force Microscopy (AFM) 190, 211–215, 240
Attenuated Total Reflectance (ATR) 186

Bessel function 91
Bimetallic Clusters 48, 281–283
Biosensors 159, 183
Brownian Dynamics 243–245, 247, 257

Capping Principle 67
Catalysts 3, 17, 19–38, 48–51, 237, 267, 283–288
Chichen Itza (metal nanoparticles in) 2
Colloid
  definition 18–19
  stabilization 21
Conjugation 56, 58
Core-shell structures 104, 151–153, 164, 169–173, 177–179
Coulomb
  barrier 35

[Coulomb]
  blockade 2
  staircase 298, 299
Current Voltage Curves (see also Coulomb Staircase) 9
  cyclic voltammograms 284, 312–315
  for nanoparticle superlattices 228–231

Dance Circles of Nanoparticles 219–220
Debor Principle 67
Dendrimers
  encapsulation of metal nanoparticles within 267–283
  properties 264–267
  synthesis 263
Dimers 90, 114–115, 153–160 (see also paired particles)
Divine Proportion 61
Discrete Dipole Approximation 90, 101–104
  solvent effects 106
DNA
  dendrimers as shuttles for 291
  metal nanoparticle assay application 159, 163

335

Drude Theory (see also Free Electron Theory) 6
Duality (see Conjugation)
Dynamic Depolarization 96

Electric Dipole 8, 93, 133, 173
Electrodeposition
　gold rods 165
　metals on graphite substrates 239
　particle pairs 124
Electron Microscopy
　scanning (SEM) 211–216, 224, 225, 252, 255, 256
　transmission (TEM) 18, 31, 32, 36, 126, 133, 135, 166, 167, 171, 172, 211–215, 218, 219, 220, 223, 225, 226, 233, 277, 281, 290
Electron Nuclear Double Resonance (ENDOR) 269, 291
Electron Paramagnetic Resonance (EPR) 269, 291
Electron Transport 8–9, 310–311
Electrophoresis 31
Electrostatic Limit (see also Long Wavelength Approximation) 1, 173
Energy Dispersive Spectroscopy (EDS) 275
Euler's Theorem 58, 68

Faraday, Michael 2
Fibonacci Series 61
Fractals 66
Free Electron Theory (see also Drude Theory) 1, 6
Frequency 56, 58, 74, 84
Fuel Cells 284

Generating Function 56, 61, 66, 67, 71, 74–76, 82, 83
Golden Section 61
Growth 20, 55
　atom by atom (ABA) 63, 64
　autocatalytic 44–48
　burst nucleation 39

[Growth]
　cluster of clusters (COC) 63, 65
　diffusive agglomerative 39
　gnomic 61
　layer by layer (LBL) 62, 73, 78
　shell by shell 62, 73
　Stransky-Krastinov 239
　Volmer-Weber 239, 242–247

Highly Ordered Pyrolytic Graphite (HOPG), 211, 213–215, 217, 226, 231, 232, 256
Hyper-Rayleigh Scattering 13, 90, 141–161
Hyperpolarizability 145

Immunosensing (see Plasmon Resonance, biosensing applications)
Infrared (IR)
　scattering, Near IR 161
　spectroscopy 29, 31

Jellium model 56, 71–73, 79

Kunckel, Johann 3

Langmuir Films
　Second Harmonic Response 156
Ligands 5, 12, 13, 284
Lithography 1, 90, 107, 319
Long Wavelength Approximation 93–96
　Modified (MLWA) 96–101
Lorentz Dispersion Theory 6
Lycurgus Chalice 2

Mackay Sequence 63, 69, 77
Magic Numbers 12, 18, 55, 72, 74–77, 82, 83, 84
Magic Sequence 55, 57, 61
Magnetic Dipole 93, 133, 233
Magnetization Hysteresis 231
Mass Spectrometry 77
　fast atom bombardment (FAB) 27–28, 29, 31

Index

[Mass Spectrometry]
    matrix assisted laser desorption ionization-time of flight (MALDI-TOF) 271
Maxwell-Garnett Theory 3, 4, 8, 119
Maxwell's Equations 89, 90, 91, 98, 99, 112
Mercaptoethylamine
    binding to gold nanoparticles 189–194, 198
Metal clusters 4
    melting temperature 56
    transition from molecular to bulk properties 56
Micelles 164, 168, 208–210
    micelle-like domains 262
Mie Theory 2, 8, 89–94, 114, 119, 173, 224
Monolayer Protected Clusters (MPCs)
    definition 297
MOSFET 9
Mossbauer Spectroscopy 210
Multipoles 8, 93, 111, 114, 133, 157–158, 161, 173

Nanoclusters
    distinction from metal colloids 17
    synthesis 29–34
Nanocrystals (see Nanoparticles)
Nanoparticles
    electrodeposition 119–128, 237–258
    nanorods 126–133, 163–181
    nonlinear Optical Properties (NLO) (see also Second Harmonic Generation and Hyper Rayleigh Scattering) 141, 142, 154
    solution phase synthesis 12, 165–173
    template synthesis 12, 119–128, 164, 268
Nonlinear Optics 141
Nuclear Magnetic Resonance (NMR) 29,

Oscillator Strength 6, 145, 157

Paired Particles 13
    optical properties 133, 135

[Paired Particles]
    synthesis 131–134
Plasmon Resonance 1, 5, 164, 273–274,
    biosensing applications 183–204
    dipole origin for metal particles 6, 93, 173–175
    interparticle interactions 114–116, 132, 134–135
    particle shape dependence and spectral maxima 101–103, 175–179
    particle size dependence and spectral maxima 93–97
    particle surface composition effects 41, 104–106, 179–181
    in thin metal films 184–189
Photoluminescence 248–250
Polarizability 4, 5, 97, 101
Polyhedra 57
    archimedean 57, 58
    platonic 56, 57, 58
    v_n polyhedral clusters 74–78
$P_2W_{15}Nb_3O_{62}^{9-}$ (see also Polyoxoanions)
    role in metal cluster growth mechanism 45–50
    synthesis and characterization 29–30
Polyoxoanions
    metal cluster stabilization 22–25, 35
Polyoxometalates (see Polyoxoanions)
Potential of Zero Charge (E_{PZC})
    definition 300
    relation to surface charging 301

Quantum Double Layer Charging (see also single electron transfer) 298–306
    effect of monolayer chain length on, 306–310
Quantum Dots 248, 288, 291
Quartz Crystal Microgravimetry 203

Radiative Damping 96
Rayleigh 3
    hyper-Rayleigh (see Hyper Rayleigh Scattering)

[Rayleigh]
  Scattering 105

Scanning Tunneling Microscope (STM) 11, 12, 31, 90, 298
Second Harmonic Generation (SHG) 13, 90, 132–133, 156, 157–158
  near field effects 110–113
Self Assembly
  molecular 40, 50, 60
  nanoparticles 207–233
Self-Similarity 61–66
Semiconductor nanoparticles 133
Silanes
  in coated metal nanoparticle synthesis 170–171
Single Electron
  charging (see also Coulomb Blockade) 9–12
  devices 2
  transfer (SET) 298–305
  tunneling 9, 11, 320
Size Distribution 18, 36–38, 47, 89, 126, 128, 238, 242–254, 277, 290
Skeletal Electron Pair Theory (SEP) 67
SPR (see Plasmon Resonance)
Stokes, George Gabriel 3
Superlattices (See Also Self-Assembly, nanoparticles) 13, 211–224
  collective properties 224–233
Surface Enhanced Raman Scattering (SERS) 4, 111, 112, 119 (see also Surface Enhanced Spectroscopy)

Surface enhanced spectroscopy 2, 4, 90, 163
  near field effects 110–113

Template Synthesis (see Nanoparticles, template synthesis)
Tetraalkylammonium cations
  misconception about surface adsorption 25
Tetrabutylammonium cation
  contribution to cluster stability 25, 35
  surfactants in nanorod synthesis 165
Thiolates
  alkanethiolates 217–219, 221, 230, 233, 297, 307, 310, 316, 319, 322
Topological Electron Counting (TEC) 66–68
Track Etch Membranes 123
Trimers 153–160
Truncation 56, 58

Ultraviolet-Visible Spectroscopy (UV/Vis)
  polarization UV/Vis 226–227

Voltage Reduction
  in porous anodic alumina synthesis 112–113

X-Ray Crystallography 24, 25, 29
X-Ray Photoelectron Spectroscopy (XPS) 275, 281